Lipid Metabolism and Health

Lipid Metabolism and Health

Robert Brown

MURPHY & MOORE
www.murphy-moorepublishing.com

Lipid Metabolism and Health
Robert Brown
ISBN: 978-1-63987-344-9 (Hardback)

MURPHY & MOORE
Published by Murphy & Moore Publishing,
1 Rockefeller Plaza,
New York City, NY 10020, USA

Cataloging-in-Publication Data

Lipid metabolism and health / Robert Brown.
p. cm.
Includes bibliographical references and index.
ISBN 978-1-63987-344-9
1. Lipids--Metabolism. 2. Lipids. 3. Metabolism. I. Brown, Robert.
QP751 .L57 2022
612.397--dc23

For more information regarding Murphy & Moore Publishing and its products, please visit the publisher's website www.murphy-moorepublishing.com

Table of Contents

Preface

Lipids are macrobiomolecules which are soluble in non-polar solvents. They are utilized for signaling, storing energy and as basic structural components of cell membranes. Lipids are hydrophilic in nature and find extensive use in food industry, cosmetic industry and nanotechnology. Lipid metabolism refers to the synthesis and degradation of lipids which are present in cell bodies. This involves the breakdown and storage of fats, and synthesis of functional and structural lipids. The processes studied under this discipline are lipid absorption, lipid transportation, lipid catabolism and lipid biosynthesis. The biosynthesis mechanisms can be further classified into membrane lipid biosynthesis, triglyceride biosynthesis and fatty acid biosynthesis. The lipids involved in these processes can be divided into four categories, namely, phospholipids, sphingolipids, glycolipids and glycerophospholipids. This book presents the complex subject of lipid metabolism in the most comprehensible and easy to understand language. Some of the diverse topics covered herein address the varied branches that fall under this category. The readers would gain knowledge that would broaden their perspective about lipid metabolism and health.

A short introduction to every chapter is written below to provide an overview of the content of the book:

Chapter 1 - A macromolecule which is insoluble in water and is soluble in non-polar solvents such as chloroform, ether and alcohol is known as lipid. The process of degradation and synthesis of lipids is known as lipid metabolism. This is an introductory chapter which will briefly introduce all the significant aspects of lipids and their metabolism; **Chapter 2 -** There are various types of lipids such as simple lipids, complex lipids, and precursor and derived lipids. Simple lipids are the fatty acid esters of different alcohols which do not contain any other substance whereas complex lipids contain three or more chemical identities. This chapter has been carefully written to provide an easy understanding of these various types of lipids; **Chapter 3 -** When the supply of dietary carbohydrates is high, the excess carbohydrate is converted into triglycerides. It involves two processes: production of fatty acids from acetyl-CoA and esterification of fatty acids during the synthesis of triglycerides. The topics elaborated in this chapter will help in gaining a better perspective about the lipid biosynthesis; **Chapter 4 -** The oxidative degradation of lipids is termed as lipid oxidation. It is a highly complex set of reactions occurring between fatty acids and oxygen resulting in the oxidative degradation of lipids. The process takes place by a free radical chain reaction mechanism. This chapter closely examines the key concepts of lipid oxidation to provide an extensive understanding of the subject; **Chapter 5 -** The rare genetic disorders in which the body is not able to metabolize the food properly are known as inborn errors of metabolism. Some of the inborn errors of metabolism are lysosomal storage diseases, glycogen storage diseases and peroxisomal disorders. This chapter discusses in detail these diseases and disorders related to metabolism.

Finally, I would like to thank my fellow scholars who gave constructive feedback and my family members who supported me at every step.

Robert Brown

1

Introduction to Lipid Metabolism

A macromolecule which is insoluble in water and is soluble in non-polar solvents such as chloroform, ether and alcohol is known as lipid. The process of degradation and synthesis of lipids is known as lipid metabolism. This is an introductory chapter which will briefly introduce all the significant aspects of lipids and their metabolism.

Lipid metabolism is a complex process that involves multiple steps involving the dietary intake of lipids (exogenous) or the production of lipids within the body (endogenous) to degradation or transformation (catabolism) into several lipid-containing structures in the body.

Lipids

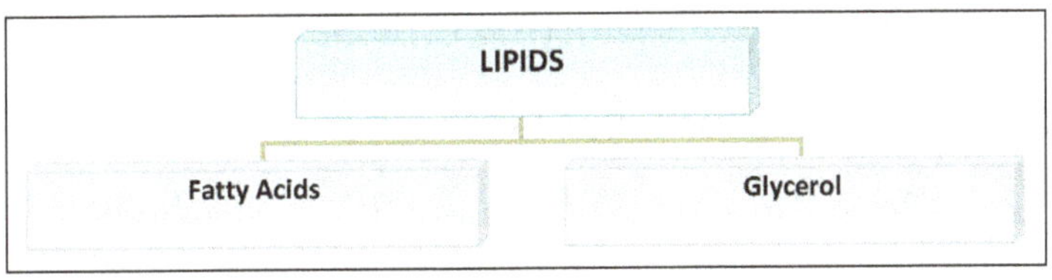

Lipids are heterogeneous group of compounds present in plants and animal tissues related either actually or potentially to the fatty acids. They are amphipathic molecules, hydrophobic in nature originated utterly or in part by thioesters (carbanion-based condensations of fatty acids and/or polyketides etc) or by isoprene units (carbocation-based condensations of prenols, sterols, etc.). Lipids have the universal property of being:

- Quite insoluble in water (polar solvent).
- Soluble in benzene, chloroform, ether (non-polar solvent).

Thus, lipids include oils, fats, waxes, steroids, vitamins (A, D, E and K) and related compounds, such as phospholipids, triglycerides, diglycerides, monoglycerides and others, which are allied more by their physical properties than by their chemical assests.

They are vital constituents related to diet because of high energy value, essential fatty acids and fat soluble vitamins present in the content of fat foods. Lipids serve as adept source of energy when stored in adipose tissue. Fat also dole out as Thermal Insulators in the subcutaneous tissues and in the region of some organs and nonpolar lipids acts as Electrical Insulators permitting quick propagation of depolarization waves by the side of myelinated nerves. Lipoproteins (combinations of lipid and protein) transport lipids in the blood and the biochemical acquaintance of lipids is obligatory in understanding atherosclerosis, diabetes mellitus, obesity etc. The function of diverse polyunsaturated fatty acids (PUFAs) in nutrition and health can also be best understood by studying its biochemical profile.

Chemically, lipids are defined as esters of glycerol and fatty acids or else refer as the triglycerides of fatty acids. They are the substances of natural origin, soluble in non-polar solvents and hence may be extorted by using organic solvents such as methanol. Lipids could be fractionated by either using adsorption chromatography (thin layer chromatography) or reverse-phase chromatography.

Fatty Acid

It may be defined as an organic acid that occurs in a natural triglyceride and is a mono carboxylic acid ranging in chain length from C_4-C_{24} carbon atoms. Fatty acids are obtained from the hydrolysis of fats. They are naturally occurring straight chain derivatives containing carbon atoms with even numbers (4-28) as they assemble from two carbon units. Typically derived from triglycerides or phospholipids, fatty acids are vital sources of fuel yielding huge quantities of ATP. Fatty acids that contain C=C are recognized as unsaturated fatty acid (UFAs) and those lacking double bonds are recognized as saturated fatty acid (SFAs). They are named after corresponding hydrocarbons and vary in length. UFAs end with suffix-enoic and SFAs ends with suffix-anoic. Based on length as short to very long, they may be categorized as under:

- SCFA: Short Chain Fatty Acids with less than six carbons (e.g. butyric acid).
- MCFA: Medium Chain Fatty Acids with 6-12 carbons.
- LCFA: Long Chain Fatty Acids with 13-21 carbons.
- VLCFA: Very Long Chain Fatty Acids with more than 22 carbons.

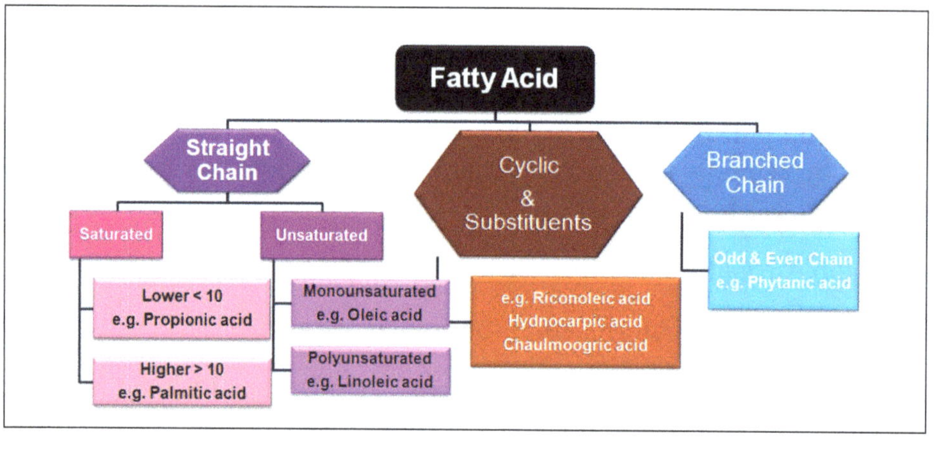

The condensation of Acetyl Co-A, a coenzyme, results in the biosynthesis of fatty acids as it carries two carbon units. That is why all FAs have even numbers of carbon atoms.

Unsaturated Fatty Acid (UFA)

UFAs contain carbon units linked by double bonds, saturated with hydrogen atoms that convert double bonds to single bonds (one or more double bonds between carbon units). The carbon atoms occur either in a cis or a trans configuration. When two hydrogen atoms nearby to the double bond fasten on the chain (same side), it is cis configuration of fatty acid (Oleic acid, Linoleic acid etc). When the neighboring two hydrogen atoms lie on the chain (opposite side), it is trans configuration of fatty acid (Elaidic acid, Vaccenic acid etc). Unsaturated fatty acid may be of following types:

- Monounsaturated: Presence of one double bond.
- Polyunsaturated: Presence of two or more double bond.
- Eicosanoid: Are signaling molecules made by the oxidation of 20-carbon fatty acids.
- Prostanoid: It includes (a) Prostaglandins e.g. PGE1, (b) Prostacyclin e.g. PCI2, (c) Thromboxanes e.g. TXA2.
- Leukotriene: Containing three double bonds sequentially e.g. LTB4, LTE4.

Table: Selected examples of Unsaturated Fatty Acids.

n-x	Name	Structure & Chemical Formula	C:D	Δ^x
n-3	α-Linolenic acid	$CH_3CH_2CH{=}CHCH_2CH{=}CHCH$-$_2CH{=}CH(CH_2)_7COOH$	18:3	cis,cis,cis Δ^9,Δ^{12},Δ^{15}
	Eicosapentaenoic acid	$CH_3CH_2CH{=}CHCH_2CH{=}CHCH_2CH{=}CHCH$-$_2CH{=}CHCH_2CH{=}CH(CH_2)_3$ COOH	20:5	cis,cis,cis,cis,cis-Δ^5, Δ^8, Δ^{11}, Δ^{14}, Δ^{17}
	Docosahexaenoi c acid	$CH_3CH_2CH{=}CHCH_2CH{=}CHCH_2CH{=}CHCH$-$_2CH{=}CHCH_2CH{=}CHCH_2C\ H{=}CH(CH_2)_2COOH$	22:6	cis,cis,cis,cis,cis,cis-Δ^4, Δ^7, Δ^{10}, Δ^{13}, Δ^{16}, Δ^{19}
n-5	Myristoleic acid	$CH_3(CH_2)_3CH{=}CH(CH_2)_7COOH$	14:1	cis-Δ^9
n-6	Linoleic acid	$CH_3(CH_2)_4CH{=}CHCH_2CH{=}CH(CH_2)_7COOH$	18:2	cis,cis-Δ^9, Δ^{12}
	Linoelaidic acid	$CH_3(CH_2)_4CH{=}CHCH_2CH{=}CH(CH_2)_7COOH$	18:2	trans,trans Δ^9, Δ^{12}
	Arachidonic acid	$CH_3(CH_2)_4CH{=}CHCH_2CH{=}CHCH_2CH{=}CHCH$-$_2CH{=}CH(CH_2)_3COOH$	20:4	cis,cis,cis,cis-$\Delta^5\Delta^8$,Δ^{11}, Δ^{14}
n-7	Palmitoleic acid	$CH_3(CH_2)_5CH{=}CH(CH_2)_7COOH$	16:1	cis-Δ^9
	Vaccenic acid	$CH_3(CH_2)_5CH{=}CH(CH_2)_9COOH$	18:1	trans-Δ^{11}
n-9	Oleic acid	$CH_3(CH_2)_7CH{=}CH(CH_2)_7COOH$	18:1	cis-Δ^9
	Elaidic acid	$CH_3(CH_2)_7CH{=}CH(CH_2)_7COOH$	18:1	trans-Δ^9
	Erucic acid	$CH_3(CH_2)_7CH{=}CH(CH_2)_{11}COOH$	22:1	cis-Δ^{13}
n-10	Sapienic acid	$CH_3(CH_2)_8CH{=}CH(CH_2)_4COOH$	16:1	cis-Δ^6

C:D = Number of carbon units and double bond ratio.

Table: List of Other Unsaturated Fatty Acids.

C:D	ω-n	Name	Structure & Chemical Formula	Δ^n	Configuration	Source
16:1	ω-7	Palmitoleic acid	$CH_3(CH_2)5CH{=}CH(CH_2)_7COOH$	Δ^9	cis	Macadamia nut

18:1	ω-7	Vaccenic acid	$CH_3(CH_2)_5CH=CH(CH_2)_9COOH$	Δ^{11}	trans	Butter, Milk, and Yogurt
	ω-9	Oleic acid	$CH_3(CH_2)_7CH=CH(CH_2)_7COOH$	Δ^{9}	cis	Canola, Olive and Pecan oil
	ω-9	Elaidic acid	$CH_3(CH_2)_7CH=CH(CH_2)_7COOH$	Δ^{9}	trans	Vegetable oil (hydrogenated)
18:2	ω-6	Linoleic acid	$CH_3(CH_2)_4CH=CHCH-_2CH=CH(CH_2)_7COOH$	$\Delta^{9,12}$	cis	Chicken fat, Olive and Peanut oil
18:3	ω-3	α-Linolenic acid	$CH_3CH_2CH=CHCH_2CH=CHCH-_2CH=CH(CH_2)_7COOH$	$\Delta^{9,12,15}$	cis	Chiaseeds, Flaxseeds and Walnuts
	ω-6	γ-Linolenic acid	$CH_3(CH_2)_4CH=CHCH_2CH=CHCH-_2CH=CH(CH_2)_4COOH$	$\Delta^{6,9,12}$	cis	Black currant oil, Borage oil, Evening primrose oil and safflower oil
18:4	ω-3	Stearidonic acid	$CH_3CH_2CH=CHCH_2CH=CHCH-_2CH=CHCH_2CH=CH(CH_2)_4COOH$	$\Delta^{6,9,12,15}$	cis	Blackcurrant, Corn gromwell and Seed oils of hemp
20:1	ω-7	Paullinic acid	$CH_3(CH_2)_5CH=CH(CH_2)_{11}COOH$	Δ^{13}	cis	Guarana
	ω-9	Gondoic acid	$CH_3(CH_2)_7CH=CH(CH_2)_9COOH$	Δ^{11}	cis	Jojoba oil (noncaloric and nondigestible but edible)
20:3	ω-6	Dihomo-γlinolenic acid	$CH_3(CH_2)_4CH=CHCH_2CH=CHCH-_2CH=CH(CH_2)_6COOH$	$\Delta^{8,11,14}$	cis	Trace amounts (Animal products)
	ω-9	Mead acid	$CH_3(CH_2)_7CH=CHCH_2CH=CHCH-_2CH=CH(CH_2)_3COOH$	$\Delta^{5,8,11}$	cis	Cartilage
20:4	ω-6	Arachidonic acid	$CH_3(CH_2)_4CH=CHCH_2CH=CHCH-_2CH=CHCH_2CH=CH(CH_2)_3COOH$	$\Delta^{5,8,11,14}$	cis	Dairy, Eggs, Meat
20:5	ω-3	Eicosapentaenoic acid	$CH_3CH_2CH=CHCH_2CH=CH-CH_2CH=CHCH_2CH=CHCH-_2CH=CH(CH_2)_3COOH$	$\Delta^{5,8,11,14,17}$	cis	Cod liver, Herring, Menhaden, Mackerel, Salmon and Sardine
22:1	ω-9	Erucic acid	$CH_3(CH_2)_7CH=CH(CH_2)_{11}COOH$	Δ^{13}	cis	Mustard oil, Wallflower seed

22:4	ω-6	Docosatetraenoic acid	$CH_3(CH_2)_4CH{=}CHCH_2CH{=}CHCH_2CH{=}CHCH_2CH{=}CH(CH_2)_5COOH$	$\Delta^{7,10,13,16}$	cis	-
22:6	ω-3	Docosahexaenoic acid	$CH_3CH_2CH{=}CHCH_2CH{=}CHCH_2CH{=}CHCH_2CH{=}CHCH_2CH{=}CHCH_2CH{=}CH(CH_2)_2COOH$	$\Delta^{4,7,10,13,16,19}$	Cis	Fish oil and Maternal milk
24:1	ω-9	Nervonic acid	$CH_3(CH_2)_7CH{=}CH(CH_2)_{13}COOH$	Δ^{15}	cis	Flaxseed, King Salmon, Macademia nuts, Sesame seed and Sockeye salmon

Saturated Fatty Acid (SFA)

They are long chain carboxylic acids without double bonds but with 12-24 carbon units. As its name indicates they are saturated with hydrogen atoms having solitary bonds with each carbon units inside the chain has 2 hydrogen atoms (except 3 hydrogens at the end for omega carbon). Examples of the SFAs are Capric acid, Palmitic acid, Stearic acid etc.

Table: List and Examples of Saturated Fatty Acids.

C: D	Name	Chemical Formula	Nomenclature
3:0	Propionic acid	CH_3CH_2COOH	Propanoic acid
4:0	Butyric acid	CH3$(CH_2)_2COOH$	Butanoic acid
5:0	Valeric acid	$CH_3(CH_2)_3COOH$	Pentanoic acid
6:0	Caproic acid	$CH_3(CH_2)_4COOH$	Hexanoic acid
7:0	Enanthic acid	$CH_3(CH_2)_5COOH$	Heptanoic acid
8:0	Caprylic acid	$CH_3(CH_2)_6COOH$	Octanoic acid
9:0	Pelargonic acid	$CH_3(CH_2)_7COOH$	Nonanoic acid
10:0	Capric acid	$CH_3(CH_2)_8COOH$	Decanoic acid
11:0	Undecylic acid	$CH_3(CH_2)_9COOH$	Undecanoic acid
12:0	Lauric acid	$CH_3(CH_2)_{10}COOH$	Dodecanoic acid
13:0	Tridecylic acid	$CH_3(CH_2)_{11}COOH$	Tridecanoic acid
14:0	Myristic acid	$CH_3(CH_2)_{12}COOH$	Tetradecanoic acid
15:0	Pentadecylic acid	$CH_3(CH_2)_{13}COOH$	Pentadecanoic acid
16:0	Palmitic acid	$CH_3(CH_2)_{14}COOH$	Hexadecanoic acid
17:0	Margaric acid	$CH_3(CH_2)_{15}COOH$	Heptadecanoic acid
18:0	Stearic acid	$CH_3(CH_2)_{16}COOH$	Octadecanoic acid
19:0	Nonadecylic acid	$CH_3(CH_2)_{17}COOH$	Nonadecanoic acid
20:0	Arachidic acid	$CH_3(CH_2)_{18}COOH$	Eicosanoic acid
21:0	Heneicosylic acid	$CH_3(CH_2)_{19}COOH$	Heneicosanoic acid
22:0	Behenic acid	$CH_3(CH_2)_{20}COOH$	Docosanoic acid
23:0	Tricosylic acid	$CH_3(CH_2)_{21}COOH$	Tricosanoic acid
24:0	Lignoceric acid	$CH_3(CH_2)_{22}COOH$	Tetracosanoic acid
25:0	Pentacosylic acid	$CH_3(CH_2)_{23}COOH$	Pentacosanoic acid

26:0	Cerotic acid	$CH_3(CH_2)_{24}COOH$	Hexacosanoic acid
27:0	Heptacosylic acid	$CH_3(CH_2)_{25}COOH$	Heptacosanoic acid
28:0	Montanic acid	$CH_3(CH_2)_{26}COOH$	Octacosanoic acid
29:0	Nonacosylic acid	$CH_3(CH_2)_{27}COOH$	Nonacosanoic acid
30:0	Melissic acid	$CH_3(CH_2)_{28}COOH$	Triacontanoic acid
31:0	Henatriacontylic acid	$CH_3(CH_2)_{29}COOH$	Henatriacontanoic acid
32:0	Lacceroic acid	$CH_3(CH_2)_{30}COOH$	Dotriacontanoic acid
33:0	Psyllic acid	$CH_3(CH_2)_{31}COOH$	Tritriacontanoic acid
34:0	Geddic acid	$CH_3(CH_2)_{32}COOH$	Tetratriacontanoic acid
35:0	Ceroplastic acid	$CH_3(CH_2)_{33}COOH$	Pentatriacontanoic acid
36:0	Hexatriacontylic acid	$CH_3(CH_2)_{34}COOH$	Hexatriacontanoic acid

Essential Fatty Acid (EFA)

They are indispensable for the human body not produced in adequate amount from substrates, and consequently ought to be obtained from diet (food). The idiom 'EFA' refers to fatty acids obligatory for biological processes and excludes those fats that merely act as fuel. Two vital series of EFAs have been reported: (i) Three carbon units with double bond removed from the methyl end and (ii) Six carbon units with double bond removed from the methyl end. Omega 3-fatty acid (α-linolenic acid = ALA) and Omega 6-fatty acid (Linoleic acid = LA) are the two EFAs widely distributed in plant oils. Humans lack the ability to synthesize these two EFAs due to absence of desaturase enzymes required for their production. In 1923, these two EFAs were preferred as Vitamin F but later (1929), studies on mice revealed that these two EFAs should be categorized under fats rather vitamins. Omega-3 fatty acid (Docosahexaenoic acid) and Omega 6-fatty acid (γ- Linoleic acid) are occasionally referred to as 'Conditionally Essential' as they happen to indispensable under disease or some developmental circumstances. In the human body, EFAs dole out numerous functions such as:

- Customized to make:
    - Eicosanoids: Distressing several cellular functions including inflammation.
    - Endocannabinoids: Upsetting mood, behavior and inflammation.
    - Lipoxins: A faction of eicosanoid derivatives produced from ω-6 EFAs via the lipoxygenase pathway and resolvins from ω-3 (down regulating inflammation in the presence of acetylsalicylic acid).
    - Epoxyeicosatrienoic acids (EETs), Hepoxilins, Isofurans, Isoprostanes, Neurofurans and Neuroprotectin D.
- EFAs affect cellular signaling by forming lipid raft.
- They either activate or inhibit transcription factors such as NF-κB and act on DNA.

Examples of the food sources with EFAs are canola (rapeseed) oil, chia seeds, fish and shellfish, flaxseed (linseed), hemp seed, leafy vegetables, pumpkin seeds, soya oil, sunflower seeds and walnuts. Nearly, all the PUFAs in the human diet are EFAs that play a vital part in the existence and loss of cardiac cells. The deficiency of EFAs results in depression, dermatitis and osteoporosis.

Free Fatty Acid

When fatty acids do not affixed to supplementary molecules, they are acknowledged as 'Free Fatty Acids' (FFAs) or 'Uncombined Fatty Acids' (UCFAs). The FFAs or UCFAs ensued from the triglyceride breakdown, insoluble in water, circulated, solubilized and transported through albumin (a plasma protein). However, their blood echelon is restricted by the accessibility of binding sites of albumin.

Table: Composition of dietary fats.

	Monounsaturated	Polyunsaturated	Saturated	Cholesterol	Vitamin E
	g per 100 g	g per 100 g	g per 100 g	mg per 100 g	mg per 100 g
Vegetables fats					
Canola (Rapeseed Oil)	64.3	24.8	5.3	0	22.21
Coconut oil	6.6	1.7	85.2	0	0.66
Corn oil	24.7	57.8	12.7	0	17.24
Cottonseed oil	21.3	48.1	25.5	0	42.77
Hemp oil	15	75	10	0	42.77
Olive oil	69.7	11.2	14.0	0	5.10
Palm kernel oil	11.4	1.6	81.5	0	3.80
Palm oil	41.6	8.3	45.3	0	33.12
Safflower oil	12.6	72.1	10.2	0	40.68
Soybean oil	23.2	56.5	14.5	0	16.29
Sunflower oil	20.2	63.0	11.9	0	49.00
Wheat germ oil	15.9	60.7	18.8	0	136.65
Animal Fats					
Butter	19.8	2.6	54.0	2.30	2.00
Duck fat	49.3	12.9	33.2	100	2.70
Lard	43.8	9.6	40.8	93	0.60

Triglycerides (TG)

Triglycerides are most abundant form of lipids and constitute about 98% of total dietary lipids. TGs are esters of glycerol with three fatty acid molecules. Glycerol (Alcohols) contains a hydroxyl (OH) moiety and Fatty acid (Organic acids) encloses a carboxyl (-COOH) group. Both join together to form esters. During each esterification one molecule of water is released. In TGs, the OH group of the glycerol connects the COOH group of the fatty acid to form ester bonds. They may be fats and oils and are also known as Triacylglycerol (TAG) or Triacylglyceride. TGs contain three moles of fatty acids which may be similar or dissimilar. Similar kind of FAs in all the three positions are called simple TGs e.g. Tripalmitin, triolein etc. Most of the TGs contains different kinds of fatty acids in position 1, 2 or 3 and are called mixed TGs e.g. Oleodipalmitin etc. There are a lot of triglycerides obtainable from the oil source, a number of them are highly unsaturated and a few are unsaturated. Saturated are those having single bonds between the carbon atoms (C-C) where hydrogen atoms bonds with carbon atoms while unsaturated compounds bears double bonds between carbon units (C=C), plummeting the integer of places wherever hydrogen atoms bonds with carbon atoms. Furthermore, at room temperature saturated have an elevated melting point and are solid while unsaturated have a lower melting point and are liquid.

TGs are the vital constituents of animal fats (saturated) including human skin oils and vegetable oil (unsaturated). In naturally occurring TGs, the chain lengths of the FAs include even number (16, 18, 20) of carbon units. However, in bacteria and ruminants fat odd number (15) carbon atoms are present. Majority of natural fats include an intricate blend of individual TGs and due to this, they deliquesce over a wide array of temperatures. In TGs form, lipids cannot be engrossed by the duodenum unless broke dowm into fatty acids, monoglycerides and a few diglycerides. In the intestine, TGs ripped into FFAs and monoacylglycerol following the secretion of bile and lipases in a process called lipolysis. TGs advances to the intestine through enterocyte cells and reinstate it from their wreckage, wrap up with proteins and cholesterol to guise chylomicrons. An array of tissues incarcerates the chylomicrons and releases the TGs to be worn as an energy source. TGs can pass through cell membranes freely via the fatty acid transporter (FAT) only after its split into fatty acid and glycerol by lipoprotein lipases. TGs being the foremost components of chylomicrons and VLDL (very low density lipoprotein) perform an imperative role (energy source) in metabolism and transporters of dietary fat (38 kJ/ g or 9 kcal/ g compared to carbohydrates: 17 kJ/ g or 4kcal/g). A high level of TGs in human body has been related to atherosclerosis, stroke risk and heart disease.

In oil paints and coating, di and triunsaturated fatty acid components present in linseed and related oil are used which apt to congeal in the presence of oxygen. Using trans-esterification phenomenon, TGs are also ripped into their components in the biodiesel manufacturing. The ensuing esters of FA be able to be worn as a fuel in diesel engines. The glycerin is used in the production of pharmaceuticals and food. Lysochromes (Fat soluble dye, Oil Red O, Sudan Black B, Sudan IV) has been employed for staining fatty acids, triglycerides, lipoproteins, and other lipids.

Lipids contain hydrocarbons which are the base for the structure and function of living cells. The biological functions of the lipids are as distinct as their chemistry. Oils and fats are the primary stored arrangement of energy in numerous organisms. Sterols and phospholipids are key structural essentials of biological membranes. Other lipids, though present in reasonably small quantities, perform crucial roles as electron carriers, anchors for proteins (hydrophobic), enzyme cofactors, light absorbing pigments, as intracellular messengers, as chaperones in protein folding of membranes and as an emulsifying agents in the digestive tract.

Being largely hydrocarbon, lipids yield huge amounts of energy on oxidation and represent exceedingly reduced forms of carbon. Based on biochemical subunits, lipids may be alienated into the following class: saccharolipids, sphingolipids, sterol lipids, polyketides (consequential of ketoacyl subunits condensation), prenol lipids (resultant of isoprene subunits condensation), glycerophospholipids, glycerolipids and fatty acids.

The major functions of lipids include energy storage, integral part of cell membrane components and signaling. Oils and fats are the main source of energy in innumerable organisms. Sterols and phospholipids are key structural elements of membranes. Other present in comparatively diminutive quantities, perform essential roles as electron carriers, enzyme cofactors, light absorbing pigments, hormones, as intracellular messenger, as an anchor for hydrophobic proteins, as chaperones in membrane proteins folding, as an emulsifying agent in the digestive tract. Lipids have a burly relevance in nanotechnology as well as in food and cosmetic industries.

In a nutshell, lipids:

- Acts as a fuel in the body.

- Yields 9.0 kcal of energy per gram.
- Exerts an insulating effect in the body.
- Provide padding and protect the internal organs like kidney.
- Supply EFAs for normal health, development and growth.
- Vital for fat soluble vitamins.
- Fundamental constituent of cell wall, cell membrane and cell organelle like mitochondria.

Foods Affect Triglyceride Levels

Eating foods high in simple sugars significantly contributes to high triglyceride levels. Follow these guidelines to limit simple sugars in your diet:

- Substitute beverages like colas, fruit drinks, iced tea, lemonade, Hi-C and Kool-Aid with artificially sweetened beverages labeled "sugar-free" or "diet."
- Limit hard candies, chocolates, candy bars, and gummy bears.
- Avoid adding table sugar and brown sugar to cereal, drinks or foods. Instead, use an artificial or herbal sweetener or nothing at all.
- Choose sugar-free gum or mints instead of the regular versions.
- Try light or low-sugar syrups on pancakes and waffles.
- Spread breads and crackers with no-sugar-added jelly or preserves.
- Snack on whole fruit instead of fruit roll-ups and other fruit-flavored treats.
- When selecting cereals, choose those with no more than 8 grams of sugar per serving.
- Try sugar-free gelatin, popsicles, yogurts, and puddings instead of the regular versions.
- Be aware that desserts labeled "fat-free" usually contain more sugar than the full-fat varieties and the same number of calories.
- Cut back on or avoid eating sweets and dessert foods, including cookies, cakes, pastries, pies, ice cream, frozen yogurt, sherbet, gelato, and flavored ices. All of these foods contain high levels of sugar.
- Read the ingredients list on food labels, and limit foods that list any of the following words (all simple sugars) in the first few ingredients:
    - Sucrose
    - Glucose
    - Fructose
    - Corn syrup
    - Maltose

- Honey
- Molasses
- High-fructose corn syrup

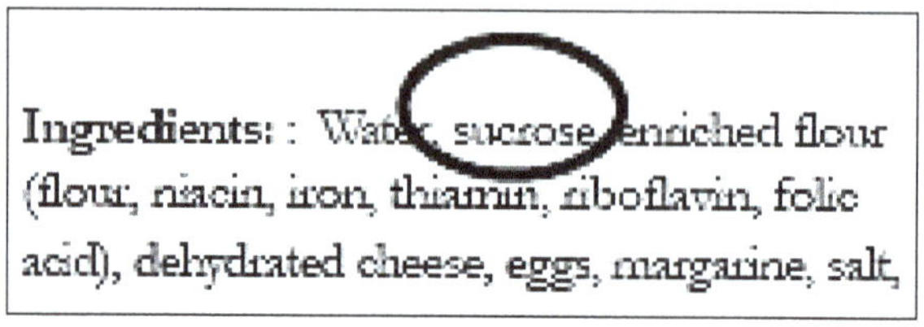

Natural sugars, when eaten in excess, can also raise your triglyceride level.

Classification of Lipids

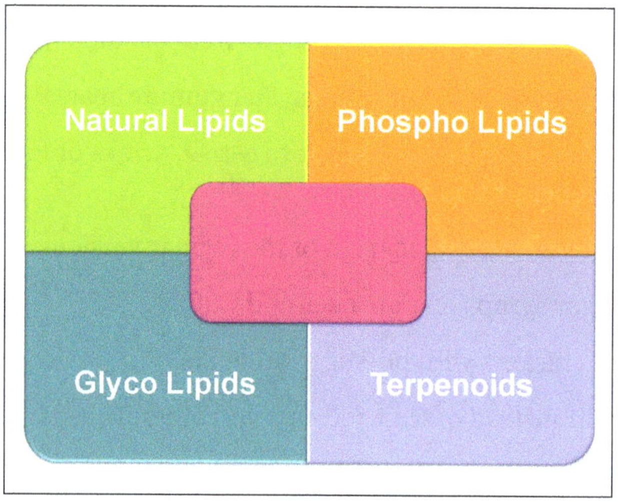

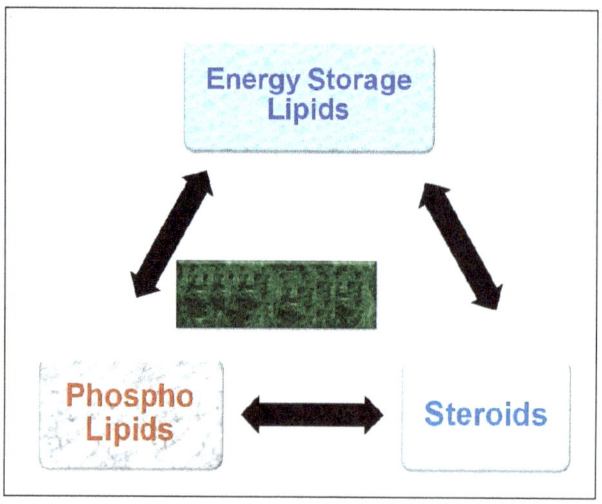

Lipids can be broadly classified into following major classes: (a) Saponifiable lipids and (b) Non-saponifiable lipids. This classification is based on the basics of their reaction with sodium or potassium hydroxide.

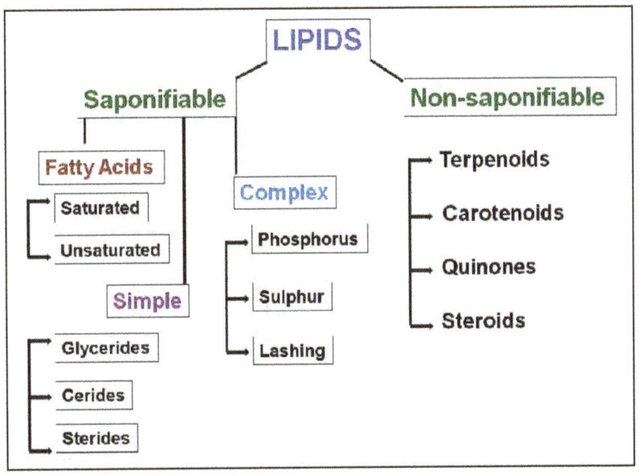

a. The saponifiable lipid has one or more ester groups allocating it to endure hydrolysis in the charisma of an acid, base or enzyme.

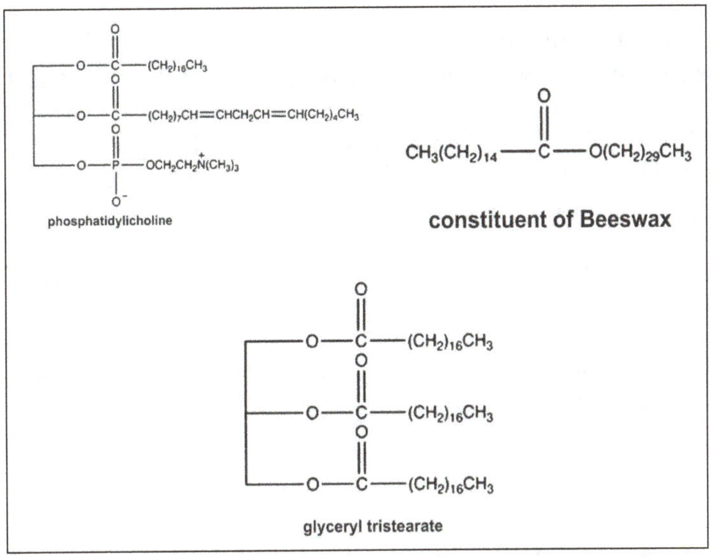

b. The non-saponifiable lipid includes phospholipids, prostaglandins, sphingolipids, steroids, terpenes, triglycerides and waxes which cannot be wrecked up into smaller molecules by hydrolysis.

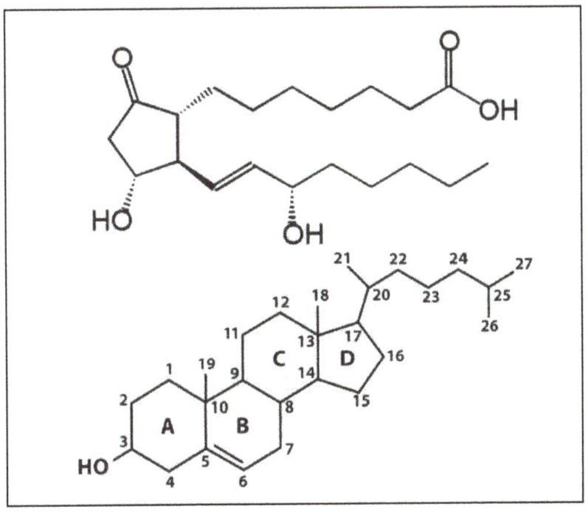

On the basis of polarity, lipids are divided into two classes: (1) Polar and (2) Non-Polar.

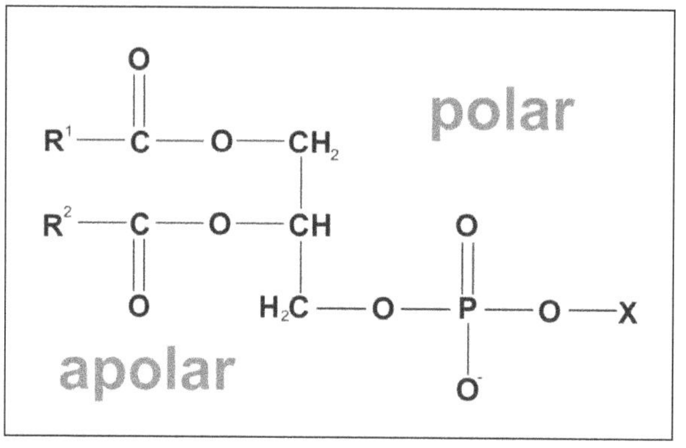

Polar (Glycerophospholipids and Sphingolipids)

Some of the characteristic of polar lipids are as under.

- It forms bilayer impulsively in water. Phosphatidyl choline (PC, a purified polar lipid from membranes) restrain up to 30 combinations of fatty acids on its chains.
- The bilayers of lipid have four states: Crystal, Gel, Liquid and Liquid crystal. It exhibits a sharp Tm defined from the gel to liquid crystal transition. The Tm is very soaring for saturated chains and very short for polyunsaturated chains.
- In the liquid crystal state, biological membranes are always just over the Tm. Examples are fish and alcoholics.
- There are three sub-classes of polar lipids: anionic, uncharged and zwitter-ionic. No cationic lipids are reported in nature.

Anionic Lipids

Examples of the anionic lipids are Cardiolipin (CL), Phosphotidyl glycerol (PG), Phosphotidyl inositol (PI), Phosphotidyl serine (PS) and many others.

- Agreement of X-ray structures of crystals with NMR conformations.
- Vile interactions improved by screening of cations from solution and H-bonds.
- pH at the surface lower that of water (2-3 units if not in salt) and the pH difference wanes off at high ionic strength.
- The anionic lipids constantly involved in fusion and the surface can bind calcium and hydrogen which eases fusion.
- Cardiolipin acts as an acid-anion provokes hydrogen ion to operate as an attractant. Fusion arises because of acid-anion formation and negative charges become striking by trapping protons.
- The entire membranes have a net negative charge on their exterior that attracts cations (H^+ & Na^{++}).

- Similarly, all biological membranes have a proton gradient sustained by a proton pump (except presence of sodium gradient in animal plasma membrane).

Uncharged Lipids

Examples of the uncharged lipids are Cerebrosides, Gangliosides and Glycolipids. The surface of gangliosides & sphingolipids are anionic but have H-bond lateral interactions. Glycolipids have H-bonds amid neighbours with each lipid has three neighbours.

Zwitter-Ionic Lipids

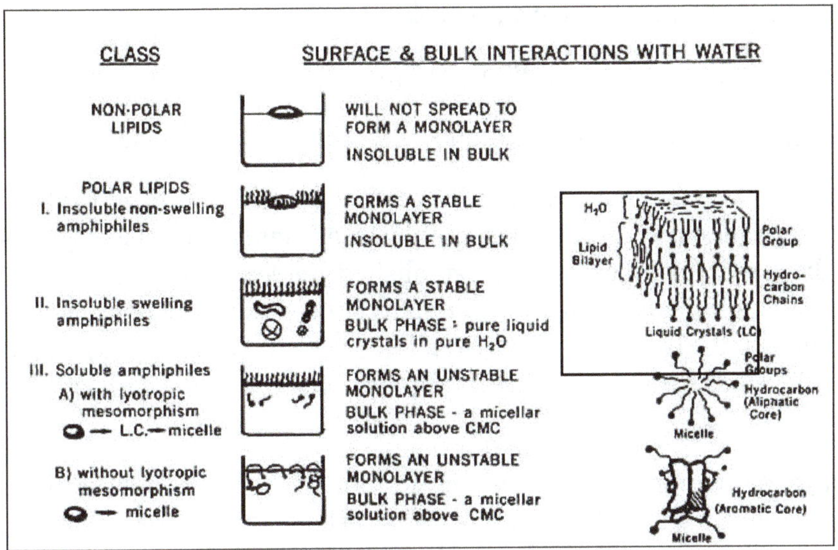

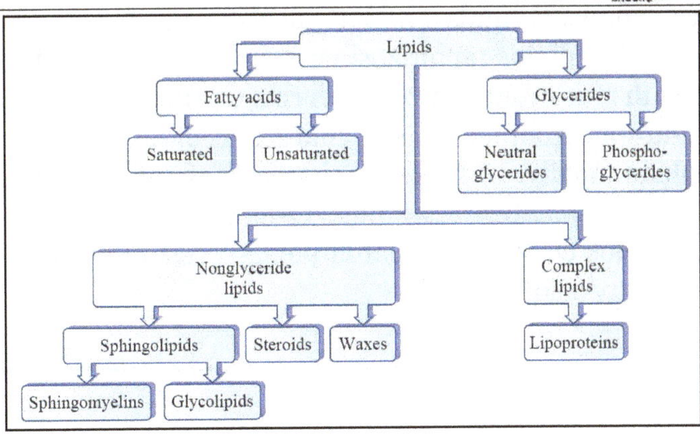

Examples of the zwitter-ionic lipids are Phosphotidyl choline (PC), Phosphatidyl ethanolamine (PE), Sphingomyelin (SM).

- Position wise phosphate is on top of the glycerol and glycerol is upright to the membrane.
- Primary ester is underneath the glycerol and secondary ester is parallel to the membrane before (at C2) the chain plunges.
- Choline/Ethanolamine (base) gyrates at an angle of ~15^{0} to the plane.
- Each headgroup of zwitter ionic lipids interact with three neighbours ionically (+,-).

Non-polar (Triglycerides)

A lot of lipids, on the other hand, are non-polar, implicating that the charge allocation is uniformly distributed. Infact, non-polar molecules do not dissolve well in water. Polar and non-polar molecules tend to deter each other in the similar way as the oil and water do not mix and will split from each other even if they are shaken robustly in an endeavor to mix them. This peculiarity between non-polar and polar molecules has imperative effect for living things, which are poised of both molecules.

On the basis of function, lipids can be classified into following:

- Storage lipids (Fatty acids, Triacylglycerols, Sterols).
- Structural lipids in membranes (Phospholipids, Saccharolipids).
- Lipids as signals, cofactors and pigments (Lipoproteins, Proteolipids, Polyketides).

In the beginning it was understood that lipids are oily substance utilized for two major reasons: (a) dole out as energy source and (b) structural blocks of membranes. George & Mildred Burr on the other hand drive out this allegory and established linoleic acid as an indispensable dietary constituent that play an imperative function in many processes in the body. Bergström et al., also revealed arachidonate (EFA) as the antecedent of the prostaglandins and explained its effect on inflammation and allied disease giving impetus to lipids that gained new-fangled significance among biochemists.

Amphipathic nature of membrane lipids and orientation of their hydrophilic and hydrophobic regions articulate their stuffing into the bilayer. Among structural membrane lipids, three types have been described: (a) Glycerophospholipids (two fatty acids connected to glycerol are present in hydrophobic regions), (b) Sphingolipids (sphingosine: a fatty amine linked to single fatty acid) and (c) sterols (compounds with four fused hydrocarbon rings arranged in an inflexible arrangement). Glycerophospholipids and sphingolipids include charged alcohols at their polar ends and a few in addition contain phosphate moeity. In these amphipathic compounds, the hydrophilic moieties could be more complex or as elementary as a single -OH group at one end of the sterol ring system. Among these membrane lipids classes, colossal miscellany results from various combinations of 'tails' and polar 'heads' of fatty acid.

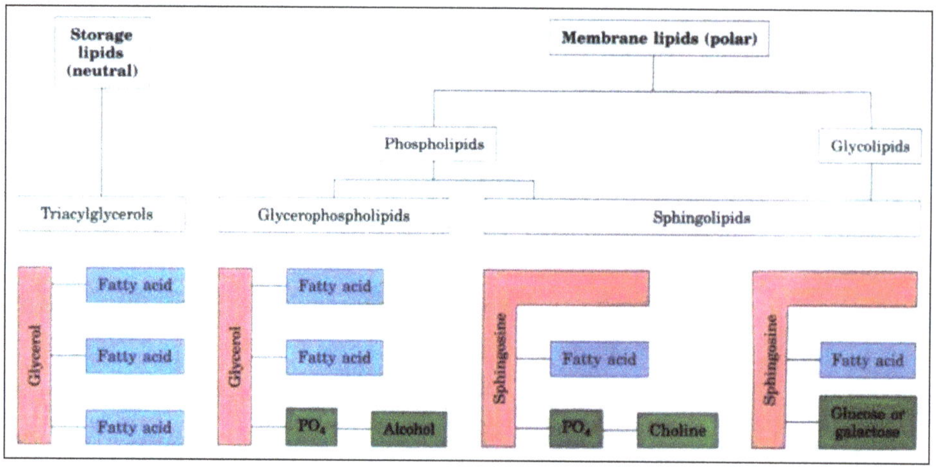

The products of hydrolysis have also been considered for classifying the lipids into following category:

1. Simple Lipids: On hydrolysis yield alcohol (monohydric or trihydric) and fatty acids:

- Simple glyceride: Include identical FAs.
- Mixed glyceride: Include dissimilar FAs.
- Oils: Glycerol with UFAs.
- Fats: Glycerol with SFAs.
- Waxes: Mono or dihydric alcohol with FAs.

2. Complex Lipids: (Compound lipids): On hydrolysis yield fatty acids, glycerol and phosphoric acid along with serine, sphingosine, ethanolamine and various sugars.

- Phospholipid: Glycerol + fatty acids + phosphoric acid + nitrogenous base, e.g. Lecithin: Glycerol + fatty acids + phosphoric acid + choline.
- Cephalin: Glycerol + fatty acids + phosphoric acid + ethanolamine.
- Glycolipids: Glycerol + fatty acid + carbohydrates (on hydrolysis).
- They are sub-classified as sulpholipids, galactosyl diglyceride and cerebrosides.
- Sphingophosphoiplds: Sphingosine + fatty acids + phosphoric acid + choline.

3. Derived Lipids: Are the hydrolytic commodities of complex and simple lipids.

- Examples are Alcohols (Sterol and glycerol), fatty acids, terpenoids etc.

Functions of Lipids

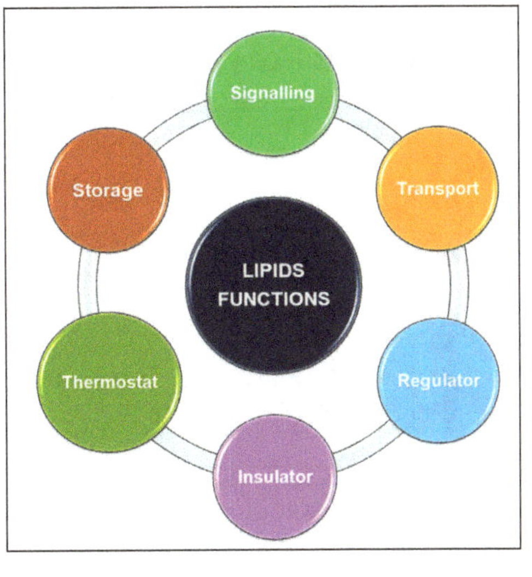

Lipids are natural hydrophobic molecules (organic compounds) occur all through the existing living organisms including plants, microbes, animals and also in each and every cell. Being a vital dietary component, lipids are a basis of high energy treasure, essential fatty acids and fat soluble vitamins. They dole out as an extremely greater cause of energy and accumulated in adipocytes. Lipids also function as an insulating material in the subcutaneous tissues and are also spotted in the region of certain organs. They participate in nerve cell functioning, cell structure built-up and supply stored fuel.

In addition to the above-mentioned characteristics, lipids execute numerous biological functions:

- Act as storage compounds, TGs serve as stockpile energy of the body.

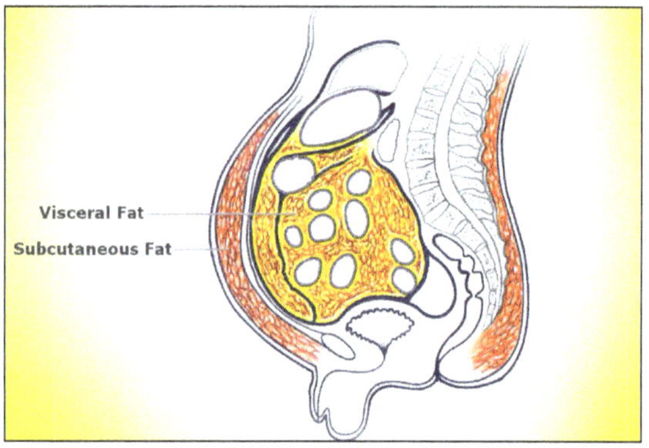

- As an essential component of membrane structure of eukaryotic cells.

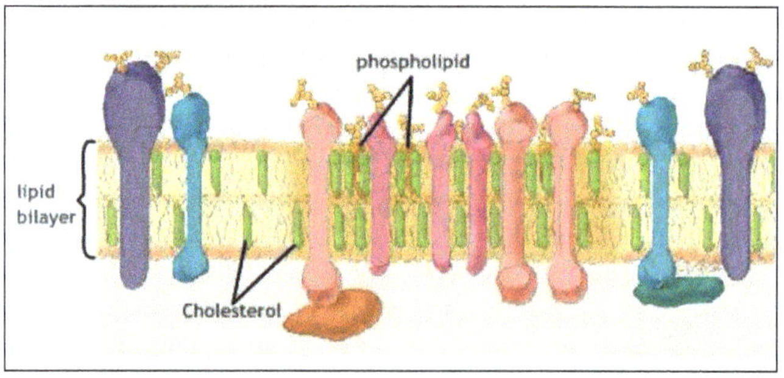

- It regulates membrane permeability.

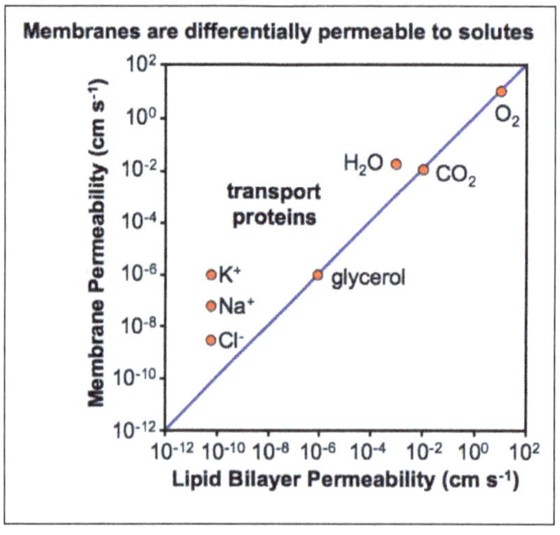

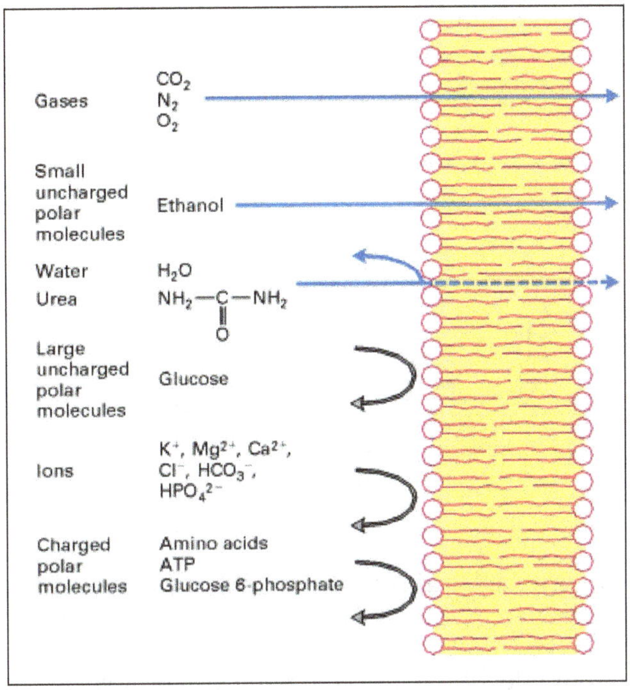

- Lipids are the source for fat soluble vitamins (A, D, E, K). Being a lipid, cholesterol is the antecedent of bile acids, steroids and Vitamin D.

- Lipids act as an electrical insulator to the nerve fibres of the myelin sheath.

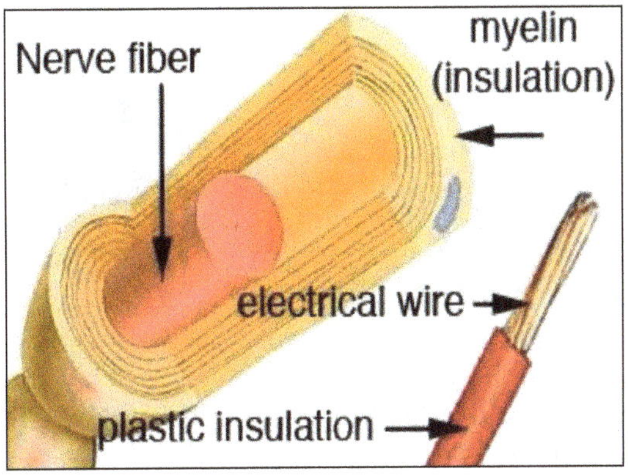

- They are components of a number of enzyme systems.

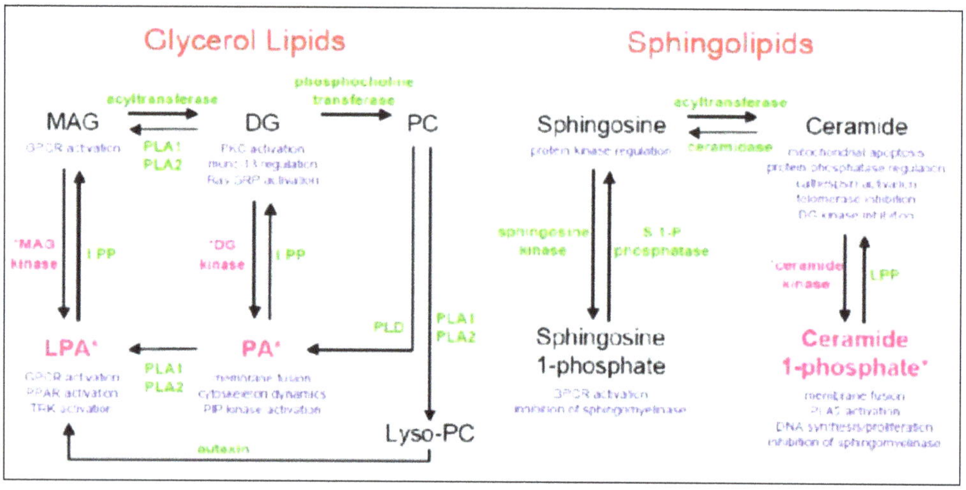

- Lipids, for example, prostaglandins and steroid hormones act as cellular metabolic regulators.

- Cholesterol is found in bile, blood and cell membranes of various organisms.

Table: National Cholestrol Education Program Cholestrol Guidelines.

	Desirable	Boderline High	High
Total Cholestrol	Less than 200	200-239	240 and Higher
LDL Cholestrol (The "Bad"cholestrol)	Less than 130	130-159	160 and Higher
HDL Cholestrol (The "Good"cholestrol)	50 and toghether	40-49	Less than 40
Triglycrides	Less than 200	200-399	400 and Higher

- Being diminutive molecules, lipids are insoluble in water and act as signalling molecules.

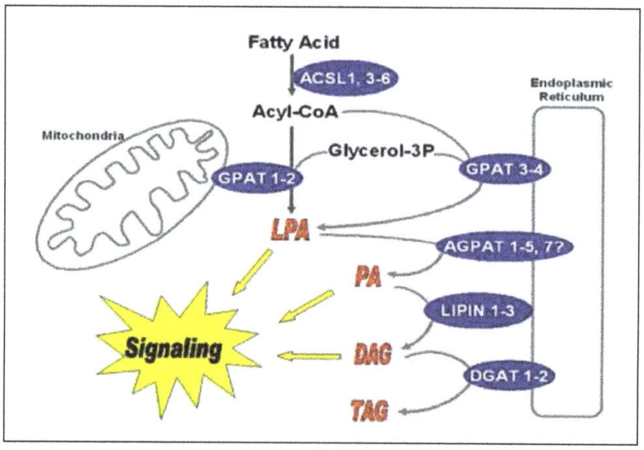

- Brown fat maintains body temperature and fats in the sub-cutaneous layer; afford insulation and protection from cold.

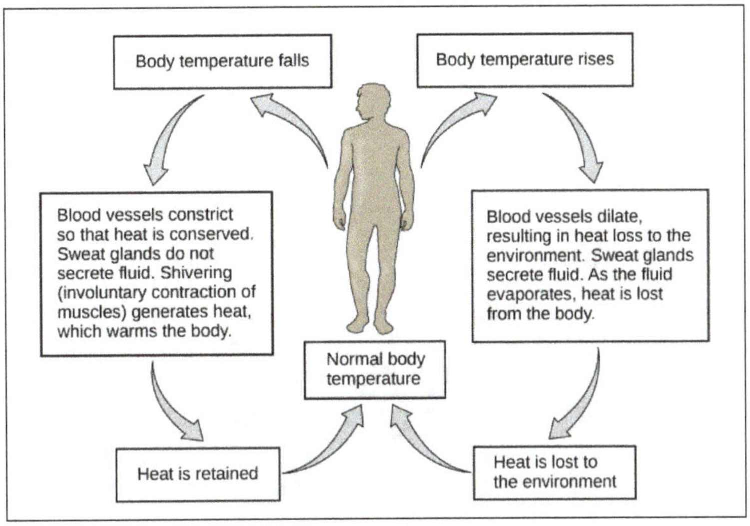

- Phospholipids provide fluidity and flexibility to the cell membranes and cholesterol sustain fluidity of membranes by intermingling with lipid complexes.

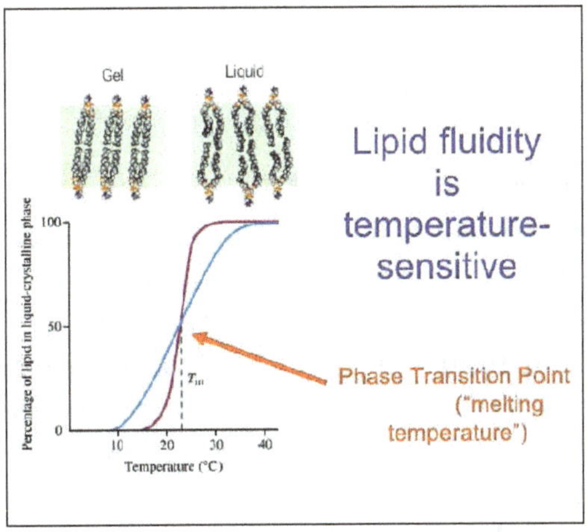

- Lipoproteins crop up in blood as plasma lipoprotein and enable transportation of lipids in aqueous environment throughout the body.
- EFAs (linolenic and linoleic acids) are predecessor of several types of ecosanoids including prostaglandins and thromboxanes which play a significant role in fever, inflammation, pain and blood clotting.

Types/Category	Functions
Glycolipids	Enhance the immune system of the body.
Lipoproteins	Transport fat in the bloodstream and subsist in HDL and LDL form.
Phospholipids	Beguile and deter water, Constitute cell membranes.
Steroids	Establish and organize the plasma membrane structure.
Triglycerides	Storage of energy.
Waxes	Protection.

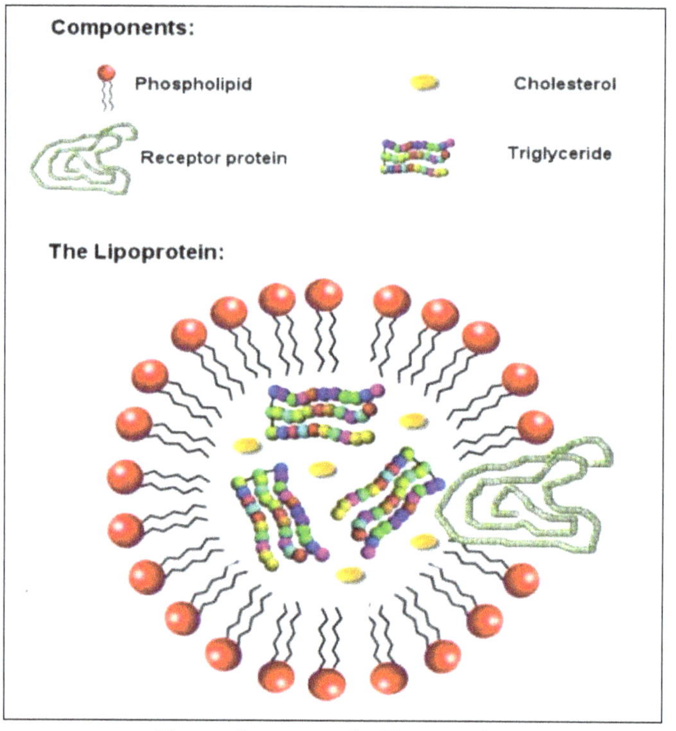

Figure: Structure of a Lipoprotein.

It is well-known that lipids perform a fundamental activity in the body (e.g. chemical messenger) than earlier described storage and cell membrane formation. Researchers have found that lipids have a pervasive activity in the body in provisos of hormonal regulation or intracellular signaling etc. They are blended through intricate biosynthetic pathways, yet deemed essential and ought to be complemented via diet. In 1929, George & Mildred Burr established that linoleic acid is an indispensable dietary component while Bergström et al reported that prostaglandins originate from arachidonate that exert their effects on inflammation and allied disease condition. In 1979, platelet activating factor (earliest bioactive phospholipid) was revealed and awareness was heaved regarding the role of phosphatidylinositol and their metabolites in signaling and cell message transfer.

Lipids as Chemical Messengers

All the multicellular organisms utilize lipids as chemical messengers to propel information among cells, tissues and organs. Since lipids are water insoluble and diminutive, they are exceptional contender for signaling. The esterified forms of lipids penetrate membranes and are elated to transmit signals to the adjacent cells. After encountering the suitable receptor, lipids remains idle till they arrive at action site and bind to certain proteins to accomplish the deed.

Energy Storage

TAGs are storage lipids, indolent and consist up of a glycerol moiety and three fatty acids. TAGs release FFAs or non-esterified form of FAs to outline the cell structural components. Dietary FAs of small chain and average size are non-esterified, oxidized swiftly in tissues to serve as a fuel source. Elongated FAs are first esterified to structural lipids or TAGs.

Temperature Safeguarding

Sub-cutaneous fat beneath the skin layers too assist in protection from cold and insulation. Brown fat as divergent to white one primarily maintains body temperature. Children boast a superior brown fat level.

Membrane Formation

Lipids of membrane are equipped with PUFAs, an essential constituent of the phospholipids that bestow numerous essential membrane properties. Fluidity and flexibility of the membrane are the most imperative assets. EFAs (Linoleic and linolenic) form arachidonic, docosahexaenoic and eicosapentaenoic acids, in part required for membrane lipid formation.

Cholesterol Formation

To a great extent, cholesterol is placed in the cell membranes; appear as plasma lipoprotein in the blood. Lipoproteins (lipids and proteins aggregates) facilitate the transport of lipids throughout the body and craft their movement feasible in an aqueous environment. Based on the relative densities, the major groups of lipoprotein are categorized as chylomicrons (CM), high density lipoproteins (HDL), low density lipoproteins (LDL) and very low density lipoproteins (VLDL). Phospholipids (Phosphatidylcholine and Sphingomyelin) interact with cholesterol and maintain the fluidity of membranes. Cholesterol, in addition, is also the predecessor of bile acids, steroidal hormones and vitamin D.

Role in Prostaglandin Formation and Inflammation

The EFAs (linolenic and linoleic acids) are predecessor of a lot of diverse types of eicosanoids such as prostanoids (thromboxanes, prostaglandins, prostacyclins), hydroxyeicosatetraenes, resolvins and leukotrienes (lipoxins) etc. that plays an imperative part in blood clotting, inflammation, fever and pain.

Regulating and Signaling

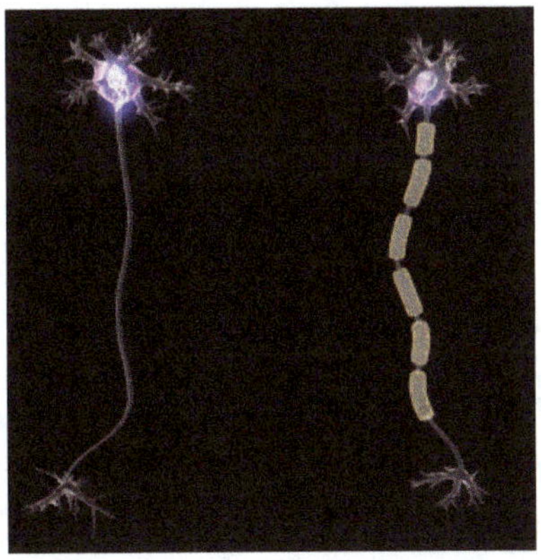

In figure: Lipids serve as signaling molecules; they are catalysts of electrical impulse activity within the brain. Nerve response in myelinated neurons (right) propagate faster than in unmyelinated neurons (left). Myelin is a mixture of proteins and phospholipids that insulates nerves. The myelin coating is ~70% lipids.

Triacylglycerols control the body's internal climate with the goal of maintaining constant temperature. Those who don't have enough fat in their bodies tend to feel cold sooner. Triacylglycerols also help the body produce and regulate hormones. For example, adipose tissue secretes the hormone leptin, which regulates appetite. In the reproductive system, fatty acids are required for proper reproductive health; women who lack proper amounts may stop menstruating and become infertile. Omega-3 and omega-6 essential fatty acids help regulate cholesterol and blood clotting and control inflammation in the joints, tissues, and bloodstream. Fats also play important functional roles in sustaining nerve impulse transmission, memory storage, and tissue structure. More specifically in the brain, lipids are focal to brain activity in structure and in function. They help form nerve cell membranes, insulate neurons, and facilitate the signaling of electrical impulses throughout the brain.

Insulating and Protecting

Within the body there are two types of fat - visceral and subcutaneous. Visceral fat surrounds vital organs such as the heart, kidneys, and liver. Subcutaneous fat, or fat underneath the skin, insulates the body from extreme temperatures and helps keep the internal climate under control. It pads our hands and buttocks and prevents friction, as these areas frequently come in contact with hard surfaces. It also gives the body the extra padding required when engaging in physically demanding activities such as ice- or roller skating, horseback riding, or snowboarding.

Transporting

Fat-soluble nutrients are important for good health and exhibit a variety of functions. Vitamins A, D, E, and K—the fat-soluble vitamins—are mainly found in foods containing fat. Fat-soluble nutrients require fat for effective absorption. These vitamins are best absorbed when combined with foods containing fat. Fats also increase the bioavailability of compounds known as phytochemicals, which are plant constituents such as lycopene (found in tomatoes) and beta-carotene (found in carrots). Phytochemicals are believed to promote health and well-being. As a result, eating tomatoes with olive oil or salad dressing will facilitate lycopene absorption. Other essential nutrients, such as essential fatty acids, are constituents of the fats themselves and serve as building blocks of a cell.

Role of Lipids in Food

High Energy Source

Fat-rich foods naturally have a high caloric density. Foods that are high in fat contain more calories than foods high in protein or carbohydrates. As a result, high-fat foods are a convenient source of energy. For example, 1 gram of fat or oil provides 9 calories compared with 4 calories found in 1 gram of carbohydrate or protein. Depending on the level of physical activity and on nutritional needs, fat requirements vary greatly from person to person. When energy needs are high, the body welcomes the high-caloric density of fats. For instance, infants and growing children require higher amounts of fat to support normal growth and development. If an infant or child is given a low-fat diet for an extended period, growth and development will not progress normally. Other individuals with high-energy needs are athletes, people who have physically demanding jobs, and those recuperating from illness.

When the body has used all of its calories from carbohydrates (which can occur after just twenty minutes of exercise), it initiates fat usage. A professional swimmer must consume large amounts of food energy to meet the demands of swimming long distances, so eating fat-rich foods makes sense. In contrast, if a person who leads a sedentary lifestyle eats the same fat-rich foods, they will likely get more fat calories than their body requires.

Smell, Taste, Texture and Satiety

Fat contains dissolved compounds that contribute to mouth-watering aromas and flavors. Fat also adds texture to food and helps keep baked foods moist. Frying foods locks in flavor and lessens cooking time.

Fat plays another valuable role in nutrition. Fat contributes to satiety, or the sensation of fullness. Fatty foods move slower through the digestive tract, thus promoting an overall sense of fullness. Oftentimes before the feeling of fullness arrives, people overindulge in fat-rich foods, finding the delectable taste irresistible. Indeed, the very things that make fat-rich foods attractive also make them a hindrance to maintaining a healthful diet.

Lipids and Vitamins

The fat soluble vitamins (A, D, E, K) are vital nutrients with plentiful purpose. The metabolism of FAs has been carried out by Acyl-carnitines involving inward and outward operation by mitochondria. Polyprenols and their derivatives (phosphorylated) facilitate the transfer of molecules through membranes. Cardiolipins (a sub-type of glycerophospholipids with three glycerol groups and four acyl chains) activate enzymes involved with oxidative phosphorylation.

Properties of Lipids

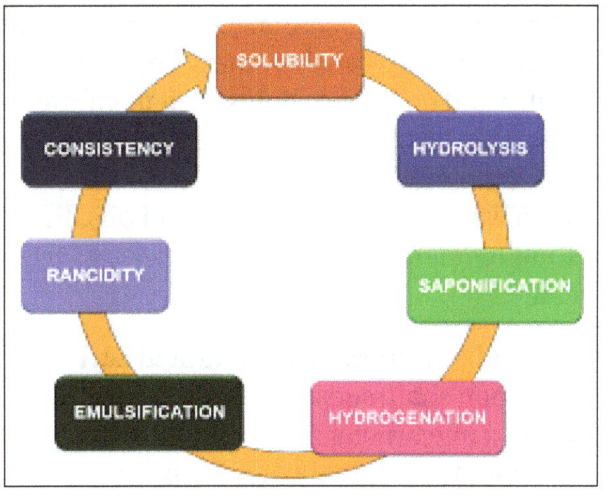

Lipids are a dissimilar group of biological compounds, but they are poised of a chain of hydrocarbons, ending with groups of bonded oxygen and hydrogen and share the general property of being at least partially not water soluble. They are essential to all forms of life and are the key constituent of every cell membrane on Earth. Some of key properties of lipids include energy storage for animals and other organisms while others are described below:

- Solubility: Lipids are soluble in non-polar solvent (ether, chloroform, benzene etc) but insoluble in water.
- Consistency: They are colorless, odorless and tasteless. Being lighter than water they have a specific gravity of 0.86. The consistency of lipids depends upon the presence of saturated and/or unsaturated fatty acids.
- Hydrolysis: It is brought about in presence of acids or alkalis under the activity of enzyme lipases. Acid hydrolysis results in the formation of glycerol and long chain of fatty acid whereas alkaline hydrolysis of fats results in the formation of sodium or potassium salts of fatty acids called as Soaps and the process is called as saponification.
- Hydrogenation: The conversion of liquid vegetable oil into solid vegetable ghee by reacting with hydrogen gas in the presence of catalyst (Ni) at 200°C.
- Emulsification: When fats or oils are rubbed with water, the large molecules of lipids breaked into smaller ones forming the emulsion and the process is called emulsification.
- Rancidity: When lipids are exposed to atmosphere (heat, light, air, moisture) for more than 30 days, an unusual and undesirable odour is developed. Such types of lipids are called rancid lipids and the phenomenon is termed as Rancidity. Rancidity is of two types: (a) Hydrolytic and (b) Oxidative.
    - Hydrolytic Rancidity: This type of phenomenon occurs due to liberation of volatile acids during the hydrolysis of fats.
    - Oxidative rancidity: When a part of unsaturated fatty acid oxidized to form aldehyde ketone, the phenomenon is termed as oxidative rancidity.

Fatty acids are basic units of lipids with hydrocarbon chains amid oxygen-hydrogen ends, making it hydrophilic at one end and hydrophobic at the other. That is why soaps made by processing accustomed lipids soften in water and emulsifies lipids. Cell membranes are made from phospholipids bilayer with hydrophobic ends facing the cell's externally and internally but the hydrophilic ends in front of each other within the membrane. In most of the lipids, polar hydrophilic areas are veiled and do not dissolve in water at all. Among fats and oils, the lone unswerving difference between fats and oils is melting temperature as both the types of triglycerides contains glycerol bonded to three fatty acid chains.

Besides the frequent fatty acids with unsaturated or saturated acyl chains, chemically several fatty acids are customized by hydrocarbon chain substituent e.g. the tidy duck gland secretes 10 carbons long fatty acid with methyl groups on carbons 2, 4, 6, and 8 replaced with one of the hydrogens. Few bacteria make FAs with a methyl group on the last but one carbon from the acidic group. Other bacteria integrate a cyclopropane ring in the vicinity of the acyl chain. Mycobacterium tuberculosis synthesizes α-mycolic acids, a member of cyclopropane-containing fatty acids. Monohydroxyl acids are another frequent component of a hydroxyl group, present in both animals and plants in minuscule quantity, but they are ubiquitous in bacteria.

Physical Properties of Lipids

The chaste FAs form crystals set up of stacked layers of molecules. The thickness of each layer of two extended molecules arranged in such a way that the water loving (hydrophilic) carboxylic acid groups shape the two faces and the water-fearing (hydrophobic) chains outline the inner core. In case of specific FAs, the specifics of the molecular packing might differ in forming different crystal configuration termed polymorphs. The SFAs of biological significance have melting temperatures exceeding 27° C (81° F) and rise with escalating hydrocarbon chain length. Polyunsaturated and monounsaturated molecules liquefy at considerably lower temperatures than their saturated moiety. In most of the biological molecules, the melting temperatures becomes lowest when the C=C are positioned in close proximity to the centre of the hydrocarbon chain. At room temperature, these molecules form viscous liquids and the hydrophobic nature of the majority of FAs surpasses the hydrophilic character of the carboxylic acid, creating the aqueous solubility of these molecules very squat. At 77° F (25° C), the solubility of FA is 3×10^{-6} per gram of solution and the aqueous solubility lessens rampantly with the accumulation of each carbon units to the hydrocarbon chain. This association imitates the energy necessary to transmit the molecule to water from a pure hydrocarbon solvent. With every CH_2 group, for example, additional energy is obligatory to organize water molecules of FA in the region of the hydrocarbon chain resulting in the hydrophobic effect.

The carboxylate group separates a positively charged hydrogen ion in pure water to a very small degree thus representing the equation: $R-COOH \rightarrow RCOO^{-} + H^{+}$.

Where R signify the hydrocarbon chain.

The negative charge bearing carboxylate ion is more polar than the undissociated acid. RCOOH can be changed to $RCOO^-$ with the addition of equal number of base molecules (e.g. NaOH). In fact, it reinstates the Na^+ with H^+ to provide the FA salt refereed as soap. The $RCOO^-$ anions in water impulsively shape stable, spherical aggregates known as micelles that dole out the useful detergent property of soaps. The micelle diameter is more or less twice the length of the unmitigated FA. Concentrated preparations of micelles-water dispersions display immense cleansing power. Usually look like pure water, these dispersions are quite stable. Foams and bubbles on the exterior surface of soap dispersions results due to spontaneous adsorption of $RCOO^-$ ions at the edge between the air and aqueous dispersion with the mechanically stretched and vigorously stabilized effect between the air-water interfaces.

Chemical Properties

Chemically, the acidic carboxyl group (COOH) of the fatty acid is the most reactive portion. It reacts with alcohols (R′OH) to yield esters (RCOOR′) releasing water molecule. In complex lipids, the ester (principal covalent) bond link FA moieties to other groups. Ether (R′–O–R, second chemical) bond also links FAs and are chemically more stable as compared to ester bond.

In fatty acids, the hydrocarbon part is relatively defiant to chemical assail except C=C and different molecules respond with similar double bond. For instance, when platinum is present as a catalyst, hydrogen append to the double bond to bestow a SFA. Halogens (Iodine, Chlorine, Bromine) and their derivatives (hydroiodic acid) also retort with the double bond to structure the SFAs, but herein the halogen (one or two atoms) swap the hydrogen atoms (one or two) usually endowed in the saturated acyl chain. C=C can also respond with oxygen in either enzymatically catalyzed oxidation reactions or non-enzymatic processes. The procedure engenders a range of products, several of which add to the stale smell in vegetable products and spoiled meat. Universally, it is known that highly unsaturated fatty acid is more easily oxidized.

Properties of lipids explain its consistency, solubility, hydrolysis, saponification, emulsification and rancidity. Based on the chemical and physical properties, lipids undergo many reactions which are described below:

Reactions of Fatty Acids

Alike other carboxylic acids, fatty acids demonstrate some of the key reactions i.e. they undergo acid-base reactions and esterification.

Acid-Base Reactions

pH, Buffers, Acids and Bases

Acids dissociate into H^+ and lower pH, while bases dissociate into OH^- and raise pH; buffers can absorb these excess ions to maintain pH.

Self-Ionization of Water

Hydrogen ions are spontaneously generated in pure water by the dissociation (ionization) of a small percentage of water molecules into equal numbers of hydrogen (H^+) ions and hydroxide (OH^-) ions. The hydroxide ions remain in solution because of their hydrogen bonds with other water molecules; the hydrogen ions, consisting of naked protons, are immediately attracted to un-ionized water molecules and form hydronium ions (H_3O^+). By convention, scientists refer to hydrogen ions and their concentration as if they were free in this state in liquid water.

$$2H_2O \leftrightharpoons H_3O^+ + OH^-$$

The concentration of hydrogen ions dissociating from pure water is 1×10^{-7} moles H^+ ions per litre of water. The pH is calculated as the negative of the base 10 logarithm of this concentration:

$$pH = -\log[H^+]$$

The negative log of 1×10^{-7} is equal to 7.0, which is also known as neutral pH. Human cells and blood each maintain near-neutral pH.

pH Scale

The pH of a solution indicates its acidity or basicity (alkalinity). The pH scale is an inverse logarithm that ranges from 0 to 14: anything below 7.0 (ranging from 0.0 to 6.9) is acidic, and anything above 7.0 (from 7.1 to 14.0) is basic (or alkaline). Extremes in pH in either direction from 7.0 are usually considered inhospitable to life. The pH in cells (6.8) and the blood (7.4) are both very close to neutral, whereas the environment in the stomach is highly acidic, with a pH of 1 to 2.

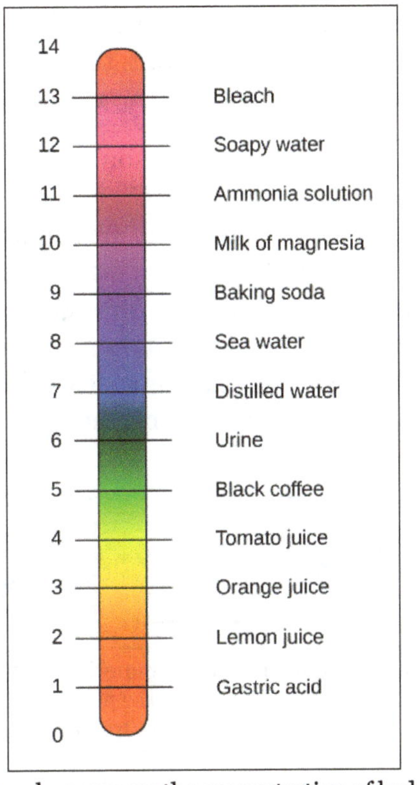

Figure: The pH scale: The pH scale measures the concentration of hydrogen ions (H^+) in a solution.

Non-neutral pH readings result from dissolving acids or bases in water. Using the negative logarithm to generate positive integers, high concentrations of hydrogen ions yield a low pH and low concentrations a high pH.

An acid is a substance that increases the concentration of hydrogen ions (H^+) in a solution, usually by dissociating one of its hydrogen atoms. A base provides either hydroxide ions (OH^-) or other negatively-charged ions that react with hydrogen ions in solution, thereby reducing the concentration of H^+ and raising the pH.

Strong Acids and Strong Bases

The stronger the acid, the more readily it donates H^+. For example, hydrochloric acid (HCl) is highly acidic and completely dissociates into hydrogen and chloride ions, whereas the acids in tomato juice or vinegar do not completely dissociate and are considered weak acids; conversely, strong bases readily donate OH^- and/or react with hydrogen ions. Sodium hydroxide (NaOH) and many household cleaners are highly basic and give up OH^- rapidly when placed in water; the OH^- ions react with H^+ in solution, creating new water molecules and lowering the amount of free H^+ in the system, thereby raising the overall pH. An example of a weak basic solution is seawater, which has a pH near 8.0, close enough to neutral that well-adapted marine organisms thrive in this alkaline environment.

Buffers

How can organisms whose bodies require a near-neutral pH ingest acidic and basic substances (a human drinking orange juice, for example) and survive? Buffers are the key. Buffers usually consist of a weak acid and its conjugate base; this enables them to readily absorb excess H^+ or OH^-, keeping the system's pH within a narrow range.

Maintaining a constant blood pH is critical to a person's well-being. The buffer that maintains the pH of human blood involves carbonic acid (H_2CO_3), bicarbonate ion (HCO_3^-), and carbon dioxide (CO_2). When bicarbonate ions combine with free hydrogen ions and become carbonic acid, hydrogen ions are removed, moderating pH changes. Similarly, excess carbonic acid can be converted into carbon dioxide gas and exhaled through the lungs; this prevents too many free hydrogen ions from building up in the blood and dangerously reducing its pH; likewise, if too much OH^- is introduced into the system, carbonic acid will combine with it to create bicarbonate, lowering the pH. Without this buffer system, the body's pH would fluctuate enough to jeopardize survival.

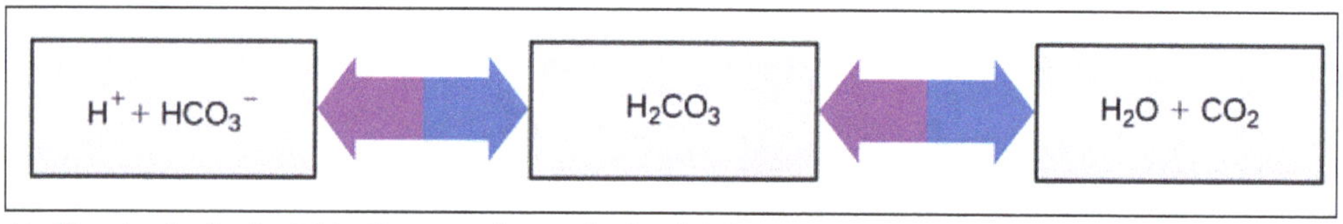

Figure: Buffers in the body: This diagram shows the body's buffering of blood pH levels: the blue arrows show the process of raising pH as more CO_2 is made; the purple arrows indicate the reverse process, lowering pH as more bicarbonate is created.

Antacids, which combat excess stomach acid, are another example of buffers. Many over-the-counter medications work similarly to blood buffers, often with at least one ion (usually carbonate) capable of absorbing hydrogen and moderating pH, bringing relief to those that suffer "heartburn" from stomach acid after eating.

Brønsted Acids and Bases

A Brønsted acid is any species capable of donating a proton; a Brønsted base is any capable of accepting a proton.

In chemistry, the Brønsted-Lowry theory is an acid-base theory, independently proposed by Johannes Nicolaus Brønsted and Thomas Martin Lowry in 1923. In this system, acids and bases are defined as follows: an acid is any species that is able to donate a hydrogen cation (H^+, a proton); a base is any species with the ability to accept a hydrogen cation (H^+). To that end, if a compound is to behave as an acid by donating a proton, there must be a base to accept that proton; the Brønsted-Lowry concept is therefore defined by the reaction:

$$\text{Acid} + \text{base} \rightleftharpoons \text{conjugate base} + \text{conjugate acid}$$

The conjugate base is the ion or molecule that remains after the acid has donated its proton, and the conjugate acid is the species created after the base accepts the proton. The reaction can proceed either forward backward; in each case, the acid donates a proton to the base.

Advantages of Brønsted-Lowry Theory

The Brønsted-Lowry acid-base theory has several advantages over the Arrhenius theory. Recall that the Arrhenius theory defines an acid as any species that increases the concentration of

H^+/H_3O^+ in solution. Consider the following reactions of acetic acid he organic acid that gives vinegar its characteristic taste:

$$CH_3COOH + H_2O \rightleftharpoons CH_3COO^- + H_3O^+$$

$$CH_3COOH + NH_3 \rightleftharpoons CH3COO^- + NH_4^+$$

Both theories easily CH_3COOH describe the first reaction: acts as an Arrhenius acid because it acts as a source of H_3O^+ when dissolved in water, and it acts as a Brønsted acid by donating a proton to water. In the second example CH_3COOH undergoes the same transformation, in this case donating a proton to ammonia (NH_3); this cannot be described using the Arrhenius definition of an acid, however, because the reaction does not produce H_3O^+.

Amphoterism of Water

Water is amphoteric, which means it can act as either an acid or a base. In the reaction between acetic acid, CH_3CO_2H, and water, H_2O, water acts as a base. The acetate ion $CH_3CO_2^-$ is the conjugate base of acetic acid, and the hydronium ion H_3O^+ is the conjugate acid of the base, water:

$$CH_3COOH + H_2O \rightleftharpoons CH_3COO^- + H_3O^+$$

Water can also act as an acid, as when it reacts with ammonia. The equation given for this reaction is:

$$H_2O + NH_3 \rightleftharpoons OH^- + NH_4^+$$

Here, H_2O donates a proton to NH_3. The hydroxide ion is the conjugate base of water, which acts as an acid, and the ammonium ion is the conjugate acid of the base, ammonia.

Acid-Base Titrations

Acid-base titration can determine the concentrations of unknown acid or base solutions.

Setting up an Acid-Base Titration

An acid-base titration is an experimental procedure used to determine the unknown concentration of an acid or base by precisely neutralizing it with an acid or base of known concentration. This lets us quantitatively analyze the concentration of the unknown solution. Acid-base titrations can also be used to quantify the purity of chemicals.

The solution in the flask contains an unknown number of equivalents of base (or acid). The burette is calibrated to show volume to the nearest 0.001 cm^3. It is filled with a solution of strong acid (or base) of known concentration. Small increments are added from the burette until, at the end point, one drop changes the indicator color permanently. (An indication of the approaching equivalence point is that the indicator changes color but changes back after stirring.) At the equivalence point, the total amount of acid (or base) is recorded from the burette readings. The number of equivalents of acid and base must be equal at the equivalence point.

Alkalimetry, or alkimetry, is the specialized analytic use of acid-base titration to determine the concentration of a basic (alkaline) substance; acidimetry, or acidometry, is the same concept applied to an acidic substance.

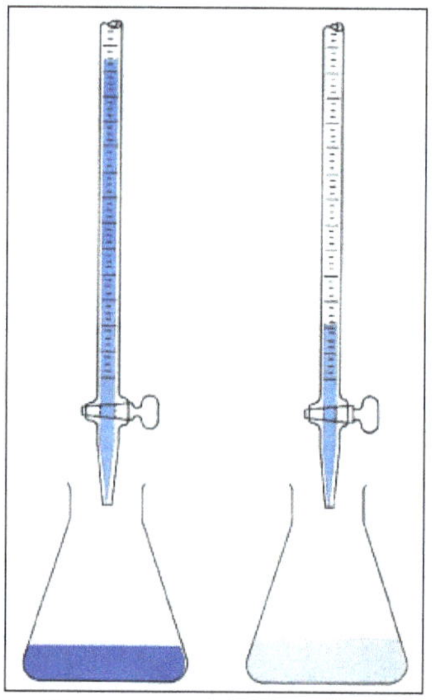

Figure: Acid-base titration. (Left): At start of titration and (Right): At equivalence point.

Materials for a Titration Procedure

- Burette.
- White tile (used to see a color change in the solution).
- Pipette.
- pH indicator (the type depends on the reactants).
- Erlenmeyer or conical flask.
- Titrant (a standard solution of known concentration; a common example is aqueous sodium carbonate).
- Analyte, or titrand (the solution of unknown concentration).

Equivalence Point Indicators

Before you begin the titration, you must choose a suitable pH indicator, preferably one that will experience a color change (known as the "end point") close to the reaction's equivalence point; this is the point at which equivalent amounts of the reactants and products have reacted. Below are some common equivalence point indicators:

- Strong acid - Strong base titration: Phenolphthalein indicator.
- Weak acid - Weak base titration: Bromthymol blue indicator.
- Strong acid-weak base titration: Methyl orange indicator the base is off the scale (e.g., pH > 13.5) and the acid has pH > 5.5: alizarine yellow indicator.

- The base is off the scale (e.g., pH > 13.5) and the acid has pH > 5.5: Alizarine yellow indicator.
- The base is off the scale (e.g., pH > 13.5) and the acid has pH > 5.5: Alizarine yellow indicator.
- The acid is off the scale (e.g., pH < 0.5) and the base has pH < 8.5: Thymol blue indicator.

Figure: Acid-base titration setup. The pink color is caused by the phenolphthalein indicator.

Estimating the Equivalence Point's pH

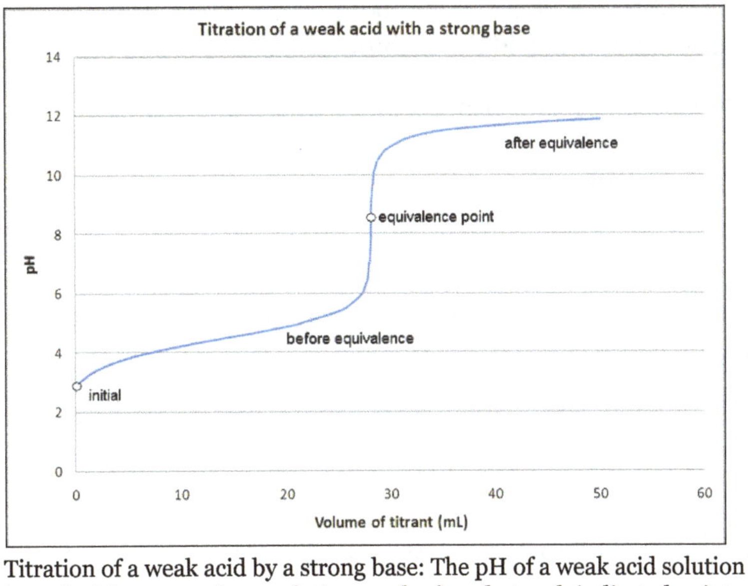

Figure: Titration of a weak acid by a strong base: The pH of a weak acid solution being titrated with a strong base solution can be found at each indicated point.

The resulting solution at the equivalence point will have a pH dependent on the acid and base's relative strengths. You can estimate the equivalence point's pH using the following rules:

- A strong acid will react with a weak base to form an acidic (pH < 7) solution.

- A strong acid will react with a strong base to form a neutral (pH = 7) solution.
- A weak acid will react with a strong base to form a basic (pH > 7) solution.

When a weak acid reacts with a weak base, the equivalence point solution will be basic if the base is stronger and acidic if the acid is stronger; if both are of equal strength, then the equivalence pH will be neutral. Weak acids are not often titrated against weak bases, however, because the color change is brief and therefore very difficult to observe.

You can determine the pH of a weak acid solution being titrated with a strong base solution at various points; these fall into four different categories: (1) initial pH; (2) pH before the equivalence point; (3) pH at the equivalence point; and (4) pH after the equivalence point.

Titration Procedure

- Rinse the burette with the standard solution, the pipette with the unknown solution, and the conical flask with distilled water.
- Place an accurately measured volume of the analyte into the Erlenmeyer flask using the pipette, along with a few drops of indicator. Place the standardized solution into the burette, and indicate its initial volume in a lab notebook. At this stage, we want a rough estimate of the amount of known solution necessary to neutralize the unknown solution. Let the solution out of the burette until the indicator changes color, and record the value on the burette. This is the first titration and it is not very precise; it should be excluded from any calculations.
- Perform at least three more titrations, this time more accurately, taking into account where the end point will roughly occur. Record the initial and final readings on the burette, prior to starting the titration and at the end point, respectively. (Subtracting the initial volume from the final volume will yield the amount of titrant used to reach the endpoint.).
- The end point is reached when the indicator permanently changes color.

Gas Evolution Reactions

A gas evolution reaction is a chemical process that produces a gas, such as oxygen or carbon dioxide. A gas evolution reaction is a chemical process that produces a gas, such as oxygen or carbon dioxide. In the following examples, an acid reacts with a carbonate, producing salt, carbon dioxide, and water, respectively.

Nitric acid reacts with sodium carbonate to form sodium nitrate, carbon dioxide, and water:

$$2HNO_3(aq) + Na_2CO_3(aq) \rightarrow 2NaNO_3(aq) + CO_2(g) + H_2O(l)$$

Sulfuric acid reacts with calcium carbonate to form calcium sulfate, carbon dioxide, and water:

$$H_2SO_4(aq) + CaCO_3(aq) \rightarrow CaSO_4(aq) + CO_2(g) + H_2O(l)$$

Hydrochloric acid reacts with calcium carbonate to form calcium chloride, carbon dioxide, and water:

$$2HCl(aq) + CaCO_3(aq) \rightarrow CaCl_2(aq) + CO_2(g) + H_2O(l)$$

The following setup demonstrates this type of reaction:

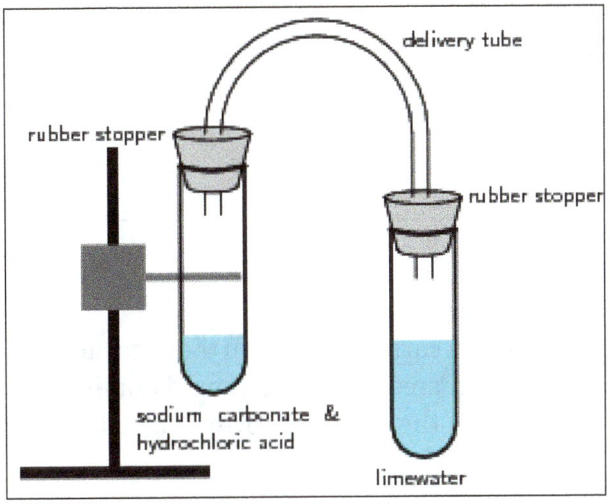

Figure: Reaction of acids with carbonates.

In this reaction setup, lime water is poured into one of the test tubes and sealed with a stopper. A small amount of hydrochloric acid is carefully poured into the remaining test tube. A small amount of sodium carbonate is added to the acid, and the tube is sealed with a rubber stopper. The two tubes are connected. As a result of the acid-carbonate reaction, carbon dioxide is produced and the lime water turns milky.

The test tube on the right contains limewater (a solution of calcium hydroxide, $Ca(OH)_2$). On the left, a solution of hydrochloric acid has been added to a solution of sodium carbonate to generate $CO_2(g)$. The test tubes are sealed with rubber stoppers and connected with a delivery tube. As the reaction proceeds, the limewater on the right turns from clear to milky; this is due to the $CO_2(g)$ reacting with the aqueous calcium hydroxide to form calcium carbonate, which is only slightly soluble in water.

Oxidation of Metals in Acidic Solution

The oxidation of a metal in acidic solution will yield a metal salt and hydrogen gas:

$$2HCl_{(aq)} + Zn_{(s)} \rightarrow ZnCl_{2(aq)} + H_{2(g)}$$

Acid and metal reaction: Hydrochloric acid oxidizes zinc to produce an aqueous metal salt and hydrogen gas bubbles.

Recall that oxidation refers to a loss of electrons, and reduction refers to the gain of electrons. In the above redox reaction, neutral zinc is oxidized to Zn^{2+}, and the acid, H^+, is reduced to $H_2(g)$. The oxidation of metals by strong acids is another common example of a gas evolution reaction.

Esterification

Esterification is the process of combining an organic acid (RCOOH) with an alcohol (ROH) to form an ester (RCOOR) and water; or a chemical reaction resulting in the formation of at least one ester product.

Ester is obtained by an esterification reaction of an alcohol and a carboxylic acid. The chemical reaction for esterification is given below.

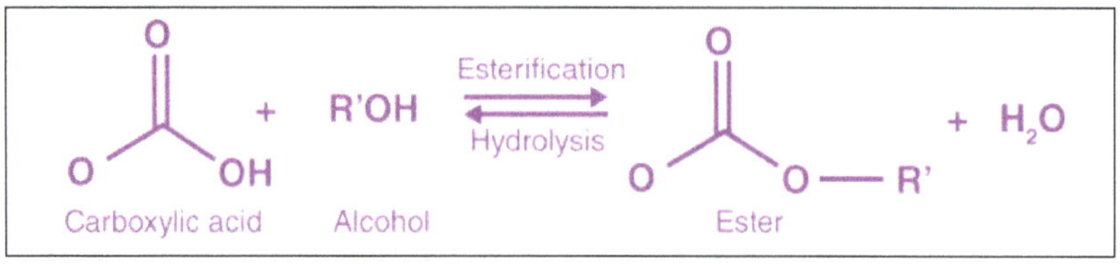

Esterification Reaction

When primary alcohol is treated with a carboxylic acid in the presence of sulphuric acid a compound is formed. This compound has a sweet smell. The compound obtained is called ester. The chemical reaction occurring in the formation of the ester is known as an esterification reaction:

$$CH_3COOH + CH_3CH_2COOH \rightarrow CH_3COOCH_2CH_3$$

Esterification Mechanism

This process involves five steps. We have discussed the steps below:

Step 1: Cation formation.

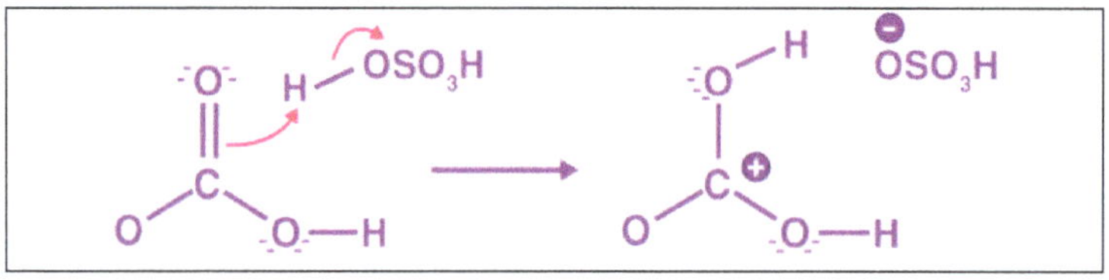

Step 2: Delocalized carbocation: Carboxyl oxygen gets protonated to give delocalized carbocation making the carbocation a better electrophile.

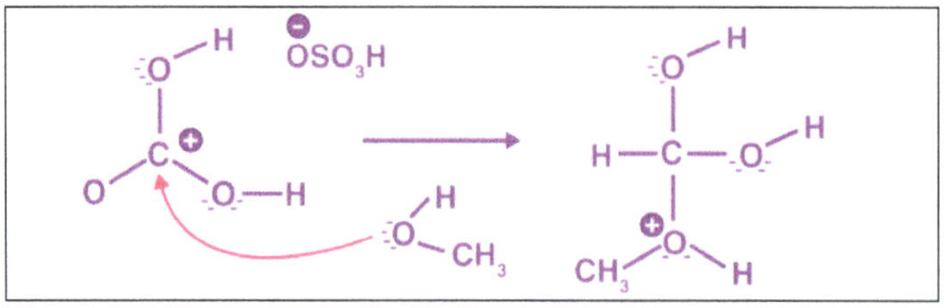

Step 3: Transfer of proton: A proton is transferred to one of the hydroxyl groups to form a good leaving group.

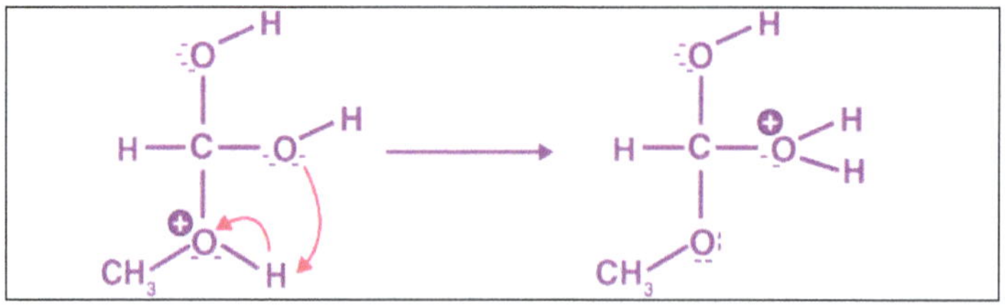

Step 4: Formation of the pi bond: The hydroxy group's alcohol oxygen atom donates a pair of

electrons to a carbon atom which makes a π bond by eliminating water. The concentration of water is less than methanol; therefore, it is not a feasible nucleophile to reverse the reaction.

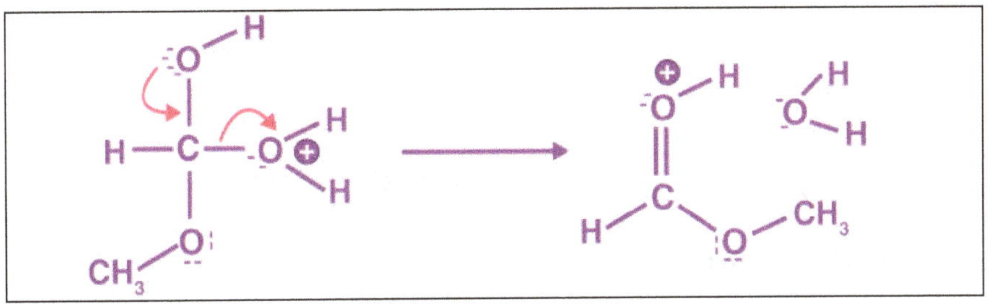

Step 5: Ester formation.

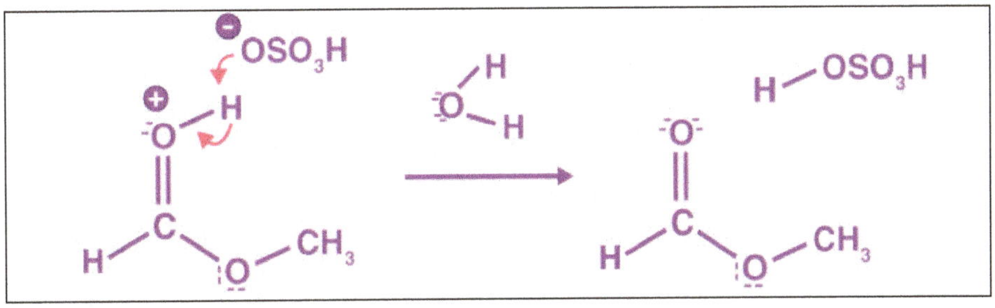

Esterification Methods

Esterification can happen in three ways. They are discussed below:

- From acid anhydride and alcohol.
- From acid chloride and alcohol.
- From carboxylic acid and alcohol.

Acid Anhydride and Alcohol

The reaction between alcohol and acid anhydride is comparatively slower when compared to acid chloride. To get a number of esters it is required to warm the mixture. For instance, consider 2,6-diiodophenol.

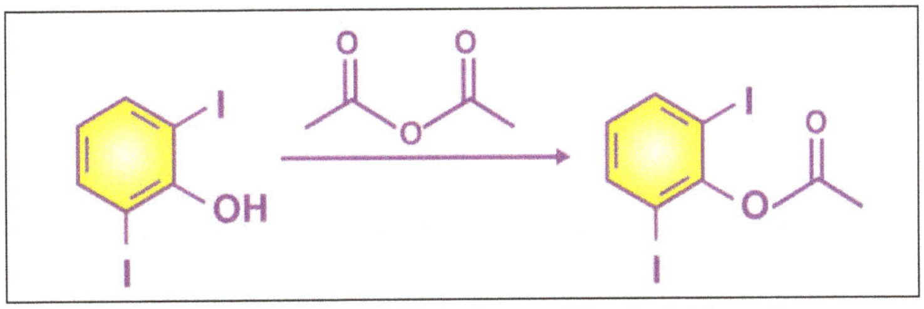

Figure: 2,6-diiodophenol reacts with an acid anhydride to form ester.

Acid Chloride and Alcohol

Esterification process can also be performed with alcohol and acid chloride at room temperature. Ester is obtained with steamy acidic fumes of hydrogen chloride. For instance, benzoyl chloride.

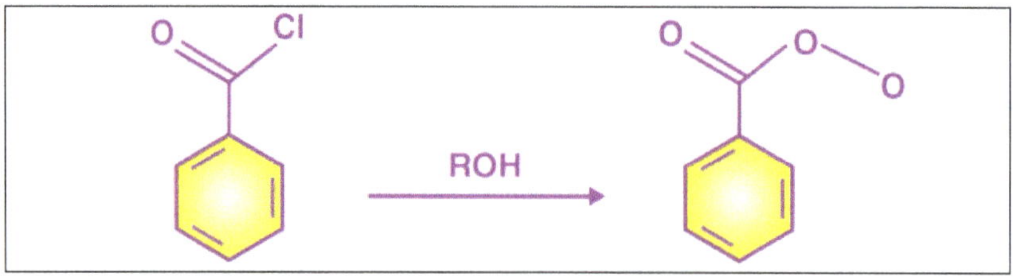

Figure: Reaction of alcohol and benzoyl chloride to form ester.

Carboxylic Acid and Alcohol

Production of esters from carboxylic acid and alcohol. Heat them in the presence of acid catalyst such as sulphuric acid (H_2SO_4) is used as a catalyst. For example, reaction of ethanoic acid and propanol to form propyl-ethanoate and water.

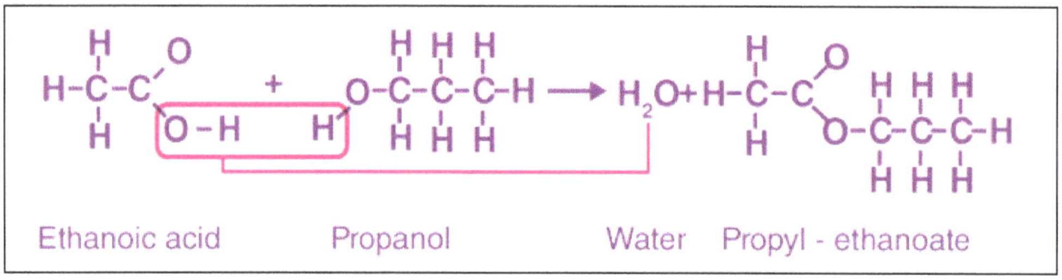

Properties of Esters

- The name of an ester is derived from its carboxylic acid that takes part in the esterification reaction.
- They have a pleasant smell.
- They have a wide application in the perfume and food industries.
- Esters are organic compounds found in oils and fats.

Uses of Esters

Some common uses of esters are mentioned below:

- Esters that have fragrance are used in perfumes, food flavourings, and cosmetics.
- Used as an organic solvent.
- Nitroglycerin is known and famous for its explosive properties.
- In the manufacturing of surfactants like detergents and soaps.
- Natural occurring esters are present in pheromones.
- The backbone of DNA molecules is formed by phospo esters.
- Polyesters are used in the production of plastics.

Acidity

The pK_a of FAs do not illustrate a vast disparity in their acidities. For instance, nonanoic acid has a pKa of 4.96, to some extent weaker than acetic acid (pK_a 4.76). In water, the solubility of the FAs decreases very rapidly due to increase in the chain length, so the longer chain fatty acids have least effect on aqueous solution pH. On the other hand water insoluble FAs liquefy in warm ethanol can be titrated by NaOH solution using phenolphthalein as an indicator with a pink endpoint. This testing is utilized to ascertain the FFA content of fats i.e. the hydrolyzed amount of the triglycerides.

Hardening and Hydrogenation

Hydrogenation is the phenomenon extensively accomplished for unsaturated fatty acids to give SFAs which are not as much of decumbent to rancidification. Since the UFAs have inferior melting than the SFAs, the procedure is called hardening. This knowledge and technology is employed to make margarine from vegetable oils. During partial hydrogenation, the isomerization of UFAs from cis to trans configuration takes place. Use of high temperature and high pressure (forcing hydrogenation) led to formation of fatty alcohols from fatty acids. However, fatty acid esters convert to fatty alcohols more simply. Certain UFAs are cleaved in molten alkali in the Varrentrapp reaction (structure elucidation from classical significance).

Rancidity and Auto Oxidation

Auto oxidation is a chemical change endured by UFAs. The process gain momentum by the charisma of trace metals and requires oxygen (air). Oils of vegetable origin defy this practice as they restrain antioxidants (tocopherol). Oils and fats and oils are repeatedly are treated with citric acid (a chelating agents) to eliminate the metal catalysts.

Ozonolysis

Ozonolysis refers to the organic chemical reaction where ozone is employed to cleave the unsaturated bonds of alkenes, alkynes, and azo compounds (compounds with the functional diazenyl functional group). It is an organic redox reaction.

- Oxidation of alkenes with the help of ozone can give alcohols, aldehydes, ketones, or carboxylic acids.
- Alkynes undergo ozonolysis to give acid anhydrides or diketones. If water is present in the reaction, the acid anhydride undergoes hydrolysis to yield two carboxylic acids.
- Ozonolysis of elastomers is also known as ozone cracking. Trace amounts of ozone gas in the atmosphere cuts the double bonds in elastomers.
- For azo compounds, the ozonolysis yields nitrosamines.

Ozonolysis Reaction

Ozone is a very reactive allotrope of oxygen. The reaction of ozone with alkenes and alkynes causes the oxidative cleaving of the alkene or alkyne. The carbon-carbon triple bonds are replaced with carbon-oxygen double bonds, giving the required carbonyl product as shown below:

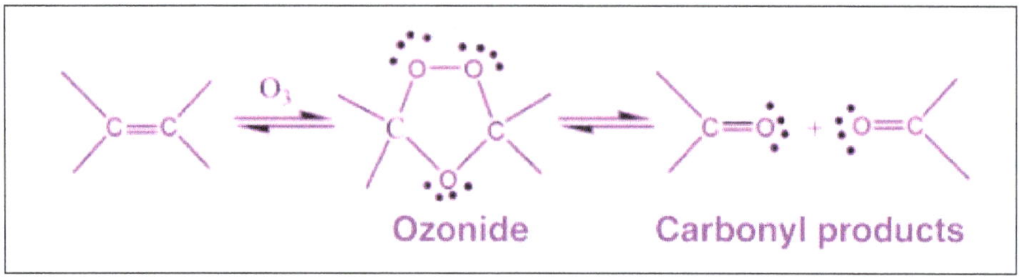

Ozonolysis Mechanism

Ozonolysis mechanism proceeds via an oxidative cleavage reaction. The ozone not only breaks the carbon pi bond but also the carbon-carbon sigma bond. It involves the attack of ozone on the given reactant to form an ozonide. To eliminate the oxygen in this intermediate stage, zinc dust is employed (since it forms zinc oxide with the oxygen). The final product varies based on the type of reactant and the workup. Given below is an illustration depicting the general mechanism of the ozonolysis reaction for alkenes and alkynes.

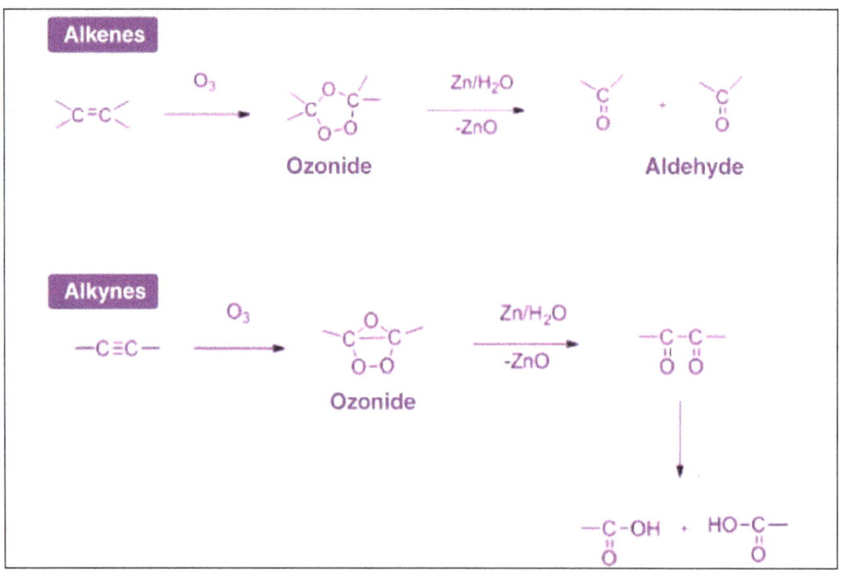

Ozonolysis of Alkenes

Alkenes can undergo ozonolysis to form alcohols, aldehydes, ketones, or carboxylic acids. The general procedure uses a solution of alkene in methanol. Ozone is bubbled through this solution at approximately 780 Celsius. When the solution turns blue, the alkene is consumed (the blue colour comes from the unreacted ozone). Other indicators of the endpoint of the reaction include potassium iodide solution. For this indicator, a stream of ozone enriched oxygen is bubbled through a mixture of the reactants. The gas that bubbles out is channeled through a solution of potassium iodide. Once the reaction is complete and there is no more alkene left to react with the ozone, the gas goes on to oxidize potassium iodide to iodine. The reaction of potassium iodide with ozone to give iodine is indicated by the distinct violet colour of iodine.

Once the ozone is added to the reaction mixture, a reagent must be added to convert the ozonide to the required carbonyl derivative. For this conversion, 2 techniques can be employed:

- Reductive Workup
- Oxidative Workup

However, reductive workup conditions see a lot more use when compared to oxidative workup conditions. In these workup conditions, triphenylphosphine, thiourea, zinc dust, and dimethyl sulfide can be used to produce aldehydes or ketones. On the other hand, hydrogen peroxide can be used to produce carboxylic acids. Ozone can also be employed to oxidize other functional groups such as benzyl ethers. Pyridine is also used to buffer the reaction since small amounts of acid may be generated at the time of reaction. A 1:1 ratio of solvent with dichloromethane co-solvent facilitates the timely cleavage of the ozonide. Ozonolysis of oleic acid is used to produce azelaic acid and pelargonic acid on an industrial scale.

Ozonolysis of Alkenes Mechanism

Step 1: The electrophilic addition of ozone to the carbon-carbon bond forms the molozonide intermediate which is quite unstable. Due to this unstable nature, the molozonide continues reacting – breaking apart to form a carbonyl molecule and a carbonyl oxide molecule as shown below:

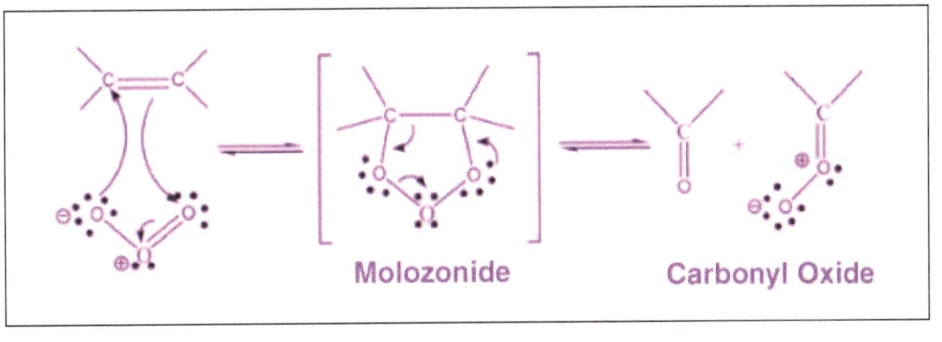

Step 2: The carbonyl molecule and the carbonyl oxide molecule formed in step 1 rearrange themselves, reforming to create a more stable ozonide intermediate. This ozonide intermediate can be subjected to either an oxidative workup or a reductive workup. The oxidative workup will give carboxylic acid as the product whereas the reductive workup will yield aldehydes or ketones.

The formation of the ozonide intermediate is illustrated below:

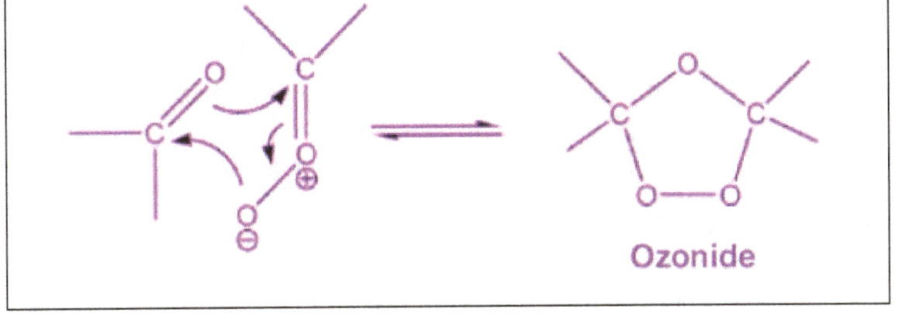

Oxidative Workup

When oxidant hydrogen peroxide is used instead of zinc or dimethyl sulfide to treat the ozonide, the aldehydes formed are oxidized to carboxylic acids. Potassium Permanganate in the presence of hot acid can also be used in the oxidative workup.

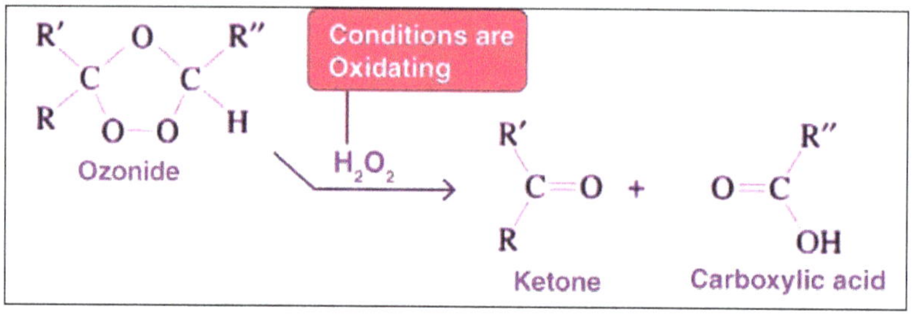

Reductive Workup

Here, the ozonide is treated with mild reducing agents such as dimethyl sulfide and zinc metal with water. The ozonide is reduced as shown below:

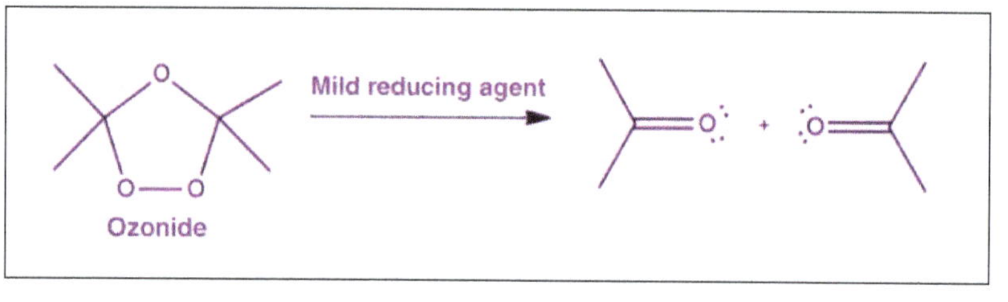

Ozonolysis of Alkynes

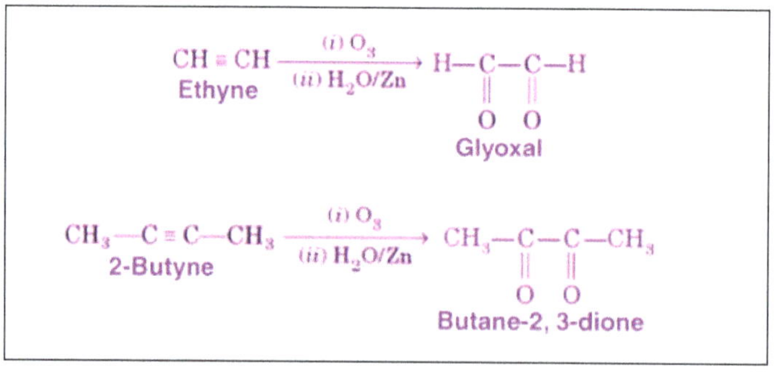

Alkynes undergo ozonolysis to give an acid anhydride or a diketone as the final product. The fragmentation is not complete in this reaction (alkenes undergo complete fragmentation). No reducing agents are required as a simple aqueous workup is followed. If the reaction happens in the presence of water, the acid anhydride undergoes hydrolyzation to give two carboxylic acids. Ozonolysis can also be used to determine the position of the triple bond in an unknown alkyne. A few examples of alkyne ozonolysis are provided below:

Ozonolysis of Alkynes Mechanism

The alkyne reacts with the ozone, causing the breakage of the alkyne. This gives rise to the ozonide intermediate. A simple aqueous workup follows with the help of zinc metal, finally yielding a dicarbonyl compound. The reaction can be written as:

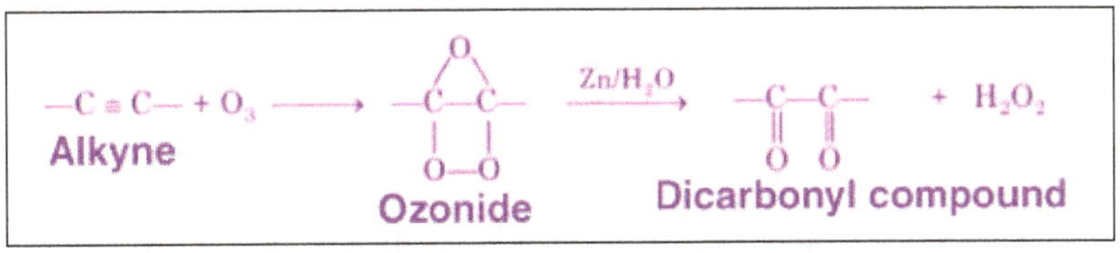

Ozonolysis of Elastomers – Ozone Cracking

The attack of ozone (present in trace amounts in the atmosphere) on the elastomer creates cracks. The double bond in the rubber chain is attacked by the ozone here. If the rubber product is under tension, ozone cracks begin to form. The cracks can appear around the circumference of a bent rubber tube since they are oriented at right angles to the axis of strain. These cracks are extremely dangerous when they occur in fuel pipes, growing inward from the exposed exterior surface. This can even lead to fuel leakage and fires. Thus, ozone can be employed to cleave unsaturated bonds in alkenes, alkynes, and azo compounds. This cleaving is used in the organic chemical reaction called ozonolysis.

Analysis

Chemically, the methyl esters of FAs are alienated by gas chromatography and furthermore, the unsaturated isomers separation is achieved by TLC. Some of the well-known examples of lipids are as under:

- Fatty acids: Palmitoleic, Oleic, Linoleic and Arachidonic acid.
- Fats & Oils: Sunflower, Peanut, Palm, Lard, Herring oil, Human fat, Coconut oil, Corn, Butter, Animal fats.
- Waxes: Beeswax, Carnauba wax, Spermacti.
- Phospholipids: Cephalins, Lecithins, Plasmalogens, Phosphatidyl inositols, Sphingomyelins.
- Glycolipids: Phrenosin, Oxynervon, Nervon, Kerasin.
- Steroids: C 21, C 24, C 27, C 28, C 29 steroids.
- Terpenes: Triterpenes, Sesquiterpenes, Diterpenes, Monoterpenes.
- Carotenoids: Xanthophylls, Lycopene, Carotenes.

References

- Triglycerides-heart-health, including-too-much-fat-in-eat-many-reduced-fat-foods-17583: my.clevelandclinic.org, Retrieved 22, March 2020
- Introduction-to-Nutrition-(Diker), functions-of-Lipids, Metropolitan-State-University-of-Denver: med.libretexts.org, Retrieved 17, July 2020
- Acid-base-reactions: courses.lumenlearning.com, Retrieved 05, April 2020
- Esterification-is-the-process-of-alcohol-and-a-carboxylic-acid: byjus.com, Retrieved 11, January 2020
- Ozonolysis-of-alkenes-alkynes-mechanism: byjus.com: Retrieved 27, May 2020

2

Types of Lipids

There are various types of lipids such as simple lipids, complex lipids, and precursor and derived lipids. Simple lipids are the fatty acid esters of different alcohols which do not contain any other substance whereas complex lipids contain three or more chemical identities. This chapter has been carefully written to provide an easy understanding of these various types of lipids.

There are numerous specific types of lipids important to live, including fatty acids, triglycerides, glycerophospholipids, sphingolipids, and steroids. These are broadly classified as simple lipids and complex lipids.

Simple Lipids

The main simple lipids are triglycerides (also known as triacylglycerols), steryl esters, and wax esters. Hydrolysis of these lipids yields glycerol and fatty acids, sterols and fatty acids, and fatty alcohols plus fatty acids, respectively. The most important of these simple lipids for food scientists are the triglycerides. They are the major components of edible oils and fats, often representing more than 95% of refined oils. Triglycerides are esters of the trihydric alcohol glycerol with three fatty acids. Many of the properties of triglycerides are dependent on the component fatty acids. Thus, the melting point of the triglyceride reflects the melting point of the component fatty acids, with three high-melting-point fatty acids yielding a high-melting triglyceride. Unsaturation in the fatty acids makes the triglyceride susceptible to autoxidation, just as the fatty acid itself would be.

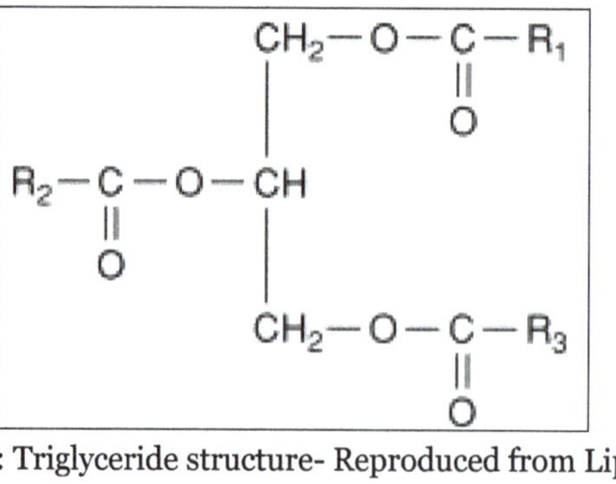

Figure: Triglyceride structure- Reproduced from Lipids.

All triglycerides are susceptible to hydrolysis in the presence of a catalyst. Acids, bases, or enzymes belonging to the hydrolase class, especially lipases, may act as the catalyst for the hydrolysis of triglycerides.

Steryl esters always occur together with sterols in plant, animal, or microbiological tissues. Wax esters may accumulate in considerable amounts in some biological tissues and this class comprises the main constituent of beeswax and jojoba oil.

Fats and Oils

Fats and oils are the most abundant lipids in nature. They provide energy for living organisms, insulate body organs, and transport fat-soluble vitamins through the blood.

Structures of Fats and Oils

Fats and oils are called triglycerides (or triacylcylgerols) because they are esters composed of three fatty acid units joined to glycerol, a trihydroxy alcohol:

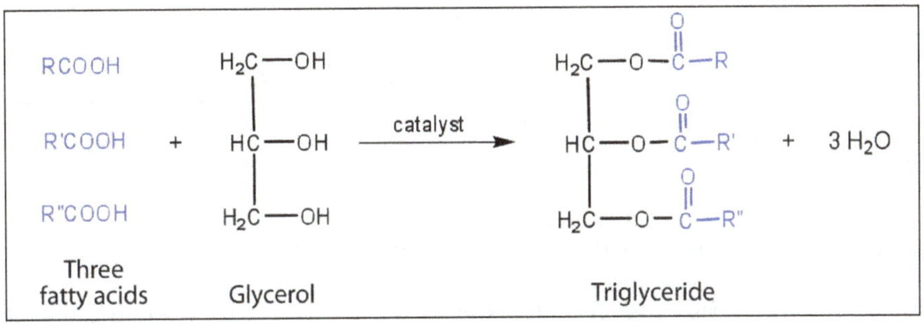

If all three OH groups on the glycerol molecule are esterified with the same fatty acid, the resulting ester is called a simple triglyceride. Although simple triglycerides have been synthesized in the laboratory, they rarely occur in nature. Instead, a typical triglyceride obtained from naturally occurring fats and oils contains two or three different fatty acid components and is thus termed a mixed triglyceride.

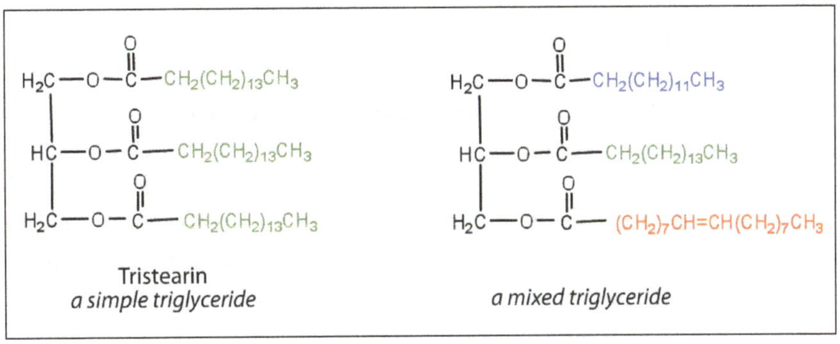

A triglyceride is called a fat if it is a solid at 25 °C; it is called an oil if it is a liquid at that temperature. These differences in melting points reflect differences in the degree of unsaturation and number of carbon atoms in the constituent fatty acids. Triglycerides obtained from animal sources are usually solids, while those of plant origin are generally oils. Therefore, we commonly speak of animal fats and vegetable oils.

No single formula can be written to represent the naturally occurring fats and oils because they are highly complex mixtures of triglycerides in which many different fatty acids are represented. Table shows the fatty acid compositions of some common fats and oils. The composition of any given fat or oil can vary depending on the plant or animal species it comes from as well as on dietetic and climatic factors. To cite just one example, lard from corn-fed hogs is more highly saturated than lard from peanut-fed hogs. Palmitic acid is the most abundant of the saturated fatty acids, while oleic acid is the most abundant unsaturated fatty acid.

Table: Average Fatty Acid Composition of Some Common Fats and Oils (%)*.

	Lauric	Myristic	Palmitic	Stearic	Oleic	Linoleic	Linolenic
Fats							
Butter (cow)	3	11	27	12	29	2	1
Tallow		3	24	19	43	3	1
Lard		2	26	14	44	10	
Oils							
Canola oil			4	2	62	22	10
Coconut oil†	47	18	9	3	6	2	
Corn oil			11	2	28	58	1
Olive oil			13	3	71	10	1
Peanut oil			11	2	48	32	
Soybean oil			11	4	24	54	7

*Totals less than 100% indicate the presence of fatty acids with fewer than 12 carbon atoms or more than 18 carbon atoms.

†Coconut oil is highly saturated. It contains an unusually high percentage of the low-melting C_8, C_{10}, and C_{12} saturated fatty acids.

Terms such as saturated fat or unsaturated oil are often used to describe the fats or oils obtained from foods. Saturated fats contain a high proportion of saturated fatty acids, while unsaturated oils contain a high proportion of unsaturated fatty acids. The high consumption of saturated fats is a factor, along with the high consumption of cholesterol, in increased risks of heart disease.

Physical Properties of Fats and Oils

Contrary to what you might expect, pure fats and oils are colorless, odorless, and tasteless. The characteristic colors, odors, and flavors that we associate with some of them are imparted by foreign substances that are lipid soluble and have been absorbed by these lipids. For example, the yellow color of butter is due to the presence of the pigment carotene; the taste of butter comes from two compounds—diacetyl and 3-hydroxy-2-butanone—produced by bacteria in the ripening cream from which the butter is made.

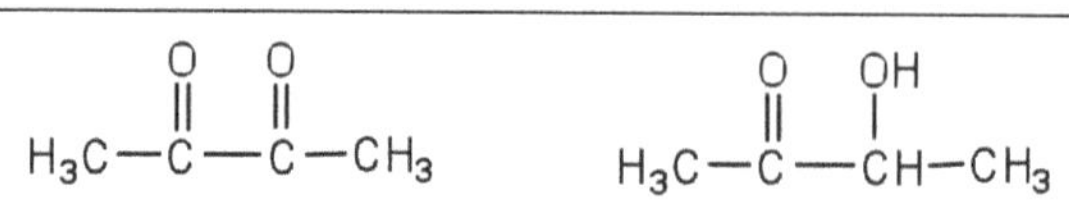

Diacetyl 3-hydroxy-2-butanone

Figure: Fats and oils are lighter than water, having densities of about 0.8 g/cm³. They are poor conductors of heat and electricity and therefore serve as excellent insulators for the body, slowing the loss of heat through the skin.

Chemical Reactions of Fats and Oils

Fats and oils can participate in a variety of chemical reactions—for example, because triglycerides are esters, they can be hydrolyzed in the presence of an acid, a base, or specific enzymes known as lipases. The hydrolysis of fats and oils in the presence of a base is used to make soap and is called saponification. Today most soaps are prepared through the hydrolysis of triglycerides (often from tallow, coconut oil, or both) using water under high pressure and temperature [700 lb/in² (~50 atm or 5,000 kPa) and 200°C]. Sodium carbonate or sodium hydroxide is then used to convert the fatty acids to their sodium salts (soap molecules):

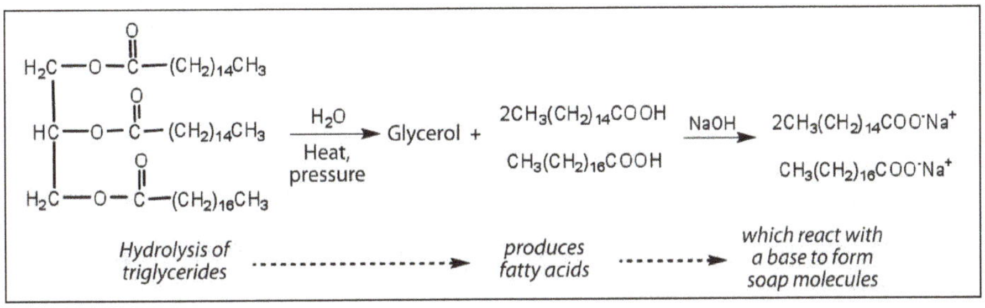

The double bonds in fats and oils can undergo hydrogenation and also oxidation. The hydrogenation of vegetable oils to produce semisolid fats is an important process in the food industry. Chemically, it is essentially identical to the catalytic hydrogenation reaction described for alkenes.

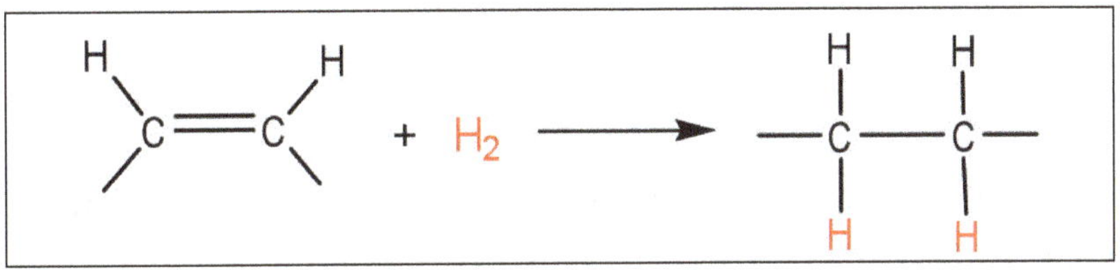

In commercial processes, the number of double bonds that are hydrogenated is carefully controlled to produce fats with the desired consistency (soft and pliable). Inexpensive and abundant vegetable oils (canola, corn, and soybean) are thus transformed into margarine and cooking fats. In the preparation of margarine, for example, partially hydrogenated oils are mixed with water, salt, and nonfat dry milk, along with flavoring agents, coloring agents, and vitamins A and D, which are added to approximate the look, taste, and nutrition of butter. (Preservatives and antioxidants are also added.) In most commercial peanut butter, the peanut oil has been partially hydrogenated to prevent it from separating out. Consumers could decrease the amount of saturated fat in their diet by using the original unprocessed oils on their foods, but most people would rather spread margarine on their toast than pour oil on it.

Many people have switched from butter to margarine or vegetable shortening because of concerns that saturated animal fats can raise blood cholesterol levels and result in clogged arteries. However, during the hydrogenation of vegetable oils, an isomerization reaction occurs that produces the trans fatty acids. However, studies have shown that trans fatty acids also raise cholesterol levels and increase the incidence of heart disease. Trans fatty acids do not have the bend in their structures, which occurs in cis fatty acids and thus pack closely together in the same way that the saturated fatty acids do. Consumers are now being advised to use polyunsaturated oils and soft or liquid margarine and reduce their total fat consumption to less than 30% of their total calorie intake each day.

Fats and oils that are in contact with moist air at room temperature eventually undergo oxidation and hydrolysis reactions that cause them to turn rancid, acquiring a characteristic disagreeable odor. One cause of the odor is the release of volatile fatty acids by hydrolysis of the ester bonds. Butter, for example, releases foul-smelling butyric, caprylic, and capric acids. Microorganisms present in the air furnish lipases that catalyze this process. Hydrolytic rancidity can easily be prevented by covering the fat or oil and keeping it in a refrigerator.

Another cause of volatile, odorous compounds is the oxidation of the unsaturated fatty acid components, particularly the readily oxidized structural unit in polyunsaturated fatty acids, such as linoleic and linolenic acids.

$$\sim CH{=}CH{-}CH_2{-}CH{=}CH\sim$$

One particularly offensive product, formed by the oxidative cleavage of both double bonds in this unit, is a compound called malonaldehyde.

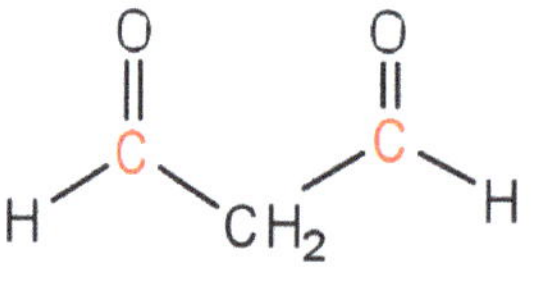

Malonaldehyde

Rancidity is a major concern of the food industry, which is why food chemists are always seeking new and better antioxidants, substances added in very small amounts (0.001%–0.01%) to prevent oxidation and thus suppress rancidity. Antioxidants are compounds whose affinity for oxygen is greater than that of the lipids in the food; thus they function by preferentially depleting the supply of oxygen absorbed into the product. Because vitamin E has antioxidant properties, it helps reduce damage to lipids in the body, particularly to unsaturated fatty acids found in cell membrane lipids.

Types of Fat

Fat is organized into two subgroups: saturated fat, and unsaturated fat. Unsaturated fat is further classified as monounsaturated fat, polyunsaturated fat, and trans-fat. These different classifications determine the effects of these fats on an organism, and the roles that they have in metabolism.

Saturated Fat

Saturated fat, or animal fat, is composed of a glycerol backbone with three fully saturated fatty acids attached. Saturated refers to all the carbons in the backbone being sp^3 hybridized, with two hydrogen atoms covalently bonded per carbon. This class of fats has higher viscosity and energy content than their unsaturated cousins. Due to poor solubility issues, this is the type of fat that is most commonly associated with heart disease.

Unsaturated Fat

Unsaturated fat, or vegetable fat, is composed of a glycerol backbone with three fatty acid chains where there is at least one sp^2 hybridized carbon. This forms a double bond somewhere in the chain. Monounsaturated fats have one double bond in the chain, while polyunsaturated fats have two or more.

Naturally occurring unsaturated fats, since they are produced by enzymes, have specific stereochemistry. Natural fats always show the cis conformation, which has a higher solubility in water, and is easily broken down by the metabolic machinery. Artificially produced fats, since they are produced using organic synthesis techniques, contain a racemic mixture of trans and cis bonds. Trans fats are less soluble – like saturated fats. However, they are not readily metabolized by cellular machinery.

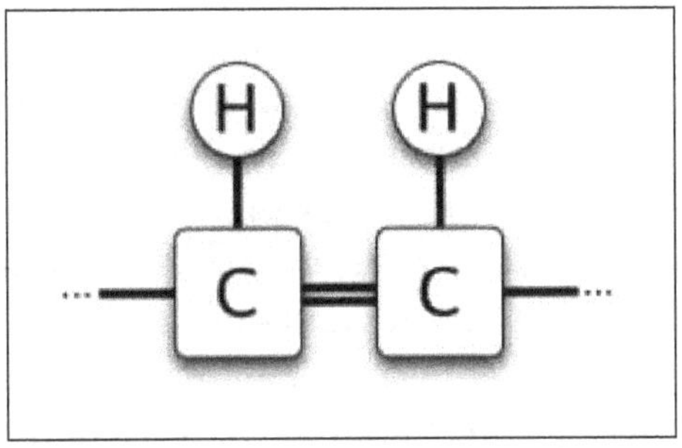

Cis bond

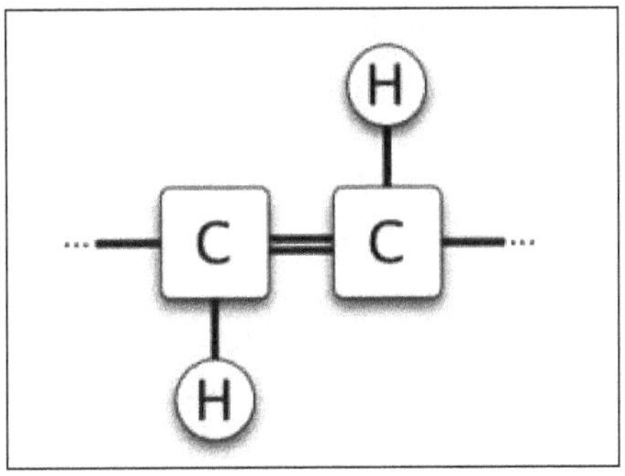

Trans bond

Types of Unsaturation

The unsaturated fats are classified by the position of the unsaturation. This designation is denoted by the ω symbol, then the number of carbon with the unsaturation. For example, ω-3 fats have an unsaturation at the third carbon position. Ω-3, 7 polyunsaturated fats have an unsaturation at the third and seventh carbon position. The position of the unsaturation determines the metabolic pathway that the fat will follow.

Examples of Fat

- Saturated Fat: Examples of saturated fat include animal fat found on beef, pork, and chicken. Since these types of fats are also called "animal fats", they are primarily found in animals. Saturated fat measurements are used in medical diagnostic tests as inverse indicators of a healthy lifestyle.
- Unsaturated Fats: Monounsaturated fats are often referred to as "good" fats. They make up the oils and fats found in avocados and olive oil. Polyunsaturated fats are found in canola oil and other less viscous plant oils.
- Trans Fats: Trans fats are chemically produced fats. They are generally considered to be

unhealthy, and are found in mass produced oil found in processed and fried foods. Trans fats have a controversial history. They have even been banned for consumer use in some countries.

Waxes

A second group of neutral lipids that are of physiological importance, though they are a minor component of biological systems, are waxes. Essentially, waxes consist of a long-chain fatty acid linked through ester oxygen to a long-chain alcohol. These molecules are completely water-insoluble and generally solid at biological temperatures. Their strongly hydrophobic nature allows them to function as water repellents on the leaves of some plants, on feathers, and on the cuticles of certain insects. Waxes also serve as energy-storage substances in plankton (microscopic aquatic plants and animals) and in higher members of the aquatic food chain. Plankton apparently uses the biosynthesis of waxes to adjust their buoyant density and thus their depth in the ocean. It has been suggested that a major source of petroleum found in deep-sea sediments originates from the deposition of wax-rich dead plankton over vast periods of time. Whales and many fishes also store large quantities of waxes.

Complex Lipids

Complex Lipids, also known as conjugated or compound lipids are fatty acids esters with groups in toting up to a fatty acid and an alcohol moeity. Being extensively distributed in animals, bacteria and plants, they are the major component of cell membranes and also found in circulating fluids. Complex lipids can be further sub divided into three main subtypes:

Phospholipids

Lipids are with a phosphate dreg, one glycerol, or an amino/fatty alcohol, with or without one or two fatty chains (remarkably one inositol group, two phosphates or four fatty chains). They frequently have bases with nitrogen and other substituents. Phospholipids are ubiquitous molecules that, due to their chemical properties, have important structural as well as metabolic roles in the cell.

Phospholipid Structure

A phospholipid is made up of two fatty acid tails and a phosphate group head. Fatty acids are long chains that are mostly made up of hydrogen and carbon, while phosphate groups consist of a phosphorus molecule with four oxygen molecules attached. These two components of the phospholipid are connected via a third molecule, glycerol.

Phospholipids are able to form cell membranes because the phosphate group head is hydrophilic (water-loving) while the fatty acid tails are hydrophobic (water-hating). They automatically arrange themselves in a certain pattern in water because of these properties, and form cell membranes. To form membranes, phospholipids line up next to each other with their heads on the outside of the cell and their tails on the inside. A second layer of phospholipids also forms with heads facing the inside of the cell and tails facing away. In this way, a double layer is formed with phosphate group heads on the outside, and fatty acid tails on the inside. This double layer, called a lipid bilayer, forms the main part of the cell membrane. The nuclear envelope, a membrane surrounding a cell's nucleus, is also made up of phospholipids arranged in a lipid bilayer, as is the membrane of mitochondria, the part of the cell that produces energy. The figure below depicts the lipid bilayer and the structure of a phospholipid:

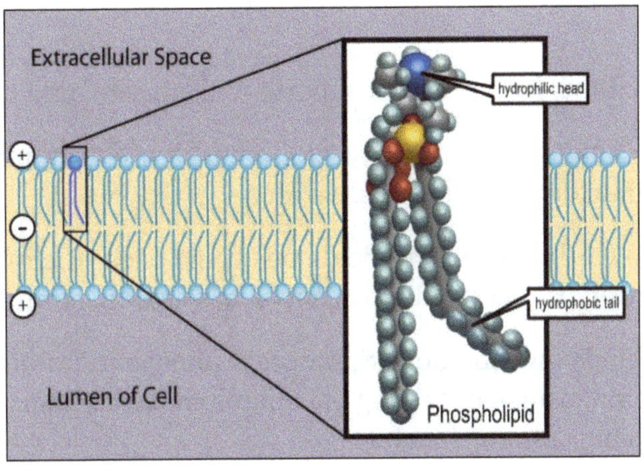

Functions of Phospholipids

As membrane components, phospholipids are selectively permeable (also called semi-permeable), meaning that only certain molecules can pass through them to enter or exit the cell. Molecules that dissolve in fat can pass through easily, while molecules that dissolve in water cannot. Oxygen, carbon dioxide, and urea are some molecules that can pass through the cell membrane easily. Large molecules like glucose or ions like sodium and potassium cannot pass through easily. This helps keep the contents of the cell working properly and separates the inside of the cell from the surrounding environment.

Phospholipids can be broken down in the cell and used for energy. They can also be split into smaller molecules called chemokines, which regulate a variety of activities in the cell such as production of certain proteins and migration of cells to different areas of the body. Additionally, they are found in areas such as the lung and in joints, where they help lubricate cells.

In pharmaceuticals, phospholipids are used as part of drug delivery systems, which are systems that help transport a drug throughout the body to the area that it is meant to affect. They have high bioavailability, meaning that they are easy for the body to absorb. Valium is an example of a medication that uses a phospholipid-based drug delivery system.

In the food industry, phospholipids can act as emulsifiers, which are substances that disperse oil droplets in water so that the oil and water do not form separate layers. For example, egg yolks contain phospholipids, and are used in mayonnaise to keep it from separating. Phospholipids are found in high concentrations in many other animal and plant sources, such as soybeans, sunflowers, cotton seeds, corn, and even cow brains.

There are three categories of phospholipids:

- Glycerophospholipids (synonymous Glycerophosphatides or Phosphoglycerides): A group for the glycerol-containing phospholipids. It implies a few imitative of sn-glycero-3-phosphoric acid that contains at least one O-alk-1'-enyl, O-alkyl, or O-acyl residue affixed to the glycerol moiety with a polar head consist of a nitrogenous base, an inositol unit or a glycerol.

  - Phosphatidic acid: When hydrolyzed yield one equivalent of each of glycerol and phosphoric acid.
  - Lecithin: It contains fatty acid, phosphoric acid, glycerol and nitrogenous base choline.
  - Cephalin: It contains fatty acid, glycerol, phosphoric acid, nitrogenous base ethanolamine (coalmine) or the amino acid serine.
  - Plasmalogens: On hydrolysis, it yields one mole each of aliphatic aldehyde, glycerol, fatty acid, phosphoric acid and nitrogenous base ethanolamine or choline.

- Sphingophospholipids (synonymous Sphingosyl phosphatides or Phosphosphingolipids): A group for the sphingosine-containing phospholipids. It contains a long chain base, glycoside moiety with a phosphorus residue.
  - Sphingomyelin: It contains fatty acid, phosphoric acid and choline. Glycerol is absent.
- Alkylphosphocholines (phospholipid like molecules): A group for the esters of phosphocholine with long chain aliphatic alcohols opposed in chain length, position and unsaturation of the cisdouble bond. Eibl et al describes its remarkable biological and therapeutic activities.

Glycolipids

Lipids with a glycosidic moiety (carbohydrates such as galactose or glucose), fatty acid, sphingosine are glycolipids. Sometimes one or more phosphate groups are present. The long chain derivatives of sugars restrain most frequently in bacteria and plants a glycerol (a diacylglycerol backbone) and in animals (a ceramide backbone). A phosphorylated polysaccharide-lipid complex or a sterol may also be found. Simple glycolipids are composed of a carbohydrate moiety linked to one fatty acid or fatty alcohol. A wide variety of glycolipids are present in bacteria where specific glycopeptidolipids or lipopolysaccharides are also available. Commensurate with the structure, glycolipids have the following categories:

Glycerol based glycolipids: It contains mono or oligosaccharide moiety connected to the glycerol (hydroxyl group), alkylated (or acylated) with one or two fatty acids. They might be uncharged and referred as neutral glycoglycerolipids having a phosphate or a sulfate group. Corresponding to their structure, glycerol based glycolipids can be sub-divided into the following:

- Neutral glycoglycerolipids
- Glycophospholipids
- Sulfoglycoglycerolipids

Distribution of Glycolipids in the Cell

Most of the glycolipids are distributed in membranous structures in the cell. Two-thirds of the total glycolipids are distributed in intracellular membranes such as golgi apparatus, endosomes, lysosomes, nuclear membrane, endoplasmic reticulum, and mitochondria. Glycolipids are synthesized in golgi apparatus by the addition of saccharides one by one to the ceramide moiety. Most glycolipids are transported between membranes as small vesicles maintaining a bilayer structure. However, the first step of the biosynthesis of most glycosphingolipids i.e. transfer of a glycosyl residue to ceramide, occurs at the cytosolic surface of cis Golgi and the other sugars are transferred at the luminal face of the golgi complex. Some glycolipids are also distributed in cytosol. The soluble fraction of brain contains 5% of the total gangliosides and their composition is similar to that of the membrane fraction. Glycolipids are distributed prudentially into exofacial leaflets of plasma membrane and the luminal side of organelles. In polarized epithelial cells, glycosphingolipid are enriched on apical side of the cells while cells with polarity, e.g. epithelial cells in intestine and kidney glycolipid composition differ at the apical and basolateral side.

Researchers observed that in plasma membrane, glycosphingolipids along with cholesterol form clusters, called rafts. This region had relatively less phospholipids than other areas of plasma membrane. Approximately 70% of total cellular glycolipids are found in rafts. Rafts are lateral assemblies of sphingolipids and cholesterol that form tight hydrophobic interactions between these molecules. Brown and London reported that sphingolipids associate laterally with one another through weak interactions between the carbohydrates heads of the glycosphingolipids and the hydrophobic interactions between their saturated side chains and any void between associated glycosphingolipids were filled by cholesterol molecules which interact with hydrophobic portions of glycolipids. The tight interactions between cholesterol and glycolipids in the membrane are the driving force that segregates them from phospholipids that remain fluid in nature.

Carbohydrate-Carbohydrate Interaction of Glycolipids

Cell-cell interactions play an important role in the development, maintenance, and host-pathogen interaction. They are highly dynamic processes, which include migration, recognition, signaling, adhesion, and finally attachment. Carbohydrates moieties of glycolipids, the most prominently exposed structures on the surface of living cells, with flexible chains and many potential binding sites are ideal to serve as important players in these events. Molecular interactions where carbohydrates are involved are usually considered as weak interactions, and therefore, there is biological relevance of carbohydrate-carbohydrate and carbohydrate-protein interactions.

Glycolipid-glycolipid interaction is a rapid process, as compared with protein-protein interaction, although the strength of the interaction is weaker than that of protein-protein interaction. A synergistic effect between cell adhesion based on glycosphingolipidsglycosphingolipids interactions and adhesion based on integrins was observed by Bucior et al. They postulated that glycosphingolipid-glycosphingolipid interaction may define the initial specificity and direction of the cell recognition and regulate the adhesion process.

One of the important properties of the bilayer is the ability to orient and cluster glycolipid species in such a way so that interactions in the biological systems are maximized. The lipid moiety of glycolipids is generally buried in the cell membrane bilayer, leaving the oligosaccharide moieties exposed but in close proximity to the bilayer surface. This represents a unique environment for carbohydrate interactions with other molecules. Glycolipids cause two types of interactions. They interact side by side within the same membrane to form clusters and by trans interactions in which two interfacing membrane interact through their surface carbohydrates. They further reported that trans interactions between the glycolipids provides the basis for glycosphingolipids-dependent cell to cell adhesion, which takes place through specific complementary structures catalyzed in many cases by calcium ions. Therefore, cell surface complex carbohydrate has emerged as key recognition molecules, mediating physiological interactions between cells. Typically glycans on one cell surface engaged by complementary carbohydrate binding proteins on opposing cells, initiate a cellular responses. The preliminary evidence for carbohydrate-carbohydrate interaction came from the studies of liposomes containing highly purified glycosphingolipids. Glycosphingolipids bind to the complementary glycosphingolipids through interaction between their carbohydrate moieties.

Cell surface carbohydrates play a major role in cell-cell or cellsubstrate recognition. They suggested that melanoma cell adhesion to endothelial cells was based on GM3/LacCer interaction, which initiates metastatic deposition, and may trigger a series of "cascade" reactions leading to activation of endothelial cells and expression of Ig family or selectin receptors, thereby promoting adhesion and migration of tumor cells, hence demonstrated dramatic changes in surface carbohydrates during oncogenesis.

Researcher suggested that galactosyl ceramide and cerebroside sulfate present in high concentrations in multilayered myelin sheath was involved in carbohydrate- carbohydrate interactions between lipid head groups. The interaction resulted in dehydration of the sulfate, changes in the intermolecular hydrogen bonding, interactions of the sugar and other oxygens, decreased intermolecular hydrogen bonding of the amide C=O of GalC and dehydration of the amide region of one or both of the lipids, and disordering of the hydrocarbon chains of both lipids. The interaction between these two lipids could be either a lateral cis interaction in the same bilayer or a trans interaction between apposed bilayers and may be involved in stabilization of the myelin sheath.

Researcher observed that human leukocytes were recognized by E-selectins present on vascular endothelium during inflammation. The recruitment of neutrophils to sites of inflammation was mediated by endothelial leukocytes adhesion molecule-1(ELAM-1) expressed on activated endothelial cells of blood vessels walls. ELAM-1 is a member of the selectin family of adhesion molecules that contain a lectin motif thought to recognize carbohydrate ligands. They observed that cell adhesion by ELAM-1 was mediated by a carbohydrate ligand, sialylLewis X (SLex; NeuAc alpha 2,3Gal beta 1,4(Fuc alpha 1,3)-GlcNAc-), a terminal structure found on cell-surface glycoprotein and glycolipid carbohydrate groups of neutrophils. The sialyl Lewis A and sialyl Lewis X, the carbohydrate determinants which are frequently expressed on human cancer cells, serve as ligands for a cell adhesion molecule of the selectin family, E-selectin, which is expressed on vascular endothelial cells. These carbohydrate determinants are involved in the adhesion of cancer cells to vascular endothelium and thus contribute to hematogenous metastasis of cancer by triggering the activation of integrin molecules through the action of several cytokines leading to extravasation of cancer cells.

Overall, it can be stated that carbohydrate-carbohydrate interactions play an important role in recognition and signaling events in a variety of biological phenomena.

Biological Functions of Membrane Glycolipids

Membrane glycolipids perform a number of functions in biological system. Glycolipids have roles in response to cell contact, as receptor components, as anchors for proteins and as markers for tumor progression and cell differentiation.

Glycolipids as Signal Transducers

Glycolipids have been known to be modulators of signal transduction. Glycosphingolipids and sphingomyelin in animal cells are clustered and organized as membrane microdomains closely associated with various signal transducer molecules such as cSrc, Src family kinases, small G-proteins (e.g. RhoA, Ras), and focal adhesion kinase. Glycosphingolipid clustering in such microdomain causes adhesion to complementary glycosphingolipids on the surface of counterpart cells through carbohydrate-carbohydrate interaction. Glycosphingolipids dependent cell adhesion in microdomain causes activation of the signal transducers, leading to cell phenotypic changes.

Glycosphingolipid microdomains mediate signal response either by associating with GPI-anchored proteins or Immuno and growth factor receptors. Kasahara et al. reported that GPI anchored proteins and acylated proteins like src-family tyrosine kinases and trimeric-G proteins are known to associate with glycosphingolipid microdomains. GPI-anchored proteins having saturated acyl chains were likely to insert preferentially into glycosphingolipid microdomains. They observed that antibody (or ligand)-mediated crosslinking of GPI-anchored proteins induces activation of src-family kinases and transient increase in tyrosine phosphorylation of several substrates. Reports showed that enzymatic removal of the carbohydrate moiety from cell-surface glycosphingolipids impairs the activation of the src-family kinase by antibody-mediated crosslinking of GPI-anchored protein.

Glycosphingolipid microdomains are also involved in signaling by immunoreceptors and growth factor receptors. Efficient T-cell activation requires one signal from a T-cell antigen receptor and a second signal from the co-stimulatory molecule. The co-stimulation leads to the recruitment of glycosphingolipid microdomains to the site of cell-cell contact between the T cell and antigen-presenting cell.

Iwabuchi et al. reported that adhesion process is based essentially on a glycosphingolipid-enriched microdomain (GEM) at the B16 cell surface. More than 90% of GM3 present in the original cells were found in GEM. GEM is also enriched in several signal transducer molecules, e.g. c-Src, Ras, Rho, and focal adhesion kinase.

The general function of glycosphingolipid microdomains in signal transduction may be to concentrate receptors and effectors on both sides of the membrane, thus speeding up binding during signaling and preventing inappropriate crosstalk between pathways.

Glycolipids as Receptors of Bacteria and Bacterial Toxins

Binding of pathogenic bacteria and bacterial toxins to host cell surfaces is an essential step in establishing infection in tissues and producing toxic effect. Glycolipids on cell surfaces are receptors for binding to cells. Many pathogenic bacteria bind to glycolipids of host cell surface for colonization and infection.

Rodighiero et al. reported that many bacterial toxins bind to gangliosides as the receptors on cell surface and invade host cells. The best known of these is the cholera toxin, an enterotoxin produced by

Vibrio cholerae, and its specific cell surface receptor was identified as ganglioside GM1. The entry of cholera toxin into target epithelial cells and the induction of toxicity depend on the cholera toxin binding to ligand-based receptor GM1 and association with the glycosphingolipid microdomain. Cholera toxin consists of a pentameric B subunit that binds to GM1 and an A subunit with direct toxic activity. The binding of B subunit to membrane GM1 induces a conformational change in the toxin, resulting in the entry of the A subunit into the cell. Wolf et al. observed that to induce toxicity cholera toxin must bind ganglioside GM1 at plasma membrane, enter the cell by endocytosis and then retrograde into endoplasmic reticulum. They proposed that GM1provides the sorting motif necessary to retrograde trafficking into biosynthetic, secreatory pathway of host cells. The entry of cholera toxin into target epithelial cells and the inductions of toxicity depend on cholera toxin binding to the lipid-based receptor ganglioside GM1 and association with detergent-insoluble membrane microdomains. Cholera toxin action depends on the stable formation of the cholera toxin B-subunit- GM1 complex.

Bock et al. studied that enterohaemorrhagic E. coli bind to neutral glycosphingolipids having an alpha-1, 4 galabiose moieties in the sugar chain, such as galabioside (Ga2Cer) and ceramide trihexoside (Gb3Cer).They suggested that E. coli bind to glycosphingolipid by recognizing not only the terminal sequence but also the internal sequence of the sugar chain. Propionibacterium, which causes skin disease, recognizes the lactosyl moiety of glycosphingolipids as a binding epitope. These bacteria bind strongly to lactosylceramide and also bind to isoreceptors such as asialo GM1 (GA1) and asialo GM2 (GA2). This bacterium could not bind to any glycosphingolipids composed of a dihydroxy base and nonhydroxy fatty acid in ceramide, even though they contain a lactosylmoiety.

Glycolipids in Modulation of Cell Proliferation

Several observations underline the role of the glycolipids in regulation of cell growth by interacting through growth factor receptors. Auge et al. observed that intracellularly produced ceramide, Lac Cer stimulated DNA synthesis in endothelial smooth muscle cells. They also potentiated the mitogenesis induced by various growth factors including platelet-derived growth factor. The lactosylceramide (LacCer, Galβ4Glcβ1Cer) activates NADPH oxidase to modulate the intercellular adhesion molecule-1 expression on human umbilical vein endothelial cells and to induce the proliferation of human aortic smooth muscle cells. Reports showed that decrease in ceramide content via increased activity of ceramidase, sphingomyelin synthase or GlcCer synthase correlated with proliferative response of smooth muscle cells.

Sphingolipid breakdown products, sphingosine and lysosphingolipids, inhibit protein kinase C, an important enzyme in cell regulation and signal transduction. Sphingolipids and lysosphingolipids affect significant cellular responses and exhibit antitumor promoter activities in various mammalian cells. These molecules may function as endogenous modulators of cell function and possibly as second messengers. Ceramides in turn activate the serine-theronine protein phosphatases PP1 and PP2A. Ceramide activated protein phosphates leads either to cell cycle arrest or to apoptosis.

Some gangliosides are apoptotic inducers. Ceramide is a mediator of programmed cell death signaling in lymphoid cells. GD3 is potent mediator of cell death. De Maria reported that the apoptotic signal triggered by CD 95 in lymphoid and myeloid tumor cells increases the ceramide levels, which is followed by an increase in ganglioside GD3 synthesis. Addition to cells in culture damaged mitochondria with consequent dissipation of mitochondrial transmembrane potential and caused DNA fragmentation in HuT78 cell. Glycosphingolipids are abundant components of external leaflet of the plasma membrane of cancer cells. Addition of anti-ganglioside GD2 monoclonal antibodies into cell medium results in apoptosis in GD2 expressing human lung cancer cells.

Glycolipids as Biosurfactant

Biosurfactants are amphiphilic compounds containing hydrophobic and hydrophilic moieties that reduce surface tension and interfacial tensions between individual molecules at the cell surface and interface respectively. Glycolipids are the most common types of biosurfactants. The polar moiety is a carbohydrate and the nonpolar moiety is a long carbon chain fatty acid. Mannosylethritol lipids (MELs) are microbial extracellular glycolipids composed of mannosylethritol and fatty acids as hydrophilic and lipophilic moieties, respectively. They are produced by Schizonella melanograma, Candida antarctica T-34 and Ustilago maydis. MEL glycolipids produced by Candida antarctica T-34 showed excellent surface and interfacial tension lowering actions and exhibited antimicrobial activity particularly against Gram-positive bacteria, thereby suggesting a strong potential for their industrial use. The excellent surface tension reducing and emulsifying characteristics of MELs glycolipids and its stability over a wide range of temperature and pHs, made MELs important for commercial application. MELs glycolipids were potent inducers of apoptosis and differentiation in mouse melanoma cells.

Role of Glycolipids in Calcium Homeostasis

The association of calcium ions with glycolipids especially gangliosides were involved in neuronal function. Gangliosides micelles bind calcium ions with high affinity. This may have some significance in the process of synaptic transmission. Levade et al. reported that sphingosine and ceramide mediated a profound release of calcium ions from intracellular stores. Gangliosides function in calcium homeostasis and signaling. He observed that gangliosides induced changes in cellular calcium, which was accomplished through modulation of calcium influx channels, calcium exchange proteins and various calcium dependent enzymes that were altered through the association with the gangliosides. Lloyd-Evans et al. demonstrated that elevation of intracellular glucosylceramide (GlcCer) levels results in increased functional Ca^{2+} stores in cultured rat neurons. Surface glycolipid galactocerebroside participate in the opening of the Ca^{2+} channels in oligodendrocytes cells.

Neutral Glycoglycerolipids

Most commonly this class contains one or two saccharide units allied glycosidically to diacylglycerol or glycerol, however, neutral glycolipids with three or four sachharide units are also reported. Being essential in algae, bacteria and higher plants, they are positioned in photosynthetic membranes and are also present in minor amount in animals. About 85% of neutral glycoglycerolipids (DGDG and MGDG) are present in photosynthetic membranes of all oxygenic photosynthetic organisms. These glycolipids represent the most abundant lipid class on Earth based on the innate loads of photosynthetic organisms. In Plasmodium falciparum (a protozoan responsible for causing malaria), the discovery of a plastid without galactoglycerolipids heaves the issue of trouncing the galactoglycerolipids during evolution. In a number of plants, galactosyl monoacylglycerol with a variety of galactose moieties has also been uttered. There are two families of Neutral glycoglycerolipids:

- Non-acylated glycoside moiety containing compounds.
- Acylated glycoside moiety containing compounds.

Monogalactosyl diacylglycerol (MGDG) and digalactosyl diacylglycerol (DGDG) Carter et al. stated the presence of Glycoglycerolipids in the wheat flour extracts followed by its explicit alliance with tylakoid membranes of chloroplasts reported by Benson et al. and their accurate structure was explicated by Carter et al. This stiff alliance is epitomized by the algaecide or herbicide commotion of Galvestine (a MGDG synthase inhibitor).

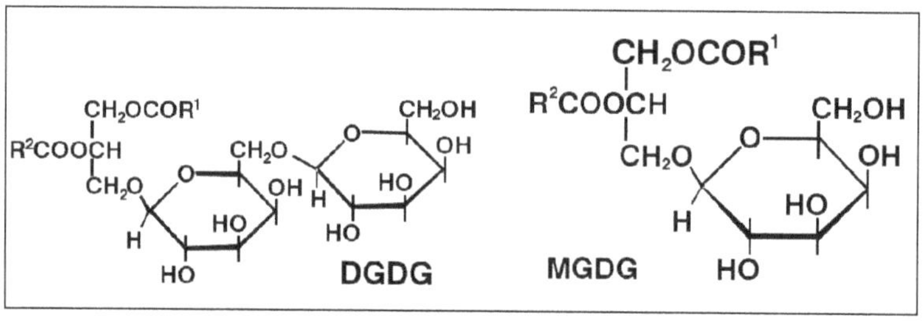

Figure: R^1 and R^2: Two fatty acid chains.

Tri and tetragalactosyl diacylglycerols (superior homologues of galactolipids) were recognized later in all plant tissues. Zhan et al. elucidated the structure of trigalactosyl monolinolenylglycerol (a new glyceroglycolipid) isolated from Premna microphylla, a plant used in Chinese folk medicine. A large amount of linolenic acid (18:3 n-3) and an explicit trienoic acid (16:3 n-3) is present in both MGDG and DGDG. In higher plants, linolenic acid is the lone fatty acid in MGDG and hence these plants are known as '18:3 plants'. In angiosperms, 16:3 n-3 is absent and linolenic acid is concerted in both sn 1 and sn 2 positions of both DGDG and MGDG. Contrastingly, lower plants (conifers, ferns, mosses, green algae) and some angiospermic families (Solanaceae, Chenopodiaceae, Brassicaceae) have 16:3 n-3 concerted in the galactolipids at sn 2 position, whereas a short proportion of 18:3 n-3 is acylated at both positions. These types of plants have a structural similarity with cyanobacteria and are referred to as 16:3 plants. In photosynthetic diatoms and red algae, a high proportion of 20:5 n-3 galactolipids have been noted. Numerous fatty acids (n-3) have been dogged in MGDG isolated from a brown alga (Sargassum thunbergii). In most fungi, glycosphingolipids is the main glycolipidsand galactolipids have been found in small amounts. In MGDG, an unusual fatty acid (18:3 n-1, ~25 %) was described from a marine diatom (Skeletonema costatum). They were revealed to be metabolized into aldehydes (short chain), octadienal (8:2 n-4) and octatrienal (8:3 n-1) that could have lethal upshot on zooplankton crustaceans. Heptadienal (7:2 n-3), another aldehyde was also known to be produced from eicosapentaenoic acid of MGDG. Two linolenic acid (18:3 n-3) acyl groups containing MGDG have been depicted in fruits of Rosa canina (rose hips) possessing anti-inflammatory properties (inhibits cell migration) which unswervingly correlates to the anti arthritis activity of rose hip herbal remedies. Other reports revealed anti tumor promoting properties, oxygen forage and virus counteracting activities of galactosyl diglycerides from different sources. In Thailand, DGDG isolated from Clinacanthus leaves demonstrated antiviral activity against herpes simplex virus. MGDG extracted from Sargassum muticum, an invasive brown alga has been revealed to display inhibition of the microbial growth including bacteria and fungi (anti-microfouling activity).

Specifics have shown that explicit galactolipids with 16:3 n-3 or 18:4 n-3 in the sn 1 position and 20:5 n-3 or 18:5 n-3 in the sn 2 position is directly involved in cytotoxic reactions induced by Phaeodactylum tricornutum, a marine diatom. It has also been discovered that Arabidopsis chloroplasts enclose an arabidopside illustrated by the existence of 12-oxo-phytodienoic acid, a phytohormone intimately allied to jasmonic acid acylated on glycerol carbon (sn 1), the 16:3 n-3 being acylated at sn 2 position. In MGDG, a 12/10 carbon oxoacids have been noticed in Arabidopsis leaves linked to six carbon aldehydes, alcohols and their esters production which are recognized as volatiles of leaf.

In the leaves of Linum usitatissimum, a novel family of oxylipin named Linolipins containing MGDG has been exposed. Esterified residues (one or two) of divinyl ether-etherolenic acid and parallel oxylipin comprising DGDG have also been alienated from the flax leaves damaged by freezing-thawing or inoculated with phytopathogenic bacteria.

In Petunia hybrid flower, it has been observed that DGDG synthesis increases and the pistilspollinic tubes include elevated quantity than any other floral organs. In bacteria, sn 3-Oglycosyldiacylglycerols possess glycoside moiety: a-D-mannopyranosyl (1->3)-O-Dmannopyranoside (Micrococcus), a-D-galactopyranosyl (1->2)-O-a-D-glucopyranoside (Lactobacillus), a-D-glucopyranoside (Pneumococcus, Staphylococcus), a-D-glucopyranosyl (1- >2)-O-a-D-glucopyranoside and b-D-galactofuranoside (Mycoplasma), b-D-glucopyranosyl (1- >6)-O-b-D-glucopyranoside (gentobiosyldiacylglycerol, Staphylococcus). In Propionibacterium propionicum, a novel glycoglycerolipid (a-D-glucopyranoside (1->3) a-D-glucopyranoside) enclosing an ether-linked alkyl chain at C-3 position of glycerol has been illustrated. Manca et al., isolated a diglucosyl diglyceride with an incredibly exceptional diglucosyl structure (1->4) from Thermotoga maritima.

Inquisitively, two moles of palmitic acid esterifies the glycerol moiety and acylated on the 6-OH of the terminal glucose by one molecule of decanoic acid. It has been well-known that the glycolipids present in flour act as surfactants and affect the texture, staling and volume of bread. Selmair et al. reviewed the role of glycolipids in breadmaking and have shown that monoalactosylglyceride and its lyso derivative perk up the baking performance of wheat flour. Hamberg reported DGDG mono estolides in the kernels of oat.

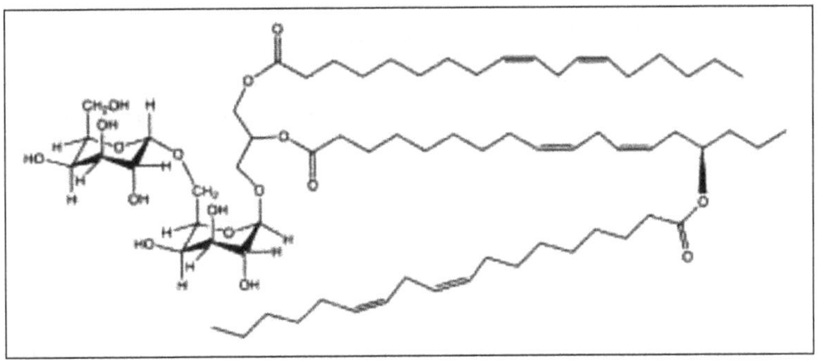

DGDG Mono Estolide from the Kernel of Oat

The estolide of DGDG include an avenolic acid (15-hydroxylated linoleic acid) and linoleic acid cluster. The di estolides of DGDG have also been isolated from the kernels of oat. A diminutive fraction of Tri and TetraGDG estolides were also noticed and overall the estolides include 10% of lipid fraction and ~29% of the total glycolipid fraction (extracted in methanol).

From the exudates of glandular trichome (secretory organ) of a cherry tree (Cerasu yedoensis) in Japan, 2-Acetoxy-1-(3-glycosyloxyoctadecanoyl) glycerol has been isolated and characterized. A 3-hydroxyoctadecanoic acid allied to disaccharide esterifies glycerol at sn 1 position, while sn 2 position being acylated.

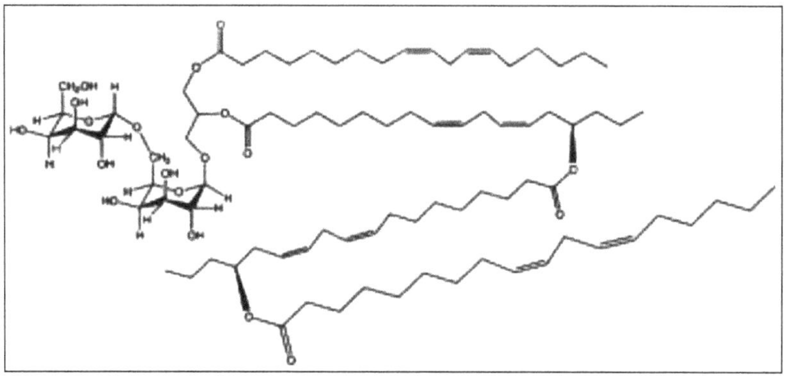

Figure: Structure of estolide of DGDG from the kernel of oat.

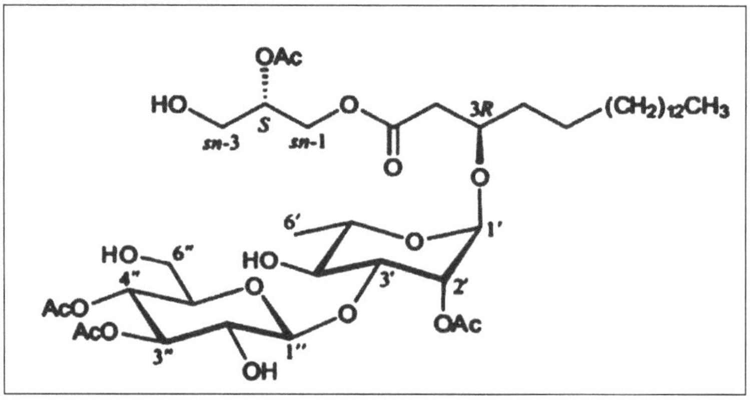

Figure: Structure of 2-Acetoxy-1-(3-glycosyloxyoctadecanoyl) glycerol.

Alkyl Acyl Glycosylglycerols

Cramerides (monoglycosyldiacylglycerol intermediate) was isolated from Pseudoceratina crassa, a sponge having a bizarre moiety of cyclitol, an alkyl chain (branched) and an acyl chain (saturated) with 14-16 carbon atoms (unbranched or branched). These compounds act as feeding deterrents in fish.

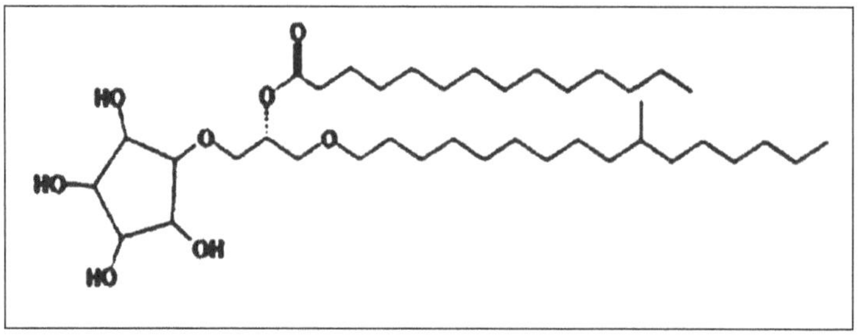

Dialkyl Glycosylglycerols

Are diether lipids from bacteria containing glycosidic headgroups noticed in carbonated chimney of the hydrothermal field. These compounds have non-isoprenoid diethers and signify an exceptional permutation of archaeal distinctiveness in lipids of bacteria. A mono or a diglycosyl moiety is present at the polar head group and alkyl chains (14-18 carbon units) is either or monomethylated, monounsaturated or saturated. Since these glycolipids exist from the creation of life on earth, it was contemplated that they were the evolutionary precursor of phospholipids and their existence is interrelated to an incredibly squat level of phosphate in the utter atmosphere.

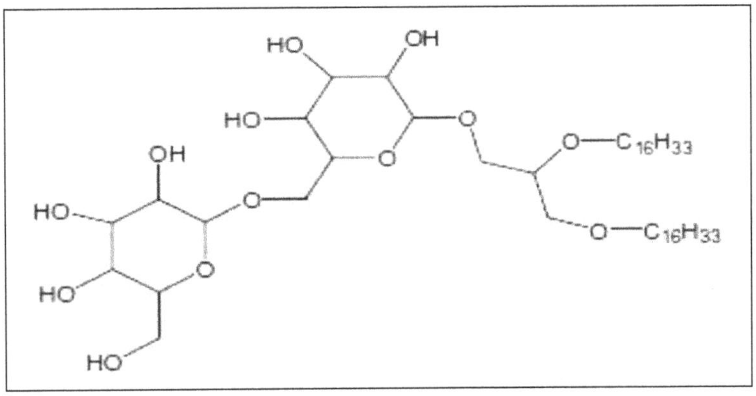

Figure: Structure of Diglycosyl alkylglycerol.

Alkyl Diglycosylglycerols

Myrmekiosides, a novel glyceroglycolipid have been reported from Myrmekioderma sp., a sponge possessing anti-tumor properties. At C1 position of glycerol, an O-alkyl ether chain (where R contains 16 carbon units/chain with or without a branched methyl group), at C2 position two moles of glucose and at C3 one mole of xylose were found.

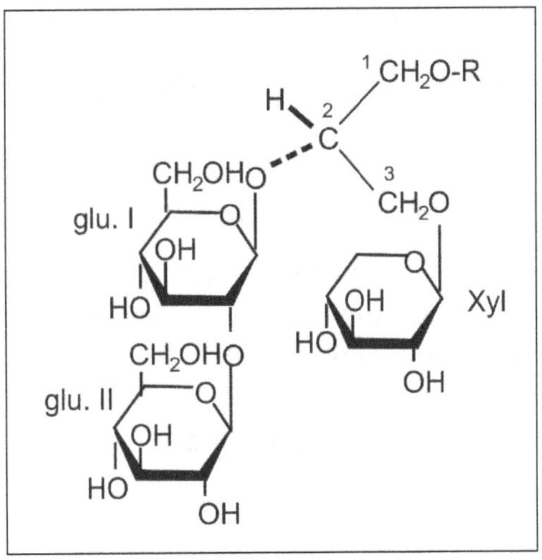

Figure: Structure of Myrmekiosides.

Trikentrion loeve, a Senegalese sponge revealed an exceptional ether lipid distinguished by the glycosylation (xylose) of two glycerol radicals and an O-alkenyl chain of ether (24 carbon units with lone double bond) at the third radical. In human brain, glyceroplasmalopsychosine (a plasmal conjugate of psychosine and glycerol), an additional ether lipid has been depicted.

Cyanobacterium, Synechocystis sp possess acylated forms of glycoglycerolipids bearing a palmitoyl group esterified at hydroxyl group of C-6 position of the terminal glycosyl moiety of either acylated DGMG or DGDG. The presence of acylated MGMG, MGDG and DGDG was also stated from nitrogen fixing cyanobacteria and leaf homogenates. Bacillus acidocaldarius, possess an extraordinary glucosamidyl glycolipids comprising ~64% of the total lipids (N-acyl derivative of Glucopyranosyl

(1-> 4) Glucosamine (1->3)-diacylglycerol along with amide- linked branched 13-cyclohexyltridecanoic acid, 11-cyclohexylundecanoic or heptadecanoic acid.

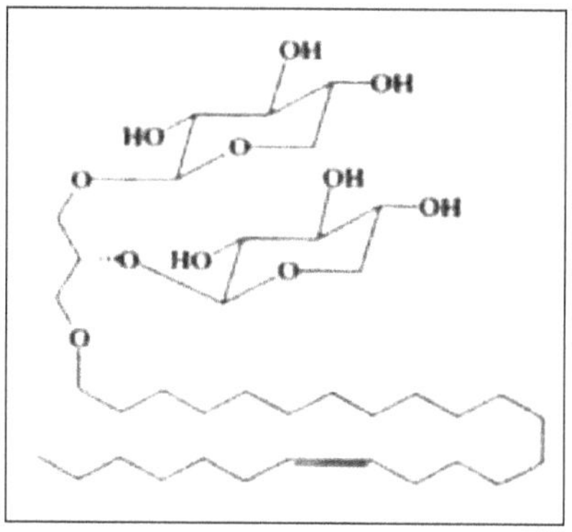

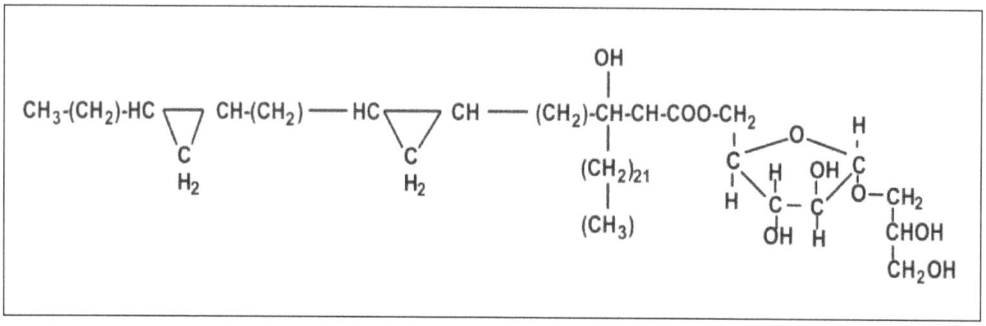

In algae and bacteria, a hefty amount of glycolipids including diverse combinations of sugar have been reported.

Mycoloyl Arabinosylglycerols

Are new glycolipids isolated from Mycobacterium avium-M. intracellulare complex, described by an arabinofuranosyl glycerol acylated on the C5 of the arabinose by an assorted mycolic acid structures (wax ester mycolic, mycolic, keto mycolic).

Alkyl Galactosylglycerols Amid Acetal Group

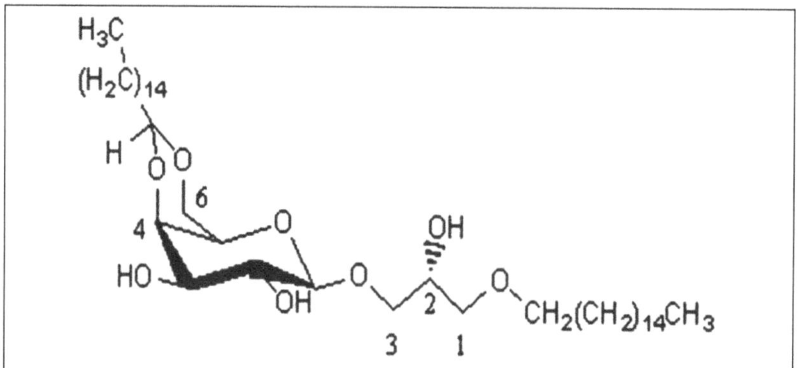

1-*O*-hexadecyl-3-*O*-(4,6-*endo*-hexadecylidene-Galβ1)-*sn*-glycerol

From the equine brain, a novel galactosylalkylglycerol named Plasmalogalactosylalkylglycerol was isolated consisting of a long chain cyclic acetal at the sugar moiety. The acetal and alkyl group chain lengths were C14 for the earlier and C16 and C18 for the last. The intact equine brain includes about 5 mg of plasmalogalactosylalkylglycerol.

Glycophospholipids

Are primarily described in bacteria, however, in other living creatures, the glycosyl phosphatidylinositol anchors are present. The simplest forms of glycophospholipid consist of phospholipid, phosphatidic acid, linked to a glycosyl group. Glucosylated phosphatidic acid is the simplest compounds present in the erythrocytes of umbilical cord, human neutrophils, human epithelium, rat brain and might be a marker for lipid rafts.

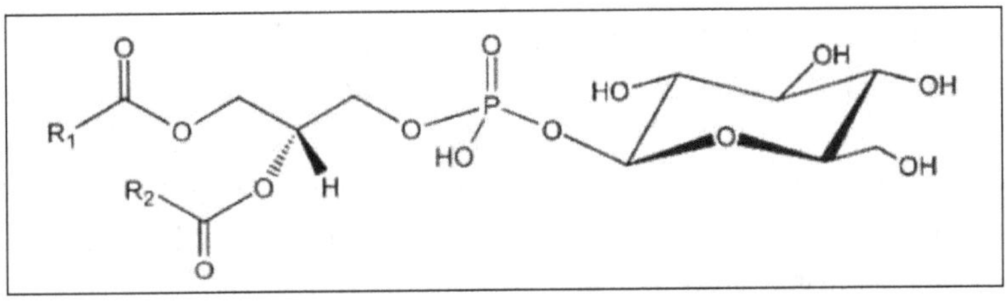

Exclusively composed of a sole pair of fatty acids (saturated, 18:0 & 20:0), its task remains indefinite, however, carbonyl compounds and sugars interrelate with proteins or amino acids in Maillard reaction. Correspondingly, it has been depicted that phosphatidyl ethanolamine undergo Amadori rearrangement after reacting with glucose leading via an unstable Schiff base, formation.

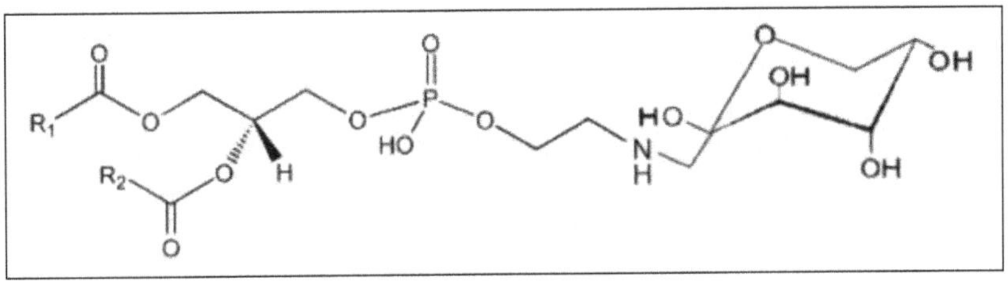

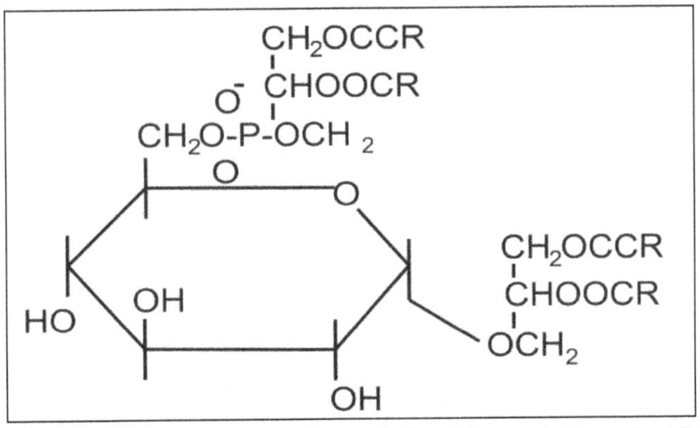

Figure: Phosphatidylmonoglucosyl diacylglycerol: Phosphatidyl glucosaminyl glycerol (PGG) is found in Bacillus megaterium.

Later, they were noticed in liver of rat, however its concentration increased in diabetic rats followed by establishment of their correct structure in human erythrocytes. It has also been observed that Amadori lipid products generate oxidative stress and accelerate membrane lipid peroxidation that amends cell survival and integrity. Concentration of Amadori PE has been found higher in erythrocytes and plasma of diabetic patients as compared to healthy individual. Phosphatidyl monoglucosyl diacylglycerol (PMDG) is found in many Gram-positive bacteria, however, its metabolism and structure elucidation was accomplished in Pseudomonas diminuta.

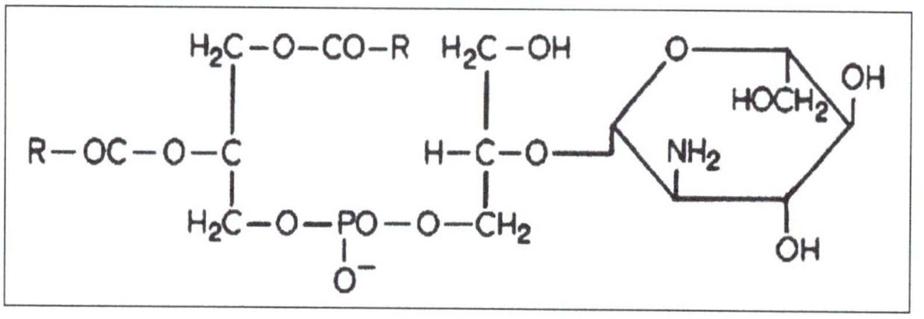

From several species of Halomonas (halophilic Gram negative bacteria), Phosphatidylglycerol moiety has been isolated. It consists of a phosphatidic acid attached to glycerol residue that is alkylated by a glucose moeity. C16:0, C19: cyclopropane, oleic acid, palmitic acids are the most plentiful acyl chains attached to the phosphatidylglycerol moiety. The tetatnus causing microbe, Clostridium tetani, produces N-acetylglucosaminyl diacylglycerol where a phosphoethanolamine moeity is annexed to the C6 position of sugar.

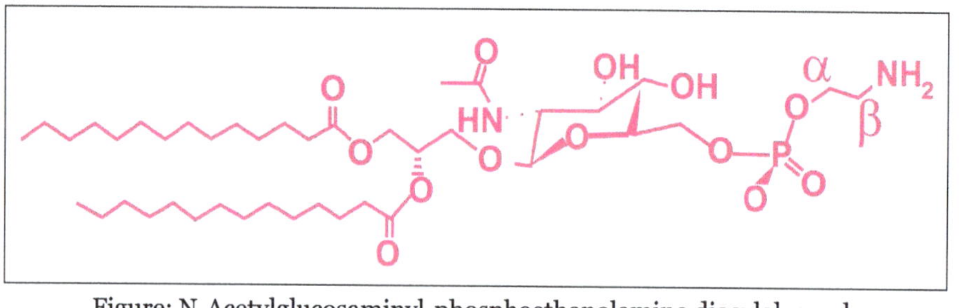

Figure: N-Acetylglucosaminyl-phosphoethanolamine diacylglycerol.

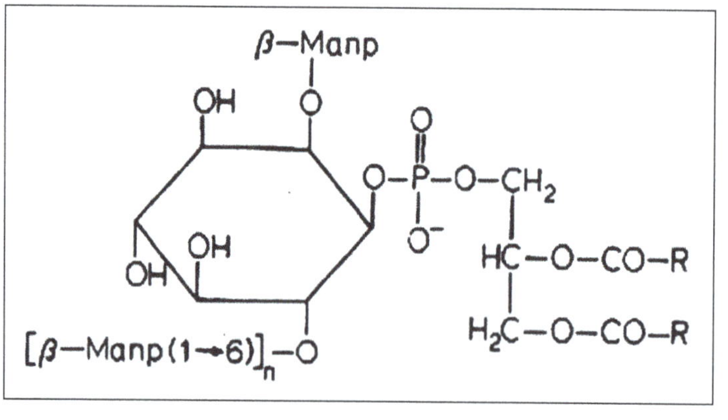

In phosphatidylinositol polymannoside (PIPM) established in Mycobacteria, at 2'position of inositol, a mannopyranoside cluster and at 6' position, one to five mannoside groups are present. It was established that phosphatidylinositol mannosides from Mycobacterium tuberculosis wield effective anti inflammatory effects and inhibit the production of chemokines, cytokines and NO both in vitro and in vivo via Toll-like receptor inhibition. Mannophosphoinositides presence in phospholipid composition is a striking feature of some bacteria and Actinomycetes. Monomannosides have been delineated in Streptomyces, Propionibacteria and Mycobacterium species. Two types of dimannosides were identified from Streptomyces griseus with one extra fatty acid fastened to a mannose residue (tri & tetra acylmannoside). In Mycobacteria, Polymannosides, one of the most complexes with a number of fatty acids acylating the mannose chain were identified.

In Corynebacterium urealyticum, C16:0 or C18:1 (R_3) acylated phosphatidylinositol dimannosides were also illustrated with acylating one of the two mannose groups.

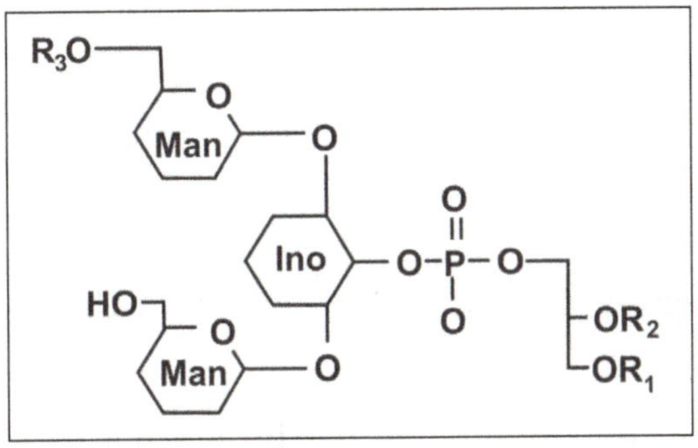

In Mycobacteria LPS, Phosphatidylinositol dimannoside (lipoarabinomannan & lipomannan) has been the anchoring domain and holds key significance in the interaction between host and pathogen. This phosphatidylinositolglycans demonstrates the earliest version from prokaryotes branded for cell surface protein anchoring and the lipoarabinomannan structural features have been used across the globe as tuberculosis vaccine.

The surface of Leishmania, Entamoeba and Crithidia contains Lipophosphoglycan, a Glycolipid serene with four domains: (1) an alkyl lyso phosphatidylinositol (2) core of glycan (3) Repeated units of Gal-Man-PO4 backbone and (4) Cap structure of an oligosachharide has been involved in a numerous functions inside the mammalian host.

In the strains of Streptococcus sp., Glucopyranosyl cardiolipin (Glycosylated cardiolipin) was initially noticed and identified (18% of the lipid phosphorus) but present as a trifling constituent in Vagococcus fluviatilis.

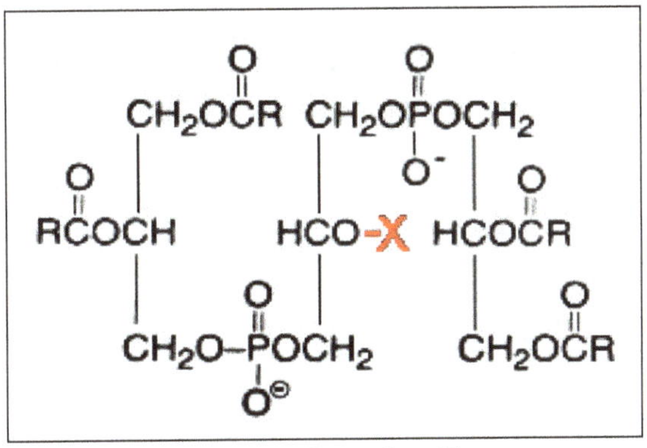

Figure: X = glucosyl residue.

This infrequent glycolipid was also reported in Geobacillus steathermophilus (4% of total phospholipids) acylated primarily by oleic acid (~25%) and palmitic acid (~33%). Revealed predominantly as iso-C15:0 and anteiso-C17:0, this composite might be implicated in the subvert effect of high temperature on the membrane organization and in the regulation of lipid composition of G. stearothermophilus. In Deinococcus radiodurans, 3-(galactopyranosyl)-2-(3- phosphatidyl) glyceroyl derivative, an exclusive glycophospholipid of a fatty alkylamine (~33% of the total lipids) has also been identified.

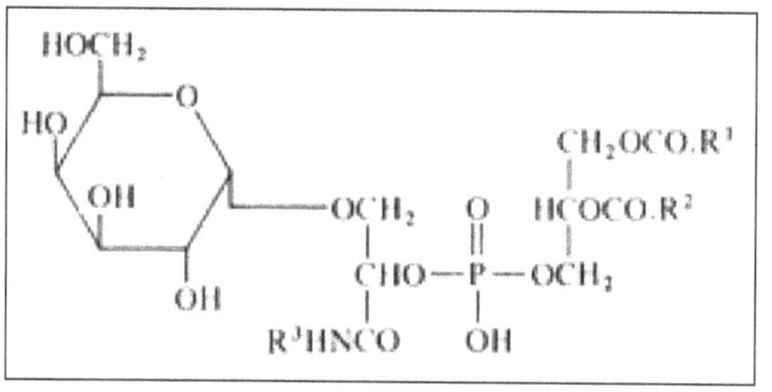

The R_3 (alkylamines) are generally straight chain compounds (C15-C17 or C16-C18), with isomeric 17:1 bases preponderant.

A phosphocholine-containing glycoglycerolipid has been reported to be produced by Mycoplasma fermentans in which the phosphocholine assembly is attached to the C6 of glucose moiety. This lipid valor in the pathogenesis of rheumatoid arthritis by eliciting cell death or inflammation through their phosphocholine dregs.

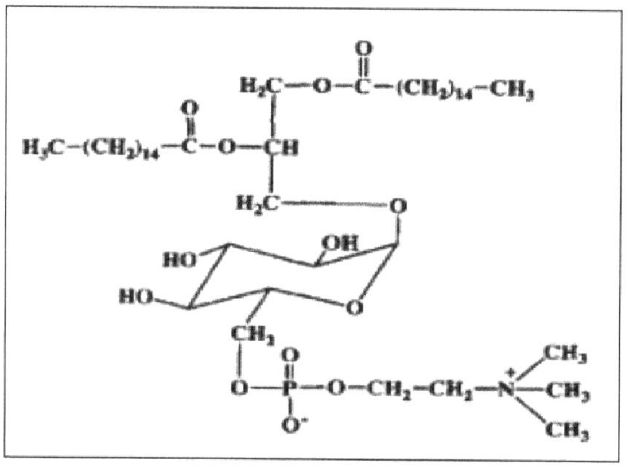

In Lactococcus and Streptococcus (Gram positive bacteria), a number of complex glycolipids deriving from glucopyranosyl-1->2-glucopyranoside-diacylglycerol (kojibiosyl diacylglycerol structure) were recognized.

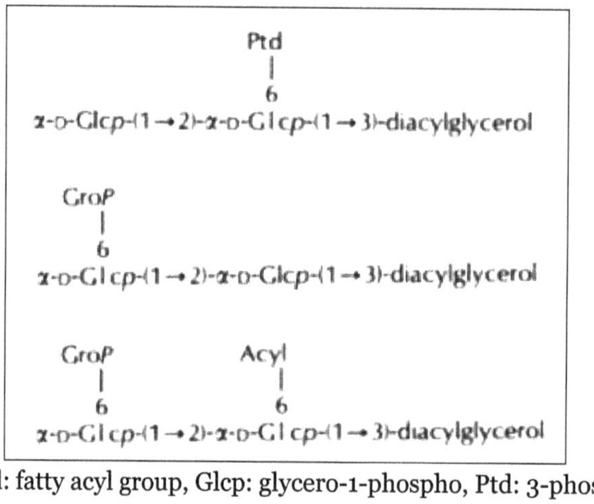

Figure: Acyl: fatty acyl group, Glcp: glycero-1-phospho, Ptd: 3-phosphatidyl.

In Staphylococcus membranes, additional complex phosphorylated compounds lipoteichoic acids contains polymer of glycerol-1-phosphate allied to a glycosyl diglyceride or a phosphatidyl glycosyl diglyceride is present. Other swapping on the glycerophosphate units with alanyl or glycosyl groups has also been reported.

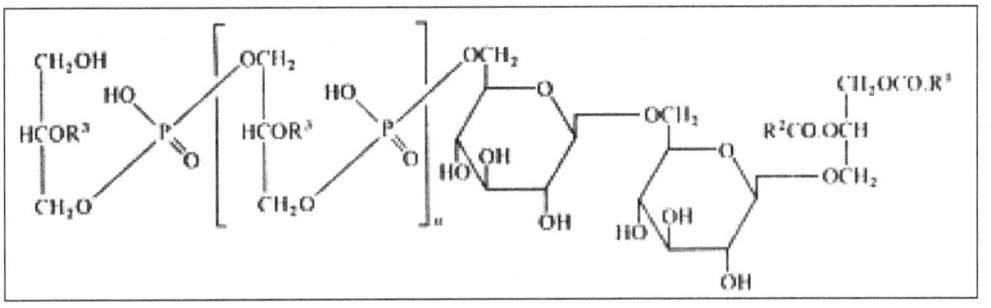

Structure of lipoteichoic acid from Bacillus subtilis where R[1] and R[2] are acyl group of fatty acid and R[3] is mainly acetylglucosaminyl, alanyl or H.

Sulfoglycoglycerolipids

Are lipids demeanoring a sulfur atom found in acidic membranes with a strong acidic pH (~2), is present in some mammalian cells, chloroplast of higher plants, bacteria, algae and in fungi. This type of lipid has also been described in Elakatothrix and Zygnema, green alga and in Fucus, Pelvetia. At very low pH, carboxylic acids occur in unionized forms, but sulfolipids subsist in the anionic form. In Chlorella, a new type of sulfolipid (sulfoquinovosyl diacylglycerol: SQDG) derived from MGDG was discovered in 1959 at the Scripps Institute of Oceanography (San Diego, CA, USA) which was afterward described in the thylakoid membranes of all higher and lower plants. Hence, sulfolipids are present algae, cyanobacteria, few basidiomycetes fungi (Clitocybe, Coprinus, Psalliota) and photosynthetic membranes of plants.

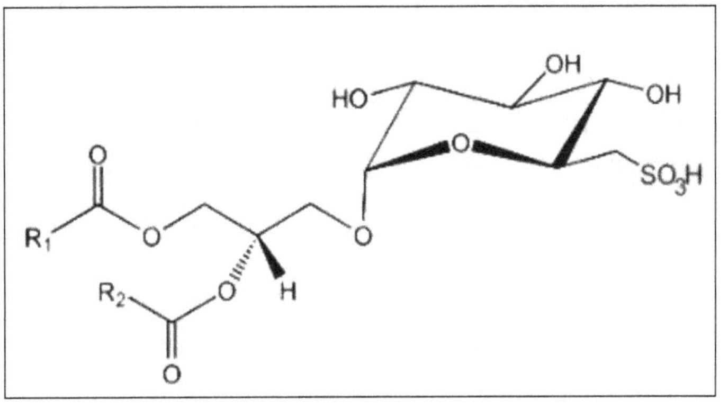

The biological function and mechanism for SQDG biosynthesis has been described in higher plants predominantly under phosphate ravenous circumstances. SQDG is the lipid with a sulfonic acid link, containing MGDG at C6 position of the monosaccharide unit, sulfoquinovosyl (quinovose : 6-deoxy-D-glucose) is the sulfonoglucosidic moiety (6-desoxy-6- sulfono-glucoside), R_1 is often the palmitic acid and R_2 is the unsaturated chain (16:3 n-3).

Heterosigma carterae (Raphidophyceae), a marine chloromonad was revealed to include a intricate mixture of SQDG. The major fatty acyl groups contain 20:5 (n-3), 16:1 (n-3), 16:1 (n5), 16:1 (n-7) and 16:0. In Rhizobium meliloti, a symbiotic bacterium as well as in other strains of Rhizobiaceae, existence of SQDG has been established. The occurrence of SQDG outside the plant kingdom suggested the role of sulfolipid in the Rhizobium/legume relationship. In Halobacterium, sulfate esters are reported with two alkyl diphytanylether chains. SQDG type bearing mono-unsaturated (20:1 to 24:1) and saturated (14:0 to 24:0) fatty acids were also extorted from a sea urchin (Scaphechinus mirabilis). Gustafson et al (1989) reported anti HIV-1 activity of SQDG from cyanobacteria containing linolenic and palmitic acids. Sulfate esters of galactoglycerolipids are reported in quite a few mammalian tissues like brain, testis, spermatozoa (3% of total lipids, >90% of total glycolipids in mammalian testis and ~0.3-1 mmole/g). The acyl groups R' and R" are most frequently saturated (16/18 carbon atoms).

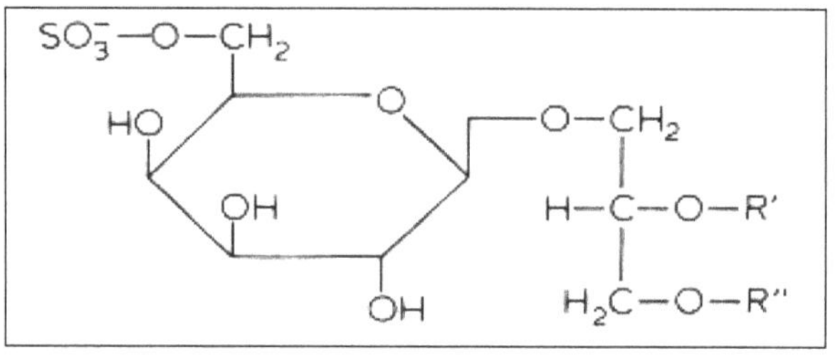

Seminolipids, mixture of alkylacyl and DAG is associated with myelination process of the nervous system. They are also essentail for mice spermatogenesis. From Hyphomonas jannaschiana, a sea water bacterium, a tauroglycolipid has been isolated lacking phospholipids. The structure determined as 1,2-diacyl-3- glucuronopyranosyl-glycerol taurine-amide with 16:0, 16:1 n-7, 18:0, 18:1 n-7 and 19:0 as major acyl chains.

Ceramides based Glycolipids

Out of 150 known varieties in this cluster of glycosphingolipids, 50 categorized into the structure of ganglio. Strewned mostly at the surface of the cell, they participate in the regulation of cell interactions and dole out as distinctive markers for cells and mediate cell to cell detection and contact. Ceramides based glycolipids are vital for the growth and development of organisms and allude to viral or microbial infections and cancer. They form molecular complex named 'Glycosynapse' and helps in explicit recognition between cells function and signaling. They are found in animals, bacteria, fungi, and plants. The grouping of glycosphingolipids is based on the structure of the carbohydrate moiety and is as under:

- Neutral glycosphingolipids (unsubstituted glycosyl group): It includes single or extra glycosyl moieties, hexoses (sometimes pentose) and an added hexosamine. As per the presence of glycosyl units they may be one glycosyl unit (monoglycosylceramides), more than one glycosyl units (oligoglycosylceramidest) or other structure containing only one glycosylated sphingoid base (deacylated glycosphingolipids), one fatty acid linked to cerebrosides (acylated glycosphingolipids), complex sphingolipids with a ceramide containing a cyclopropane chain and prenylated residues on galactose (plakosides).
- Acidic glycosphingolipids: Are zwitterionic molecules with glycosyl and carboxyl group, sulfate or phosphate. Examples are gangliosides (a carboxyl group on sialic acid and several glysosyl residues), acidic monoglycosylceramides, phosphonoglycosphingolipids, sulfosphingolipids (ceramide di & tri hexosyl sulfate with one sulfate moeity on the glycosyl unit, sulfatides), phosphosphingolipids (phosphorylceramide with oligosaccharide or glycosyl head moeity).

Lipopolysaccharide

Lipopolysaccharide (LPS) is an important cell wall constituent of Gram-negative bacteria and is vital for bacterial cell integrity, viability, and defense against environmental stress. Lipopolysaccharide is highly conserved among almost all Gram-negative bacteria and is a potent inducer of inflammatory responses. It exists ubiquitously in the environment and, thus, LPS can directly modulate the immune system and susceptibility to disease. Lipopolysaccharide is a heat stable amphiphilic molecule composed of three regions: lipophilic lipid A, hydrophilic polysaccharides or oligosaccharide core, and O-antigen. The lipid A portion of LPS has been demonstrated to be the immunostimulatory moiety of LPS. Studies of various Gram-negative bacteria show that the structure of lipid A is diverse among bacterial species, and the number of acyl chains (fatty acid) determines the immunostimulatory capacity of LPS. Nonetheless, highly immunostimulatory LPS comprises comparatively similar lipid A with six acyl chains. For example, both Escherichia coli and Salmonella enterica serovar Typhimurium comprise hexa-acylated lipid A which are highly immunostimulatory.

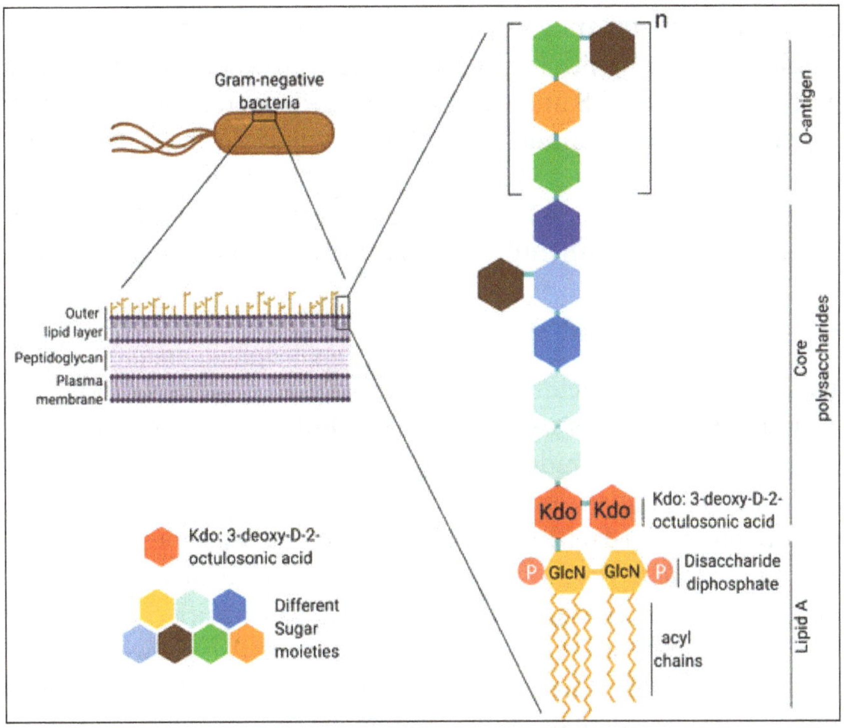

Figure: Structural details of lipopolysaccharide from a Gram-negative bacterium. Lipopolysaccharide (LPS) provides structural and functional integrity to outer membrane of Gram-negative bacteria.

LPS is an amphipathic molecule with a general structure consisting of three different regions: hydrophobic lipid A, core polysaccharide, and O-antigen (repeats of polysaccharide chain, where n can be up to 40 repeats). Lipid A consist of bisphosphorylated diglucosamine backbone substituted with six acyl chains that are attached by ester or amide linkage.

The discovery of LPS dates back to the eighteenth century with the search for a substance found in putrid matter that was believed to cause fever. Later, Robert Koch showed that various diseases were caused by bacteria. Richard Friedrich Johannes Pfeiffer demon-strated that some bacteria consisted of a heat stable, non-volatile pyrogenic substance that caused disease and termed it "endotoxin" to distinguish it from exotoxins that are released by bacteria. Subsequently, endotoxin was shown to characterize Gram-negative bacteria. By the 1940s, pure extracts of endotoxin were prepared and demonstrated to be made of a small portion of lipid A and polysaccharide, hence named lipopolysaccharide. In the 1980s, Tetsuo Shiba et al. synthesized free lipid A molecule and proved it to be the endotoxic center of LPS.

Intensive research in the field of innate immunity has led to the identification of a myriad of pattern recognition receptors (PRRs) on host immune cells that recognize non-self-molecules, i.e., pathogen-associated molecular patterns (PAMPs) derived from various pathogens including LPS. Extracellular LPS is a potent PAMP recognized by toll-like receptor-4 (TLR4) which is a PRR present on the surface of phagocytic cells like macrophages, neutrophils, and dendritic cells. Recognition of LPS by TLR4 induces a signaling cascade that eventually induces inflammation and production of the pro-inflammatory cytokines that help eliminate invading pathogens. Conversely, excessive production of pro-inflammatory cytokines leads to life-threatening pathological consequences such as septic shock.

While TLR4 was thought to be the only sensor for LPS, recent studies have provided insight into two TLR4-independent LPS recognition systems: transient receptor potential (TRP) channel-dependent sensing of extracellular LPS and caspase-4/5/11-dependent sensing of intracellular LPS. Extracellular LPS sensed by TRP channels present on the neuronal cells drives neurogenic inflammation and pain in mice. Caspase-4/ 5 in humans and caspase-11 in mice sense intracellular LPS within the cytoplasm of innate immune cells, such as macrophages, to drive production of pro-inflammatory cytokines IL-1β and IL-18 and inflammatory cell death, termed pyroptosis.

LPS Structure and Immunogenicity

Studies of various Gram-negative bacteria suggest a common general structure of LPS. The membrane-embedded lipophilic lipid A is usually composed of a dimer of glucosamine (D-GlcN) attached to acyl chains by ester or amide linkages. Lipid A is covalently attached to hydrophilic anionic groups, 3-deoxy-d-manno-2-octulosonic acid (Kdo) in the core region together with L-glycero-D-manno-heptose (l,d-Hep) and hexoses and hexosamines. In most Gram-negative strains, the core region is attached to the repeated units of saccharides called O-polysaccharides or O-antigens. O-antigens vary among bacterial strains and give bacteria a rough (R-type) or smooth (S-type) phenotype. O-antigens are truncated or lacking in R-type when compared to S-type Gram-negative bacteria. As the outermost part of LPS, O-antigens are responsible for bacteria evading the immune system, particularly the complement system of the host (e.g., Salmonella enterica serovar Typhimurium). Lipid A as well as the polysaccharide regions are able to induce potent immune responses. The lipid A component, independent of the polysaccharide portion, is responsible for various pathophysiological effects including toxicity, mitogenicity, complement reactivity, and Limulus lysate gelation. Researcher demonstrated that solubilized lipid A administered to mice and rabbits induced toxicity and pyrogenicity. Furthermore, soluble lipid A demonstrated the properties of intact LPS, i.e., lethal toxicity and pyrogenicity induced by lipid A was comparable to their corresponding parent LPS. Furthermore, chemically synthesized E. coli lipid A has demonstrated identical endotoxic activities as natural free lipid A, further supporting lipid A as the endotoxic center of LPS.

The immunogenic potential of LPS varies among bacterial species and is mostly determined by the structure of lipid A. Based on the potency of the pro-inflammatory reaction, LPS can be agonistic, weakly agonistic or antagonistic. Generally, the LPS present in major Gram-negative bacteria are agonistic and hexa-acylated (six acyl chains are esterified with disaccharide backbone) and are therefore potent immunostimulants. The structure of Lipid A can vary in number, length, position of esterified acyl chain, and number of phosphate groups, resulting in various degrees of immunogenicity. The tetra-acylated lipid A structure found in Yersinia pestis grown at 37 °C has very low immunogenic activity and has been shown to be a TLR4 antagonist. Similarly, Francisella tularensis expresses lpxD1 lipid A biosynthase acyl transferase at 37 °C which synthesizes the tetra-acylated lipid A of a longer acyl chain length with lower activity.

Studies have demonstrated that, in addition to lipid A, charged molecules, such as phosphate groups, have a direct impact on the interaction between phagocytic cells and cytokine production. For example, both hexa-acyl groups and two phosphate groups of E. coli are required for full activation of the TLR4/MD2 complex. Similarly, Francisella tularensis demonstrates weak endotoxic activity when one or both phosphate groups are lacking, compared to hexa-acylated Escherichia coli lipid A with two phosphate groups at the 1 and 4 positions. In addition, several studies have demonstrated the contribution of inner core sugars (two Kdo moieties) of LPS in inflammatory responses. Studies with N. meningitids LPS variants and the synthetic lipid A variants showed the important role of two Kdo with lipid A; in fact, the response of human monocyte-derived dendritic cells to N. meningitids Re-LPS was higher compared to lipid A alone. Similarly, the production of proinflammatory cytokines, like TNFα and IFN-β, were significantly higher in the murine macrophages when stimulated with synthetic N. meningitids Re-LPS compared to lipid A, and similar results were observed with E. coli and Salmonella LPS. Thus, the structural features of these components are directly related to the virulence of Gram-negative bacteria and their ability to evade immune detection. Understanding the intricate process of LPS recognition by the host immune system can provide important insight into the nature and magnitude of inflammatory responses.

LPS Recognition by TLR4

The critical role of the innate immune system is to detect and recognize the presence of foreign invaders and promote phagocytic killing of microorganisms. The primary targets of LPS are the phagocytic cells of the immune system including tissue macrophages, peripheral monocytes, neutrophils, and dendritic cells. Lipopolysaccharide activates macrophages and induces the production of several pro-inflammatory cytokines, some of which are responsible for inflammation-induced fever. Although, the controlled production of these inflammatory mediators is required for clearance of invading pathogens, uncontrolled production of inflammatory cytokines can provoke fatal consequences such as septic shock. Therefore, fine tuning of pro- and anti-inflammatory mediators is required for proper immune function and homeostasis maintenance. In normal physiological conditions, several molecules, including sCD14, lysozyme, BPI protein, collectins, lactoferrin, tissue factor pathway inhibitors, and cationic antimicrobial protein CAP18, actively regulate low levels of LPS and remove it from circulation to limit inflammatory responses.

TLR4, present on the surface of various phagocytic cells, is the specialized cellular sensor for extracellular LPS. Efficient LPS recognition and production of inflammatory mediators by TLR4 requires an orchestrated action of various accessory proteins such as LPS-binding protein (LBP), CD14, and MD-2. Although the exact nature of the interaction between LPS and TLR4/MD-2 is somewhat obscure (structure–activity relationship), various studies have been able to unravel the complex process to some extent. Initiation of LPS recognition begins with dissociation of the LPS monomer from the aggregates by LPS-binding protein (LBP) which traffics it to CD14 (glycosylphosphatidylinositol-anchored protein), present on most phagocytic cells (except dendritic cells) that make use of soluble CD14 (sCD14). CD14 then carries and loads LPS to the TLR4/MD-2 receptor complex.

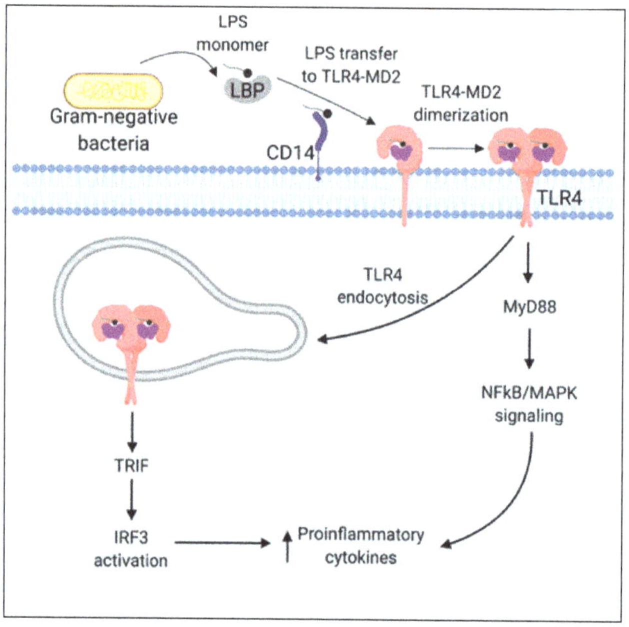

Figure: LPS recognition by toll-like receptor 4 (TLR4).

Serum protein LBP (LPS-binding protein) binds monomer of LPS from Gram-negative bacteria and delivers it to a CD14 molecule that can be either soluble or membrane-bound (glycosylphosphatidylinositol-anchored protein). CD14 transfers LPS to the ectodomain of the TLR4/MD-2 receptor complex which leads to homodimerization of TLR4. This change in conformation leads to dimerization of the cytoplasmic TIR-domain (Toll-interleukin-1 receptor) that provides a binding site for MyD88. This activates the transcription factor nuclear factor-κB (NF-κB) and MAPK (mitogen-activated protein kinase) and transcription of various proinflammatory cytokines. In addition, the endocytosis of the LPS-TLR4/MD-2 complex leads to the TRIF-dependent signaling pathway that mediates the induction of interferon regulatory factor 3 (IRF3) and type-1 interferons.

Various studies have been conducted to understand the sequential molecular and structural basis of LPS recognition by TLR4. Lipopolysaccharide aggregates (i.e., micelles) are often secreted by Gram-negative bacteria in the form of outer membrane vesicles (OMVs). Extraction of LPS monomer from the LPS micelles requires concerted action of both LBP and CD14. LBP is a glycoprotein with elongated structure consisting of N-terminal and C-terminal domains connected by a central domain. Using negative-stain TEM imaging and mutagenesis experiments, A study showed that the N-terminal domain of LBP binds to LPS aggregates with high affinity. Mutations at the N-terminal conserved basic amino acids (Arg73, Lys67, Lys69) and basic amino acids (Arg119, Lys120, and Lys124) previously shown to facilitate LPS binding at N-terminal abolished binding of LBP to LPS. In a reaction mixture of LPS micelles with LBP or CD14, only LBP molecules bound to the surface of thread-like LPS micelles, suggesting CD14 cannot directly interact with LPS aggregates. However,

CD14 can bind to LBP-LPS aggregate and extract LPS monomers from the LBP–LPS aggregate. Interestingly, CD14 bound to the C-terminal end of LBP separated from the LBP N-terminal end bound to LPS micelles. Furthermore, using total internal reflection fluorescence (TIRF), Ryu et al. demonstrated the requirement of the physical interaction between C-terminal of LBP and CD14 for the transfer of LPS monomer to CD14. These experiments support the hypothesis that LPS monomers are somehow transported from the N-terminal end of LBP to the C-terminal end whereby CD14 can then extract these LPS monomers. CD14 carrying an LPS monomer then dissociates from LBP and subsequently facilitates LPS transfer to MD2/TLR4 complex. When TLR4–MD2 complex or MD2 alone was introduced to surface immobilized CD14 complexed with Cy5 labelled LPS, Cy5 fluorescence intensity drastically decreased only with the addition of TLR4–MD2, suggesting the requirement of the TLR4–MD2 complex for LPS extraction from the CD14–LPS complex. Furthermore, LPS binds with TLR4–MD2 with higher affinity compared to MD2 alone. Thus, LPS transfer to the TLR4–MD2 complex is a one-step process in that MD2 alone cannot receive/extract LPS from CD14. Researcher demonstrated that LPS binding leads to the dimerization of the TLR4/MD2–LPS complex, and LPS interacts with the hydrophobic pocket in MD-2 where five of six lipid chains are buried deep inside the pocket and one is exposed to the surface which interacts with a second TLR4 molecule. This leads to further structural changes in MD-2 that supports the dimerization of the TLR4–MD-2 complex. Binding of LPS to TLR4–MD2 depends on the structure of LPS as LPS from R. sphaeroides does not lead to dimerization of TLR4 compared to LPS from E. coli. Recently, Hubert el al. developed a computational model to study the thermodynamics and affinity of the receptor complex in the process of LPS transfer to TLR4 and suggested that the lipid affinity increases along the TLR4 signaling cascade. The affinity for lipid A binding to CD14 (136 ± 13 kJ mol^{-1}) is less than the affinity for MD2 (156 ± 11 kJ mol^{-1}) and lipid A binding to the MD2/TLR4 complex (~400 kJ mol^{-1}) is the most thermodynamically favored event. Furthermore, Phe126 residue in MD2 is important for the transfer of LPS from CD14 to the TLR4–MD2 complex which acts as a hydrophobic switch to mediate the signaling at the TLR4 complex dimerization interface.

TLR4–LPS binding promote homodimerization of the ectodomains of TLR4, and the subsequent structural and conformational changes induce dimerization of cytoplasmic TIR (Toll-interleukin-1 receptor) domains. The dimerized TIR domain structure is recognized by the downstream adaptor proteins MyD88 (myeloid differentiation primary response gene 88) and TIRAP (TIR domain-containing adaptor protein), leading to the formation of a protein complex, Myddosome, along with several other serine-threonine kinases of the IRAK family. Myddosome mediates the signaling cascade that leads to the activation and subsequent translocation of transcription factor NF-κB and expression of pro-inflammatory cytokines such as TNF-α, IL-1β, and IL-6. Furthermore, the MyD88-dependent pathway also activates the downstream MAP kinase pathway that leads to AP-1 transcription and expression of other pro-inflammatory cytokines.

In addition to signaling at the cell surface, TLR4 signaling within the endosomes begins with the assembly of TRIF (TIR domain-containing adaptor inducing IFN-β) and TRAM (TRIF-related adaptor molecule), leading to the assembly of triffosome which promotes activation of IRF3 (interferon regulatory factor 3) and induction of type-1 interferons. Thus, signaling cascades downstream of TLR4 in distinct subcellular sites leads to the production of different pro-inflammatory cytokines.

Sugar-dependent receptors such as C-type lectins on macrophages and dendritic cells play important roles in innate immunity. They recognize microbial polysaccharides and enhance signaling through TLRs; for instance, Neisseria gonorrhea lipooligosaccharide (lacks O-antigen) lipid A is detected by the TLR4/MD2 complex, and the oligosaccharide (phenotype C) is detected by sugar-dependent receptor C-type lectin DC-SIGN on dendritic cells (in human). Neisseria has several variants that differ in the terminal carbohydrate moieties of their lipooligosaccharide which play a vital role in immunomodulatory properties. These variants induce similar DC maturation, but the differences are seen in the cytokine profile and subsequent T-cell polarization depending on the differential C-type lectin usage. C-type lectin, SIGNR1 in macrophages of mice, enhances TLR4 oligomerization when macrophages are stimulated with E. coli or S. typhimurium. Using Ba/F3 cells expressing SIGNR1 and LPS from rough mutants of S. typhimurium, Nagaoka et al. showed that the SIGNR1-recognized polysaccharide part of LPS enhanced TLR4–MD2 complex oligomerization and subsequent production of proinflammatory cytokines. Similarly, Saunders et al. demonstrated that mice deficient in SIGNR1 were resistant to LPS shock and DSS-induced colitis with reduced colon damage and production of proinflammatory cytokine. Signr1-/- mice are partially protected from LPS-induced mortality; however, both Tlr4-/- and Tlr4-/- × Signr1-/- DKO mice are equally resistant to LPS shock. Thus, it is not clear whether LPS-induced SIGNR1 signaling can occur independently of TLR4, and additional comparative studies should address this in the future.

For decades, TLR4 remained the sole immune receptor for LPS, and the importance of TLR4 in pathogen recognition and homeostasis maintenance have been studied extensively. TLR4-deficient mice are highly susceptible to different strains of Gram-negative bacteria because of the inability to recognize and initiate a proper immune response. Conversely, TLR4-deficient mice are highly resistant to LPS shock. These studies highlight the importance of TLR4 in LPS recognition. Recent advances have identified TLR4-independent LPS recognition systems that include the transient receptor potential (TRP) channels and caspase-4/5/11 sensors.

LPS Recognition by TRP Channels

Transient receptor potential (TRP) cation channels present in sensory neurons and epithelial cells have been demonstrated as novel LPS sensors that are sensitized and activated before the initiation of immune responses through TLR4.

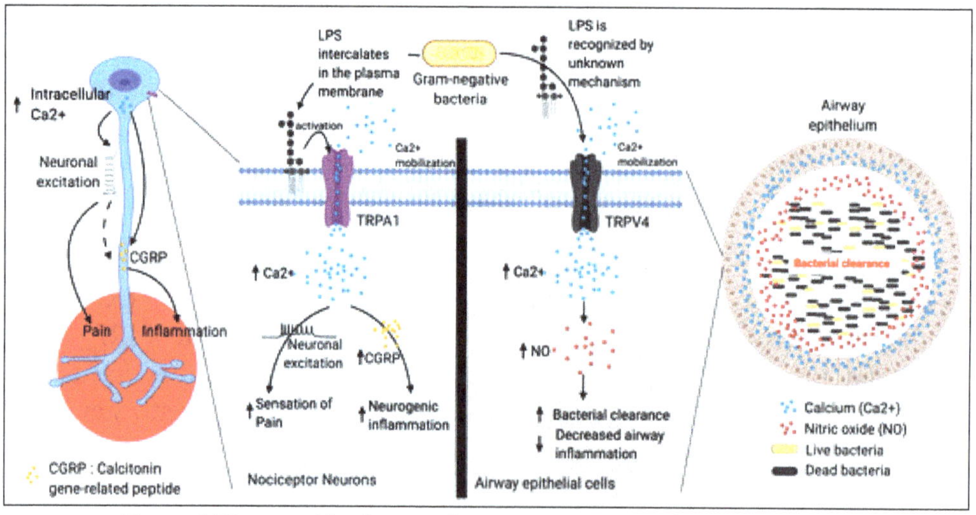

Figure: LPS recognition by transient receptor potential (TRP) channels.

Transient receptor potential (TRP) cation channels are novel sensors of LPS that are present in the sensory neurons and epithelial cells. TRP channels can recognize LPS before initiation of TLR4 signaling during inflammation. TRPA1 in nociceptive neuron and TRPV4 in non-neuronal cells, such as airway epithelium in mice, are the most studied. LPS recognition by these TRP cation channels leads to an influx of Ca^{2+} ions that generate action potential firing (neuronal excitation) which activates intracellular signaling cascade and the release of signaling peptides (CGRPs) that generate pain during inflammation. In the airway epithelium, LPS can be sensed by TRPV4 and the subsequent increase in Ca^{2+} influx leads to the production of nitric oxide (NO) which facilitates the pathogen clearance from the airways.

Transient receptor potential (TRP) channels are a large superfamily of ion channel proteins. They are divided into two groups and seven subfamilies according to the amino acid sequence homology. Group 1 consists of five subfamilies: TRPC (canonical), TRPV (vanilloid), TRPM (melastatin), TRPA (ankyrin), and TRPN (NOMPC-like). Group 2 consists of two subfamilies: TRPP (polycystin) and TRPML (mucolipin). Transient receptor potential channels share common structural features composed of homologous tetramers, each with intracellular carboxy (C–) termini, six transmembrane (TM) domains, and an amino (N–) termini with regulatory motifs. Transient receptor potential ion channels are expressed in a wide variety of cells and are activated by diverse stimuli. They are cellular sensors of various environmental and intracellular stimuli and play a critical role in cellular homeostasis. Most TRP channels are non-selective cation channels, and their activation leads to depolarization of resting membrane potential and activation of voltage-dependent ion channels which, in turn, change intracellular Ca^{2+} concentration. Recognition of pathogenic bacteria by these TRP cation channels in nociceptive neurons results in Ca^{2+} influx and action potential firing, leading to the intracellular signaling cascade responsible for release of signaling peptides that generate pain during inflammation. The role of TRPA1 and TRPV1 in neurogenic inflammation is the most extensively studied and well established.

Gram-negative bacterial infections leading to inflammation are accompanied by somatic or visceral pain which is attributed to activation of nociceptors by inflammatory mediators released from immune cells and, until recently, was thought to be secondary to TLR-dependent immune activation. However, various studies have shown that the neuronal activity in response to LPS occurs rapidly before the initiation of a pro-inflammatory immune response. Meseguer et al. showed that the neurogenic inflammation and pain produced by LPS was dependent on the activation of functional TRPA1 channels and was independent of TLR4 activation. Lipopolysaccharide activated the sensory neurons (as demonstrated by intracellular calcium measurement) isolated from the nodose and trigeminal ganglia of WT mice which was not affected by TLR4 deficiency. Importantly, LPS-induced calcium influx and activation of these neurons were dependent on TRPA1 channels as demonstrated by lack of LPS-induced calcium influx in Trpa1-/- mice. Lipopolysaccharide-induced mobilization of calcium in the neurons triggered neuronal excitation and sensation of pain. Calcitonin-gene-related peptide (CGRP, a neurogenic peptide) is produced in response to several neuronal stimuli and promotes neurogenic inflammation. Lipopolysaccharide induced the production of CGRP in a TRPA1-dependent and TLR4-independent manner. To address whether Ca2+ directly promotes CGRP production or neuronal excitation are precursors to CGRP will require further investigation. TRPA1-mediated sensing of LPS is an evolutionarily conserved mechanism, as a role for TRPA1 in LPS sensing has been observed in Drosophila melanogaster. Researcher demonstrated that TRPA1 functions as a chemosensor of pathogenic cues and is responsible for aversive responses and gustatory-mediated avoidance to LPS in Drosophila melanogaster. TRPA1 activation by LPS in neuronal cells was completely inhibited after polymyxin B treatment (a cationic cyclic antibiotic that binds to lipid A), suggesting an important role for the lipid A moiety in this process. The efficiency of TRPA1 activation and the inflammatory outcomes have been shown to vary according to the lipid A structure of LPS, i.e., LPS with asymmetrical hexa-acyl lipid A induced the strongest activation of TRPA1 as measured by calcium influx. Furthermore, LPS treatment-induced TRPA1 currents in isolated membrane patches suggest a role for TRPA1 in sensing membrane perturbations. These observations have been strengthened by a recent study demonstrating that LPS is able to be inserted into the cell membrane of TRPA1-expressing CHO cells. Importantly, when LPS was added to lipid vesicles made from 1,2-dipalmitoyl-sn-glycero-3-phosphocholine (DPPC), decreased permeability of the vesicles was noted as measured by the fluorescent probe Laurdan (6-lauroyl,1-2-dimethylamino naphthalene) with an emission shift from a 400–525 nm wavelength. Thus, these studies directly support the hypothesis that LPS intercalates into cell membranes and promotes mechanical perturbations to activate TRPA1.

In addition to TRPA1, Researcher demonstrated that LPS activates TRPV1 to promote an intracellular calcium influx in HEK293T cells transfected with recombinant human TRPV1. In the mouse dorsal root ganglion (DRG) neuron, the LPS response that was only partially blunted in Trpa1-/- neurons was completely abrogated in Trpa1-/-Trpv1-/- DKO neurons. This suggests involvement of TRPV1 in sensing LPS and neurogenic inflammation. Moreover, using recombinant TRP expression in HEK293T cells, Alenmyr and colleagues demonstrated that, in addition to TRPA1 and TRPV1, LPS weakly activated TRPM3 and TRPM8 to promote calcium influx. It should be noted that other studies have demonstrated expression of TLR4 in a subclass of TRPV1 containing trigeminal nociceptors and that LPS sensitized TRPV1 via TLR4-mediated signaling. Thus, more extensive studies are needed to determine whether TRPV1, TRPM3, and TRPM8 are directly activated by LPS-induced membrane disruption or require TLR4 for their activation.

In addition to sensory neuronal TRP, a recent study has shown the presence of a non-neuronal TRP channel that can detect LPS; specifically, the TRPV4 channel in airway epithelium in mice can detect LPS. TRPV4 is expressed in both murine and human airway epithelial cells. Researcher demonstrated that murine airway tracheobronchial epithelial cells (mTECs), when stimulated with LPS, induced calcium influx that was dependent on TRPV4 and independent of TLR4. LPS-induced activation of TRPV4 channels in mTEC led to nitric oxide (NO) production that was subsequent to calcium influx. Importantly, LPS-induced calcium influx and NO production was intact in mTEC from TLR4-deficient mice. TRPV4-mediated LPS sensing is functionally important, as WT mTEC-mediated killing of non-pathogenic E. coli in vitro was blunted in Trpv4-/- mTEC. In vivo, LPS aerosol exposure induced significantly stronger ventilator responses (as measured by Penh value) and increased neutrophil and monocyte infiltration in the lungs of Trpv4-deficient mice when compared to WT controls. Thus, TRPV4-mediated early responses to LPS may be important in curbing unwanted inflammation in the lungs.

More recently, a role for TRPV4 has also been shown in macrophages using siRNA knockdown of TRPV4 or via chemical inhibition using HC-067047 or GSK2193874 (inhibitors of TRPV4). With both approaches, macrophages in the absence of functional TRPV4 had an attenuated response to LPS as demonstrated by reduced TNF, IL-6, and NO production. Further studies are needed to clarify whether LPS-induced activation of TRPV4 in macrophages are also TLR4 independent.

Altogether, these studies provide solid evidence that TRP ion-channels, such as TRPA1 and TRPV4, can quickly detect the presence of LPS to initiate acute inflammatory responses and provide a robust defense against invading pathogens.

LPS Recognition by Caspase-11

TRP channels and TLR4 can both recognize LPS independently to initiate non-redundant signaling pathways; however, both of these receptors are membrane bound and, as a result, recognize LPS that are either extracellular or within the endosomes. So, what happens to the LPS that may find its way into the cytoplasm of the cell? This was a question that remained unanswered until recently. We know now that, in the cytoplasm, LPS are sensed by caspase-11 in mice and caspase-4 and -5 in humans.

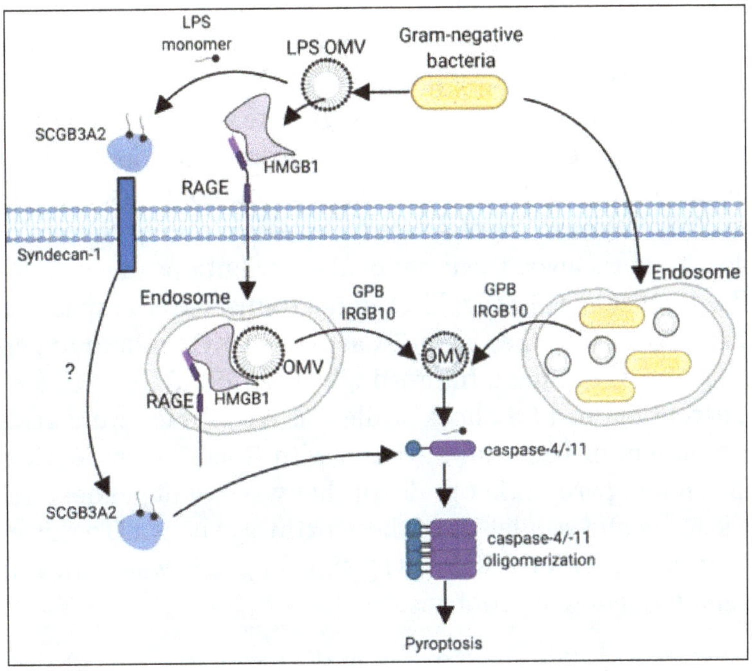

Figure: LPS recognition by caspase-4/-11.

LPS can access cytosol when Gram-negative bacteria disrupt the phagolysosome or by the uptake of LPS-containing outer membrane vesicles (OMVs) released by live bacteria and their subsequent lysis. The uptake of LPS/OMV is facilitated by HMGB1. The HMGB1–LPS complexes are internalized into endosomes through a receptor for advanced glycation end-products (RAGEs). LPS is released into cytosol after HMGB1 permeabilizes the phospholipid bilayer under acidic conditions. In addition, host GBPs and IRGB10 can facilitate the release of LPS from endosomes. Secretoglobin (SCGB)3A2 binds LPS, and the SCGB3A2–LPS complex can access cytosol (via an unknown mechanism indicate by "?" in the figure) through binding with the cell surface receptor syndecan-1. Once the LPS is delivered into the cytosol, caspase-4/-11 senses LPS, leading to its oligomerization and induction of pyroptosis.

Caspases are a family of conserved endoproteases with cysteine protease activity that specifically cleave the peptide bond after aspartic acid residue in the substrate and play central roles in driving apoptosis and inflammation. Inflammatory procaspases (caspase-1, -4, -5, -12 in humans and caspase-1, -11, and -12 in mice) require catalytic activation through the stimulated PRRs and subsequent formation of inflammasomes. Upon activation, inflammatory caspases process and produce a mature form of inflammatory cytokines (IL-1β and IL-18) and induce pyroptotic cell death. Caspase-11 in mice corresponds to caspase-4 and caspase-5 in humans and induces activation of non-canonical inflammasome.

Over a decade ago, caspase-1 and caspase-11 were thought to contribute similar signaling pathways in inflammation due to the fact of their highly homologous sequences. Further supporting this hypothesis, in vivo studies with caspase-1 (Casp1-/-) and caspase-11 (Casp11-/-) knockout mice demonstrated that both were resistant to endotoxin shock induced by LPS when compared to wild-type mice. However, in 2011, A study discovered that the original Casp1-/- mice generated using 129 mouse strain embryonic cells were actually deficient in both caspase-1 and caspase-11. Sequencing of 129S1 genomic DNA revealed a 5 bp deletion within caspase-11 locus, generating highly unstable caspase-11 transcripts and, thus, mice on a 129 background essentially have no functional caspase-11. Because of the close proximity of the caspase-1 and caspase-11 loci (~1500 bp) on the chromosome, it is nearly impossible to segregate these genes by recombination; hence, the resultant Casp1-/- mice generated on a 129 background are doubly deficient for caspase-1 and caspase-11. Indeed, these Casp1-/- mice are now commonly used in research as Casp1-/-Casp11-/- DKO mice. This discovery necessitated a re-evaluation of the studies that utilized the original Casp1-/- mice. To tease apart the functional contributions of caspase-1 versus caspase-11 during LPS-induced endotoxemia, Kayagaki et al.microinjected a caspase-11 bacterial artificial chromosome (BAC) transgene into the Casp1-/-Casp11-/- DKO embryo to generate mice lacking just caspase-1 (Casp1-/-Casp11Tg mice). Interestingly, they found that WT and Casp1-/-Casp11Tg mice were equally susceptible to LPS shock while Casp11-/- mice were resistant. Importantly, during CTB+LPS stimulation or bacterial infection with E. coli, C. rodentium, and V. cholerae, caspase-11 directly promoted pyroptotic cell death that was not dependent on caspase-1. Thus, a search for upstream sensors and regulators of these pathways began. To this end, multiple subsequent studies showed that the TLR4–TRIF–IRF3 signaling axis was critical for activation of the caspase-11 pathway and induction of proptosis.

Furthermore, TLR/IFNAR signaling-mediated complement pathway activation (Cbp1-C3-C3aR) was shown to be required for priming and activation of caspase-11. However, the first breakthrough came from two independent studies showing that intracellular LPS activated caspase-11 and subsequent pyroptotic cell death. Given that CTB promoted intracellular delivery of virulence factors in the host cell cytoplasm, Kayagaki and colleagues hypothesized that delivery of LPS into the cytoplasm of the cell drove non-canonical inflammasome activation and cell death independently of TLR4. Indeed, transfection of LPS or lipid A (active portion of LPS) was sufficient to promote caspase-11-dependent IL-1β production and cell death in BMDMs. Importantly, LPS or lipid A transfection-induced IL-1β and cell death were still observed in TLR4-deficient BMDMs, suggesting a TLR4-independent recognition of cytoplasmic LPS. In line with these observations, A researcher independently showed that transfection of Gram-negative bacterial lysates or LPS induced cell death that was dependent on caspase-11. This group demonstrated that specific lipid A structures were required to activate caspase-11; penta- and hexa-acylated lipid A activated caspase-11, whereas tetra-acylated lipid A was not detected, i.e., LPS from Francisella novicidia and Yersinia pestis that comprise the tetra-acylated lipid A structure do not activate non-canonical inflammasomes, and many pathogens seem to exploit this feature to escape caspase -11 detection. In addition to in vitro studies, LPS-induced shock in TLR3-primed mice was demonstrated to be driven by caspase-11 independently of TLR-4, suggesting the presence of a TLR-4 independent LPS sensing mechanism. Altogether, these results clearly demonstrate a novel cytoplasmic sensor that activates caspase-11 and induces subsequent non-canonical NLRP3 inflammasome activation and pyroptotic cell death.

Bringing the search for the elusive cytoplasmic LPS sensor to an end, an elegant series of studies by Shi et al. showed that caspase-4 in humans and caspase-11 in mice directly sense LPS. This was a ground-breaking study given that it identified caspase-4/caspase-11 (inflammatory caspases) as novel cytoplasmic LPS sensors. The discovery of caspase-4/11 as a potential direct LPS sensor was fueled by a serendipitous observation while attempting to generate recombinant caspase-4/11 for biochemical studies. Recombinant caspase-4/11 purified from Sf21 insect cells eluted at a size of ~100 kDa while recombinant caspase-4/11 was eluted from E. coli at a size of ~600 kDa. Analytical ultracentrifugation and static light scattering of the ~100 kDa elute showed that recombinant caspase-4/11 from Sf21 were monomers and recombinant caspase-4/11 from E. coli were oligomers. These observations suggested that some factors present in the bacteria were causing caspase-4/11 oligomerization. Given that LPS is the most abundant component in E. coli, Shi et al. hypothesized that caspase-4/11 may directly bind to LPS, resulting in its oligomerization. Indeed, biotinylated LPS or lipid A efficiently precipitated caspase-4 and caspase-11 in transfected HEK293T or immortalized BMDM cells. These results were further corroborated by surface plasmon resonance experiments where both LPS and lipid A showed a strong dose-dependent association with caspase-4 and caspase-11. Further biochemical studies demonstrated that the CARD domain of caspase-11 was sufficient to bind to LPS. This series of studies demonstrated that the CARD domain of caspase-4/11 directly binds to the lipid A portion of LPS which results in caspase-4/11 oligomerization and subsequent activation of NLRP3 inflammasome and pyroptotic cell death.

While the molecular mechanisms leading to LPS sensing by caspase-11 and subsequent inflammatory processes have been well-established, how LPS makes its way into cytosol remained unknown until recently. High mobility group box 1 (HMGB1), when added with extracellular LPS onto peritoneal macrophages or human THP1 monocytes, induced significant pyroptotic cell death that was dependent on caspase-11 and caspase-4, respectively. High mobility group box 1 directly bound LPS as shown by surface plasmon resonance and pull-down experiments. The HMGB1 A box motif binds to the LPS polysaccharide and B box motif with lipid A, and both motifs are important for HMGB1's interaction with LPS. Given that the requirement for HMGB1 can be bypassed by LPS delivery via cholera toxin B or lipofectamine, Deng and coworkers hypothesized that HMGB1 plays a role in delivering LPS into the cytoplasm. Indeed, they reported that LPS was detected in the cytosol of macrophages only when stimulated with LPS+HMGB1 but not LPS alone. Furthermore, proximity ligation assay experiments showed that the intracellular LPS interacted with caspase-11 in macrophages treated with LPS+HMGB1 but not LPS alone. Because HMGB1 internalization into cells is mediated by its receptor RAGE, genetic deletion of RAGE or RAGE inhibition by blocking antibodies both impaired HMGB1-mediated delivery of LPS into the cytoplasm of macrophages. In vivo, TLR4 stimulation of liver hepatocytes promotes HMGB1 production that can then bind extracellular LPS to be delivered into innate immune cells via RAGE.

In addition to HMGB1, a recent study implicated secretoglobin 3A2 (SCGB3A2), a protein primarily secreted by airway cells, in promoting LPS delivery into macrophages and tumor cells to induce pyroptotic cell death. Biotinylated LPS interacted with recombinant SCGB3A2 as demonstrated by pull-down assays. The SCGB3A2 binding resulted in de-aggregation of LPS micelles as demonstrated by crude LPS migration in gels and dynamic light scattering in the presence or absence of SCGB3A2. The SCGB3A2 required syndecan-1 to be present on the cell surface for mobilization of LPS into the cytoplasm. These studies clearly highlight how different host-derived proteins play important roles in binding LPS and delivering them into the cytoplasm. A study also showed that bacterial outer membrane vesicles (OMVs) containing LPS are endocytosed, resulting in subsequent release of LPS into the cytoplasm. Studies have shown that LPS, once inside cells (presumably in endosome/lysosome vacuoles), requires an additional set of regulatory molecules to be released into the cell cytoplasm to be detected by caspase-11. The family of guanylate binding proteins (GBPs) has been implicated in this process of LPS release from the intracellular vacuoles to the cytoplasm. During an active bacterial infection, GBPs and IRGB10 are recruited to the pathogen-containing vacuole and induce lysis of the vacuole to release pathogen products including LPS to the cytoplasm. Specifically, one study showed that caspase-11-dependent cell death induced by LPS transfection or LPS+CTB treatment was attenuated in Gbpchr3-/- (deficient in GBP on chromosome 3 which includes GPB1, GBP2, GBP3, GBP5, and GBP7) macrophages. Interestingly, the reduction of cell death in the absence of GBPs was only partial, suggesting that GBPchr3-independent mechanisms are also involved.

Sterol based Glycolipids

They occur as free sterols, esters (acylated), alkyl ethers (alkylated), sulfated or linked to a steryl glycoside moiety. The biosynthesis of sterol is everpresent amid eukaryotes but almost wholly deficient in prokaryotes. Being absent in Archaea, they are sparingly distributed in bacteria and Methylococcus capsulatus (proteobacterium), Gemmata obscuriglobus (planctomycete) along with myxobacteria members. As a consequence, the existence of varied steranes (saturated 4 cycle skeleton) in primordial rocks is used as proof for eukaryotic evolution somewhere 2.7 billion years ago.

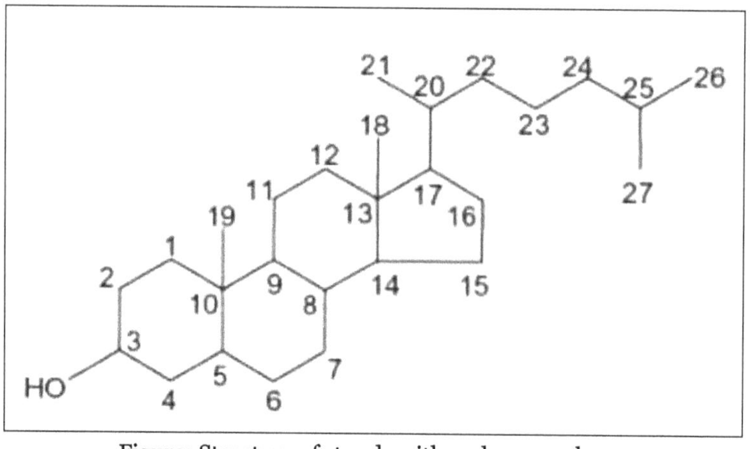

Figure: Structure of sterols with carbon numbers.

The sterols have hopanoids as precursor and an oxygen-dependent biosynthesis starting with the creation of 2, 3-oxidosqualene (primary intermediate). Hence, steroid and triterpenoid hydrocarbon detection in the history of Earth has been used to surmise the relic of oxygenic photosynthesis. It has been conjecture that levels of oxygen increased in the atmosphere made the sterols evolution feasible and make possible the eukaryotes birth giving them with an untimely protection mechanism against oxygen.

Free Sterols

Sterols shape an imperative cluster amid steroids. The unsaturated steroids involve a 3β-hydroxyl group (cholestane skeleton) and carbon atoms (8 or more) of an aliphatic side chain at position 17 figuring the group of sterols. Being saponification resistant, they are endowed in all animal and vegetal tissues important in the appraisal of marine sediment maturity as a biological marker compound, yet past hundreds millions years. They also comprise 3-hydroxysteroids (C27 to C30) crystalline alcohols and are classed as triterpenes (being derived from squalene) directly by unsaturation, cyclization and 3β-hydroxylation, cycloartenol in plants or lanosterol in animals.

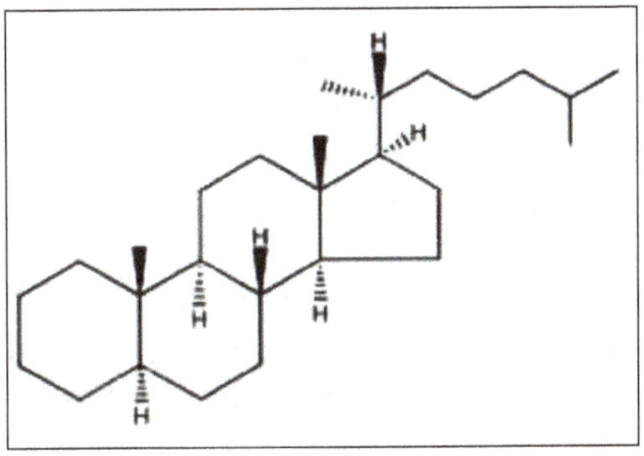

Figure: 5 α -cholestane.

Precursor and Derived Lipids

These include fatty acids, glycerol, steroids, other alcohols, fatty aldehydes, and ketone bodies, hydrocarbons, lipidsoluble vitamins, and hormones. Because they are uncharged, acylglycerols (glycerides), cholesterol, and cholesteryl esters are termed neutral lipids.

Sterols

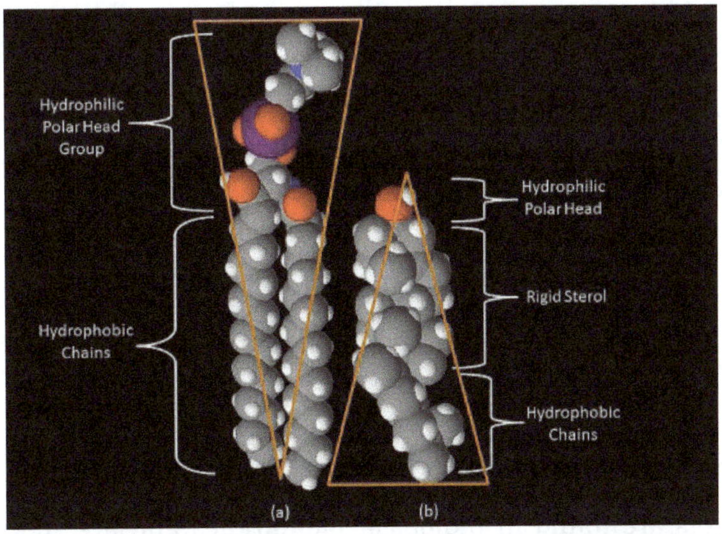

Figure: Space-filling diagram of (a) sphingomyelin and (b) cholesterol depicting relative size and depth in membrane.

Sterols are a class of hydrophobic ringed lipid molecules found in biological membranes throughout eukarya. Sterols can comprise greater than 50% of the membrane lipid content in cells, and are known to alter membrane fluidity and structure. Though the mechanisms of these effects are still under debate, sterols have been shown to induce altered lipid phases in bilayer systems, generally increasing the order of acyl chains while maintaining lipid translational fluidity. Their actions are thought to be essential to the formation of lipid rafts with various effects on cell biology and serve as precursors to steroid hormones in animals.

Sterol Structure

All naturally occurring sterols are comprised of a hydrocarbon tetrameric fused ring structure and acquire a small degree of polarity from a 3-hydroxyl in the β configuration; α-hydroxyl sterols are not found in natural membranes. The most common sterols also have a short aliphatic tail on the opposite end on C17. In lipid bilayers the hydroxyl acts as the polar head group and orients towards the phospholipid head groups of the bilayer, while the rings and short hydrophobic tail extend into the hydrophobic core of the membrane, aligning with the acyl chains of the phospholipids. The topography of sterols is such that the α side of the relatively rigid ring structure is flat while the β side has protruding methyl groups; this structure constrains their interactions with lipids allowing significant contact on the flat surface.

Common Sterols

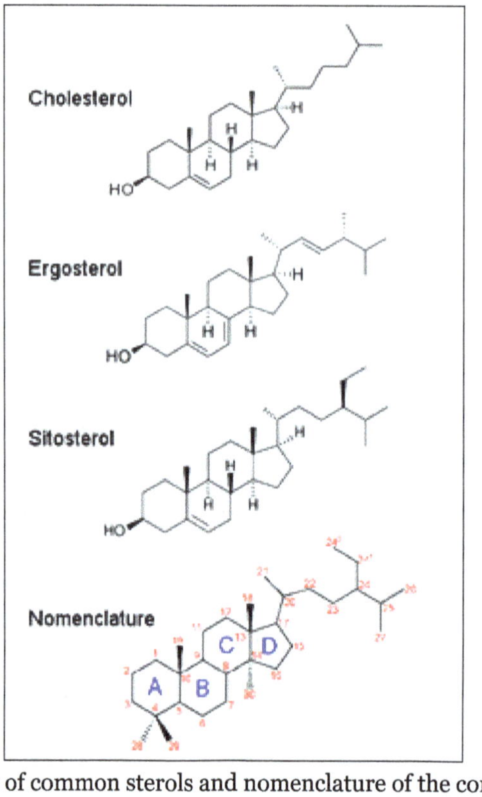

Figure: Backbone structures of common sterols and nomenclature of the conserved backbone carbons.

Many different sterols are found in biological membranes and are specific to certain phyla. Subtle differences in their structure can change their effect on membrane structure and dynamics. Cholesterol is the most common sterol in animal membranes accounting for approximately 20-30% of sterols found therein. It is characterized by a single double bond in ring B (C5-C6) and the lack of a branching carbon at position 24. Ergosterol is found in the membranes of fungi and protists and differs from cholesterol by containing two more carbon-carbon double bonds (one in ring B, another between carbons 22 and 23 in the tail region) and a methyl group at position 24.

Sterols from plants are known as phytosterols. Common phytosterols include campesterol, stigmasterol and sitosterol, though hundreds of different compounds have been isolated from plants. Hopanoids are pentacyclic compounds that share a similar structure with sterols and are often found in prokaryotic membranes, where they are thought to have an effect on membrane stability and structure similar to sterols.

Liquid-Ordered Phase (l_o)

Sterols have pronounced effects on the phase behavior of lipid membranes. When added to lipids in the gel phase ($L_{\beta'}$) sterols tend to liquefy the membrane structure by disrupting the highly ordered tilted structure of the hydrocarbon chains. At low temperatures this induces the liquid crystal phase ($L_{c'}$) where the lipid tails are fully extended, but not tilted. Conversely, cholesterol tends to solidify lipids in the liquid disordered phase (L_α). In both cases, higher cholesterol concentration leads to the formation of an intermediate phase, known as the liquid ordered (L_o) phase. In the l_o phase, the phospholipid tails have higher order similar to the gel phase, but maintain lateral fluidity similar to the Lα phase. The area per molecule is also reduced.

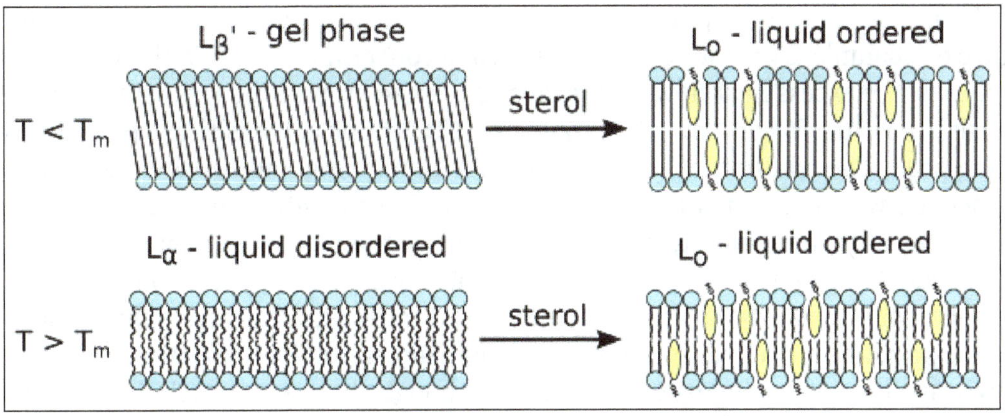

Figure: Sterol-induced liquid-ordered phase.

Lipid phase diagrams show the phase coexistence of L_o and either L_α or the ripple phase P_β (sometimes called pre-transition phase), depending on temperature in lower cholesterol content. As the relative concentration of sterol in the membrane increases, the l_o phase dominates as all phospholipids come in near contact with sterols. The ordering of saturated acyl chains in the presence of cholesterol has been studied by many different methods including 2H-NMR, EPR, and MD simulations, which show that sterols induce order in neighboring fatty acid tails. Headgroup type, tail length and saturation, and whether the lipids are glycerol or sphingosine-based all influence the effects of sterols. Notably, sterols have the most pronounced effect on the phase behavior of long saturated phospholipids like dipalmitoylphosphatidylcholine

(DPPC); however, the presence of one saturated tail is sufficient for the effect. Likewise, sterols with α conformation or methyl groups in the trans conformation are incapable of affecting the phase behavior of lipids.

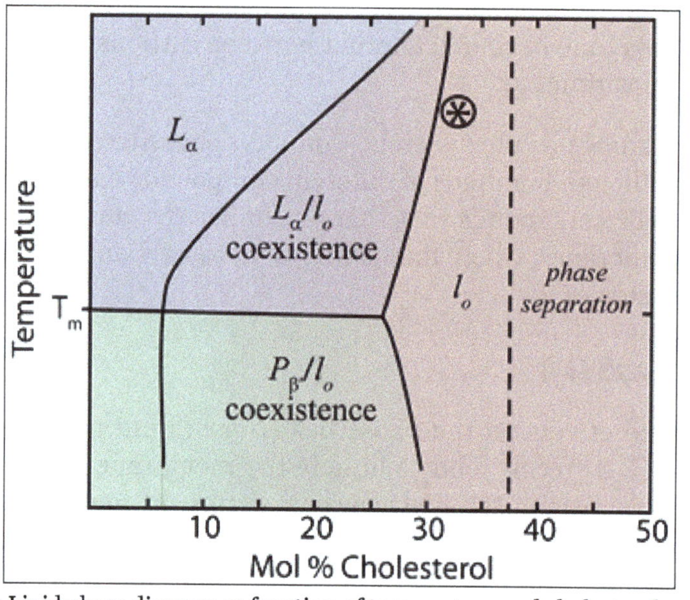

Figure: Lipid phase diagram as function of temperature and cholesterol content.

Cholesterol and the Main Phase Transition

The effect of cholesterol on lipid phases has been extensively studied with a method known as differential scanning calorimetry (DSC). With DSC the rate of heat flow can be measured as the temperature of a sample is increased or decreased at a constant rate. When the sample transitions between phases, the excess heat flow is recorded as a spike around the temperature at which that phase transition occurs. When heated or cooled past their phase transitions, pure lipids will display a sharp peak. The addition of sterols leads to a dampening and broadening of the enthalpy spikes associated with pre-transition and the main phase transition, effectively eliminating the phase transitions at higher cholesterol concentrations as the membrane enters the l_o phase. This effect is dependent on the type and amount of sterol as well as the type of phospholipid. For example, a precursor to cholesterol known as lathosterol differs in structure only by the placement of one double bond. This small difference means the lathosterol decreases the temperature of the main phase transition while cholesterol increases it, though cholesterol is eliminates the transition at high molar percent while lathosterol does not.

The Condensation Effect

When cholesterol is added to a membrane, the 2-dimensional area per molecule decreases while the membrane thickness increases; this is known as the condensation effect. The mechanism for this effect is unclear, but the rigid structure of cholesterol means that the reduction in surface area must be due to changes in the order of the lipid tails. In DMPC membranes, increasing cholesterol from 0 to 50 mol% can lead to 25% reduction in area per DMPC molecules. This condensation augments membrane strength and decreases permeability, and is strongest when paired with phosphatidylcholine-based lipids.

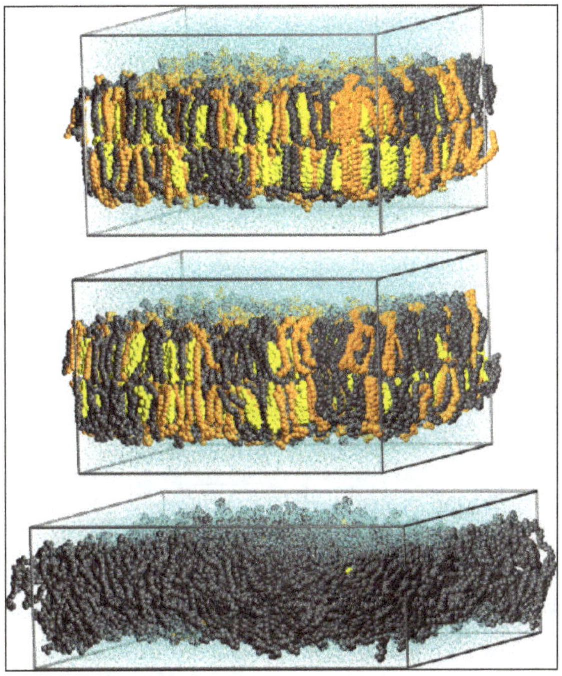

Figure shows simulation of sterol-induced phases. Palmitoyloleoylphosphatidylcholine (POPC) appears in grey, palmitoylsphingomelin in orange, and cholesterol in yellow. Note the increased order in the L_o phase and the change in dimensions of the membrane when cholesterol and PSM are present. Cholesterol also tends to interact with PSM more than POPC.

Sterols in Rafts

In biological membranes sterols are essential components of lipid rafts. Cholesterol makes a hydrogen bond with the sphingosine head group and preferentially associates with the fatty acid tails of sphingomyelin over other lipid tails. Cholesterol and sphingomyelin together solidify the membrane and decrease the lateral diffusion of lipids leading to long-lasting L_o domains amid L_α lipids. The lipid characteristics of rafts are thought to be important in determining membrane protein function.

Steroid Hormones

In animals, cholesterol serves as a precursor for all steroid hormones such as estradiol, aldosterone and cortisol, which act via intracellular receptors to regulate gene transcription in target tissues. These hormones are vital to nearly all physiological function including growth and reproduction, the stress response, water balance and metamorphosis (in invertebrates).

Cholesterol

Cholesterol is the main sterol (C27 alcohol) in the tissues of vertebrates rich in adrenals (10% w/w), gall stones, liver (0.2% w/w) and nervous tissues (2% w/w). Brain is principally cholesterol rich covering on-fourth of total free cholesterol avaialble in the vertebrate cadaver.

Cyclopentanoperhydrophenanthrene ring (sterane) is a carbon structure of cholesterol and it is the first sterol isolated by FP Poulletier de La Salle from gall stones. In 1815, ME Chevreul isolated

Cholesterine (Greek: khole-bile, stereos-solid) from the unsaponifiable fraction of animal fats. F Reinitzer proposed the accurate formula ($C_{27}H_{46}O$) but exact steric depiction of cholesterol structural came after the work of HO Wieland who won Nobel Prize in Chemistry for his work on the establishment of the bile acids and related substances and AOR Windaus who got the Nobel Prize in 1928 for establishing the connection of sterols with vitamins. RK Callow and FG Young have voted that steroids chemically allied to cholesterol. In 1913, D Steinberg proposed the central role of cholesterol in atherogenesis.

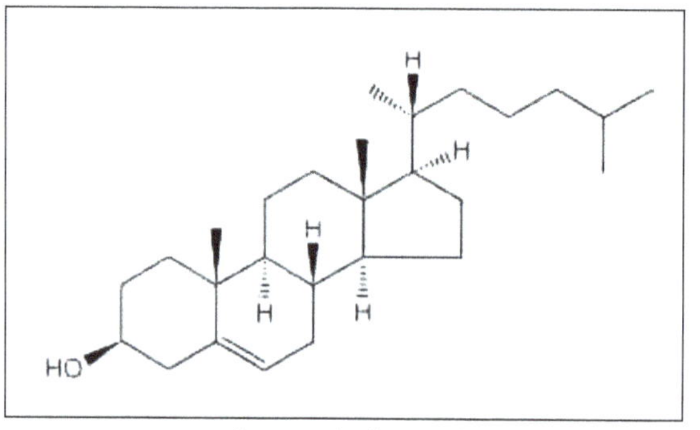

Figure: Cholesterol

In animal cell membrane, the concentration of cholesterol is usually high ranging from 30 to 50% (molar percentage of total lipids) in erythrocytes and as elevated as 80% in the membranes of ocular lens. Accordingly, cholesterol has numerous task in membranes varying from the control of phase behavior, offering mechanical strength, to precursor of hormones and vitamins and in scheming raft arrangement and membrane protein activity. Minor modifications of the side chain led to the replacement of cholesterol by sterols such as campesterol, β-sitosterol at cellular level. In the femoral gland of Acanthodactylus boskianus (male lizard) cholesterol is present in abundance and used as a scent blotching pheromone to ascertain dominance. In addition, cholesterol can also silhouette ester connections with secreted polypeptide and signaling molecules encoded by the hedgehog gene family that function in some molding actions amid metazoan evolvement. Sponges signify the richest source of peculiar sterols i.e. cholesterol and sterols, but bearing 1-3 extra carbon atoms at C24 with unusual features as quaternary alkyl groups, acetylenes, allenes, cyclopropene and cyclopropane rings. In class Demospongiae, 24-Isopropylcholesterol (with its analogue unsaturated at C22-C23) is abundant and characteristic of Neoproterozoic era (542-1000 million years) sediments and is the primeval proof of fossil record for animals but absent the eumetazoans (bilaterian and cnidarians). In Calyx nicaensis, Nicasterol was identified.

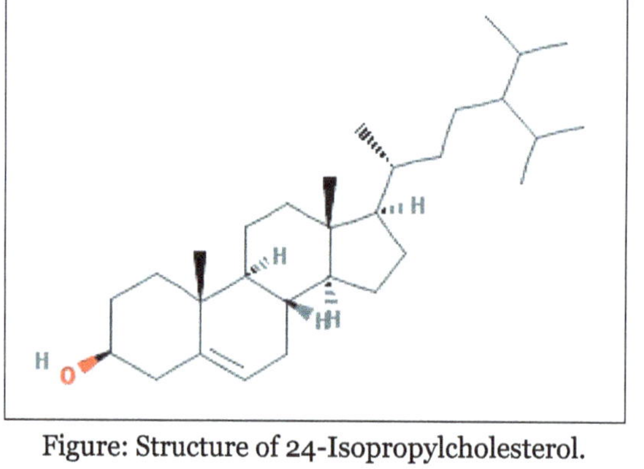

Figure: Structure of 24-Isopropylcholesterol.

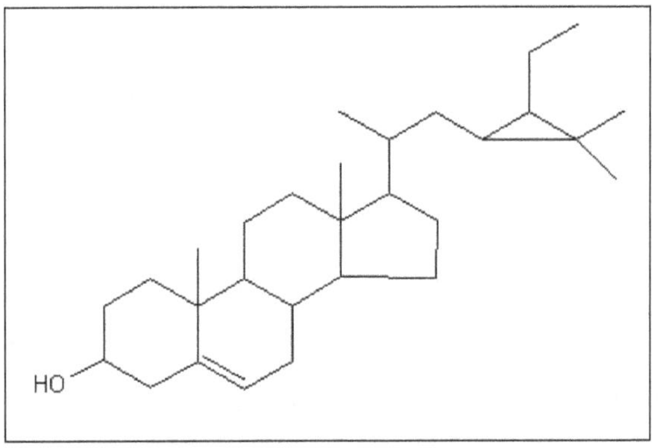

Figure: Structure of Nicasterol.

In higher plants, the presence of cholesterol is mostly accepted and detected in vegetal oils in a diminutive percentage (5% of the total sterols) however, a high cholesterol content has been reported in the oil of Camelina (~200 mg per kg). Nevertheless, numerous investigation have divulged the presence of cholesterol in chloroplasts, pollens, higher plant leaves (~72% of the total sterols in that fraction). It is the only sterol of Laurencia paniculata and also foremost in most Rhodophyceae algae.

Lanosterol, the common precursor in the synthesis of cholesterol, is also found as a foremost component of the unsaponifiable fraction of wool fat (lanoline, ~15%). Gemmata obscuriglobus, a bacterium of Planctomycete, is capable to produce lanosterol and its unusual isomer, parkeol. Derivatives of lanosterol have been reported in methanotrophic bacteria. 4- methylcholestan-8 (14), 24-dien-3β-ol, an abundant derivative was first reported in Methylococcus capsulatus and afterward in other analogous bacteria.

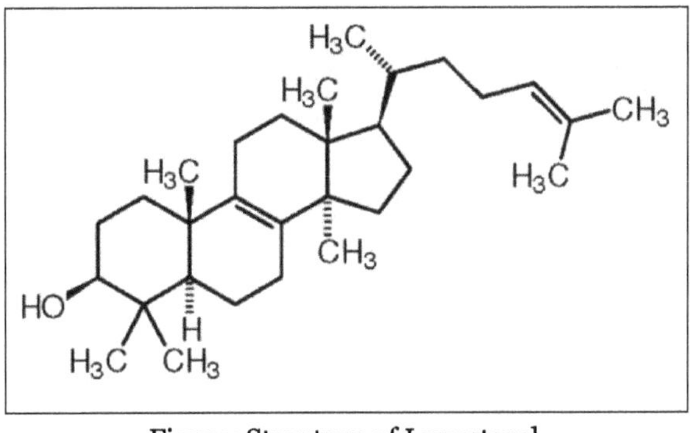

Figure: Structure of Lanosterol.

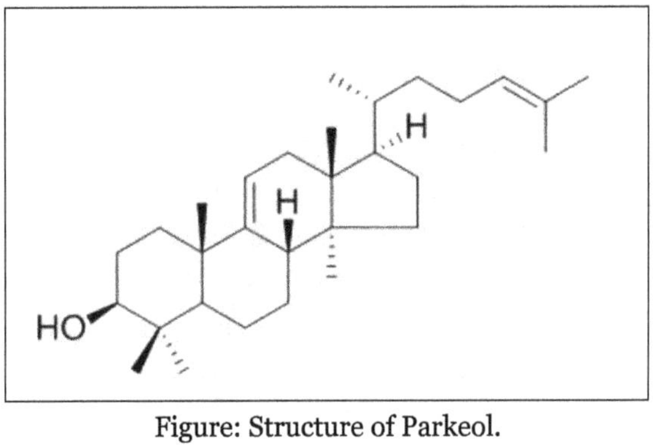

Figure: Structure of Parkeol.

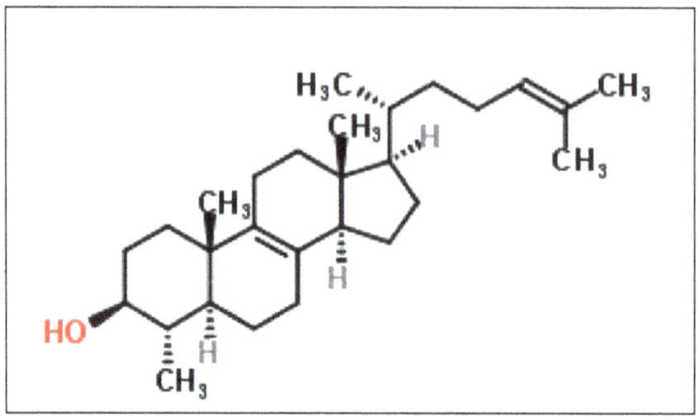

Figure: Structure of 4-Methylcholestan-8(14), 24-dien-3β-ol.

Compound with lanostane core (lucidenates, ganoderates) have been successfully isolated from a mushroom (Ganoderma lucidum). The presence of these sterols have led to the utilization of mushroom for curing bronchitis, cardiovascular, diabetes, gastritis, hepatitis, hypercholesterolemia, hypertension related pathologies in traditional Chinese and Japanese medicine.

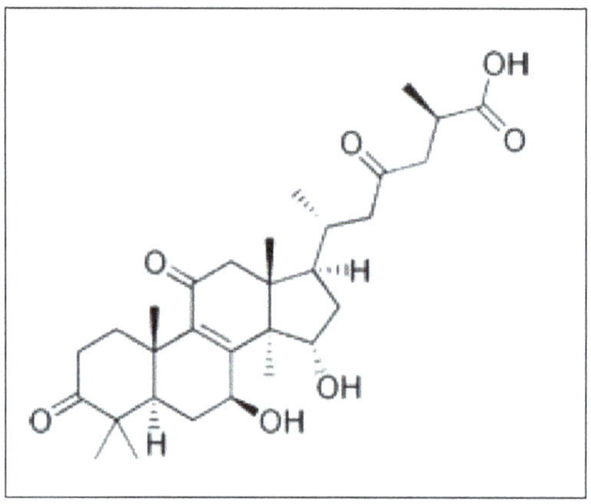

Figure: Structure of Ganoderic acids from Ganoderma lucidum.

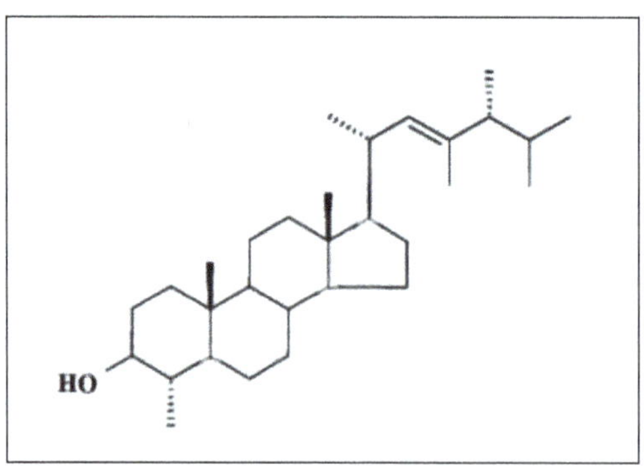

Figure: Structure of Gorgosterol.

In animal tissues, 7-dehydrocholesterol is present in minute amounts which get transformed to cholecalciferol (vitamin D3) on UV exposure. 24-dehydrocholesterol (Desmosterol), an intermediate product amid cholesterol and lanosterol is associated with myelination processes with an elevated echelon noticed in the brain of young animals and no desmosterol in adults. In astrocytes and spermatozoa of mammalian cells, it is an abundant membrane component. In red algae, 22-dehydrocholesterol and desmosterol are present in high concentrations. Desmosterolosis is a severe cognitive impairment and developmental defect in humans where desmosterol fails to get converted to cholesterol. In 1943, Gorgosterol was revealed by Bergmann from coral like animals (intracellular photosynthetic dinoflagellate symbionts belonging to zooxanthellae). The inventive structure abides a cyclopropane ring in the side chain and unusual C-23 methyl groups refurbishing interest in marine sterols. Dinosterol (4α, 23, 24-trimethyl-5α-cholest-22E-en-3β-ol) is a unswerving biomarker usually found in dinoflagellates.

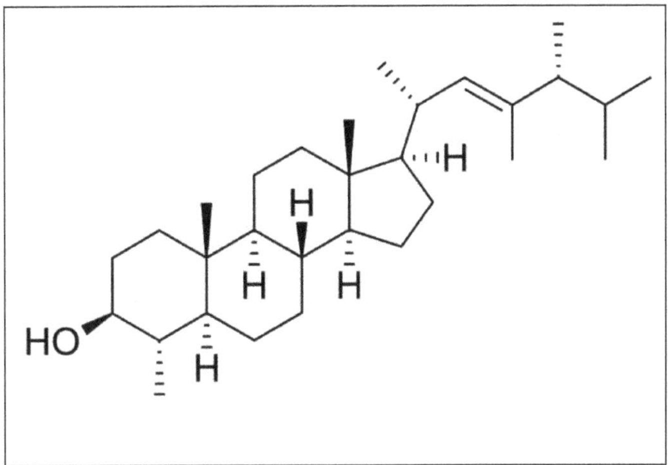

Figure: Structure of Dinosterol.

In diatoms, 23-methyl sterols and their sterane counterpart were too explicitly recognized from the Jurassic era onwards.

Oxysterols

Oxysterols are defined as oxygenated derivatives of cholesterol (or precursors to cholesterol) that may be formed directly by autoxidation or by the action of a specific monooxygenase or that may be secondary to enzymatic or nonenzymatic lipid peroxidation. The various structures that will be referred here are depicted in Figure and are referred to by trivial names.

Table: Structure of Cholestrol and Oxysterols.

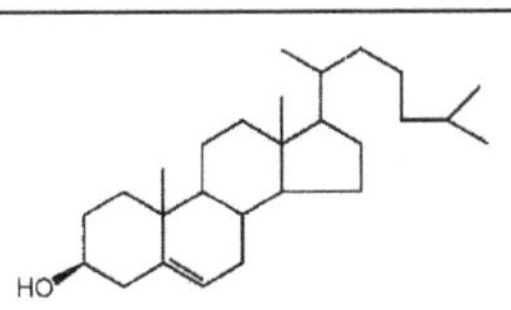 cholesterol	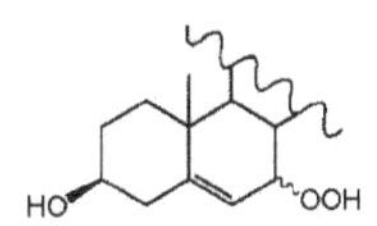 7-hydroperoxycholesterol
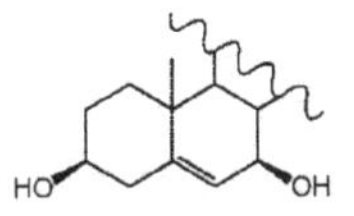 7-hydroxycholesterol	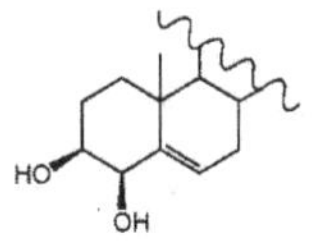 7-hydroxycholesterol
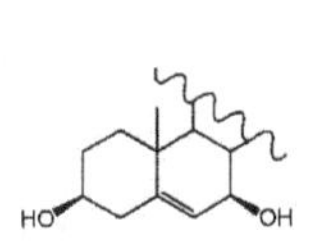 7-oxocholesterol	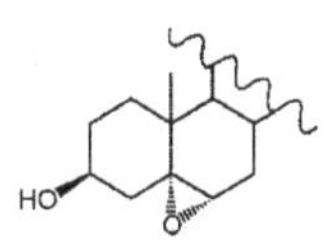 cholesterol5α,6α-epoxide

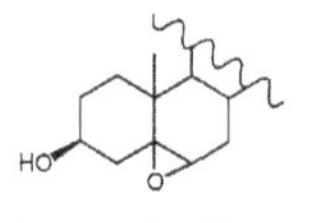 cholesterol-5β,6β -epoxide	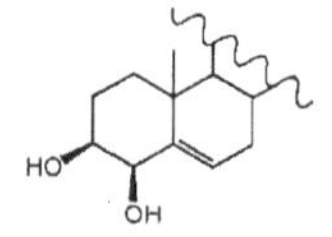 4β -hydroxycholesterol
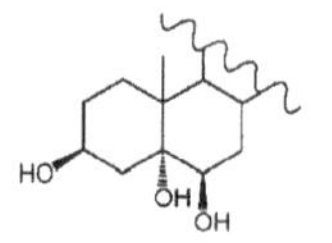 cholestane-3β,5α,6β -triol	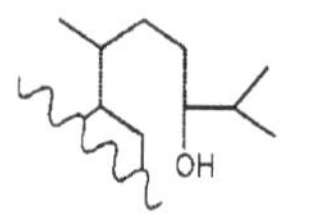 24-hydroxycholesterol
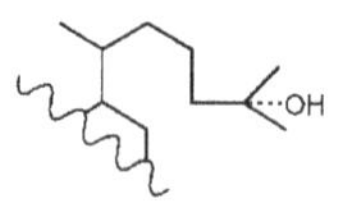 25-hydroxycholesterol	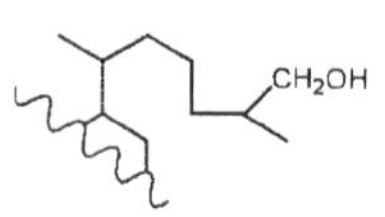 27-hydroxycholesterol
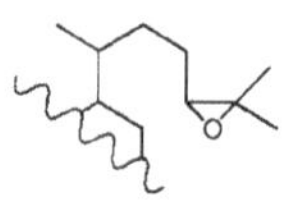 24,25-epoxycholesterol	

Analysis of Oxysterols

With few exceptions (eg, atheromas), oxysterols are present in trace amounts; excess of cholesterol is seldom 1000-fold. Because of the great excess of cholesterol and the possibility of artifactual formation of several oxysterols during isolation and workup procedures, accurate analyses are difficult. In the past, different groups have reported levels of some oxysterols that differ by 2 orders of magnitude. Even when all possible precautions are taken to avoid autoxidation during the isolation, there is always some uncertainty with respect to the origin of, for example, 7β-hydroxycholesterol, 7-oxocholesterol, 25-hydroxycholesterol, the cholesterol-5, 6-epoxides, and cholestane-3,5,6-triols. That at least the first 3 of the above oxysterols may be formed in vivo has been convincingly shown by incorporation of O_2 in oxysterol after the exposure of rats to O_2.

The most accurate technique available for assay of the oxysterols involves isotope dilution–mass spectrometry, with the use of individual deuterium-labeled analytes as internal standards.

Oxysterols as Intermediates in Cholesterol Excretion Pathways

From a quantitative point of view, the most important oxygenation reactions are those involved in the early steps in the conversion of cholesterol into bile acids. An overview of these reactions is given in Figure. The products of these cytochrome P-450–mediated oxygenations are found in the circulation as dominating oxysterols. In addition to the critical hydroxylations of cholesterol leading to the normal bile acids, there is a cytochrome P-450–mediated oxygenation of cholesterol leading to a quantitatively important oxysterol in the circulation, 4-hydroxycholesterol. The metabolic end product of this steroid has not yet been defined.

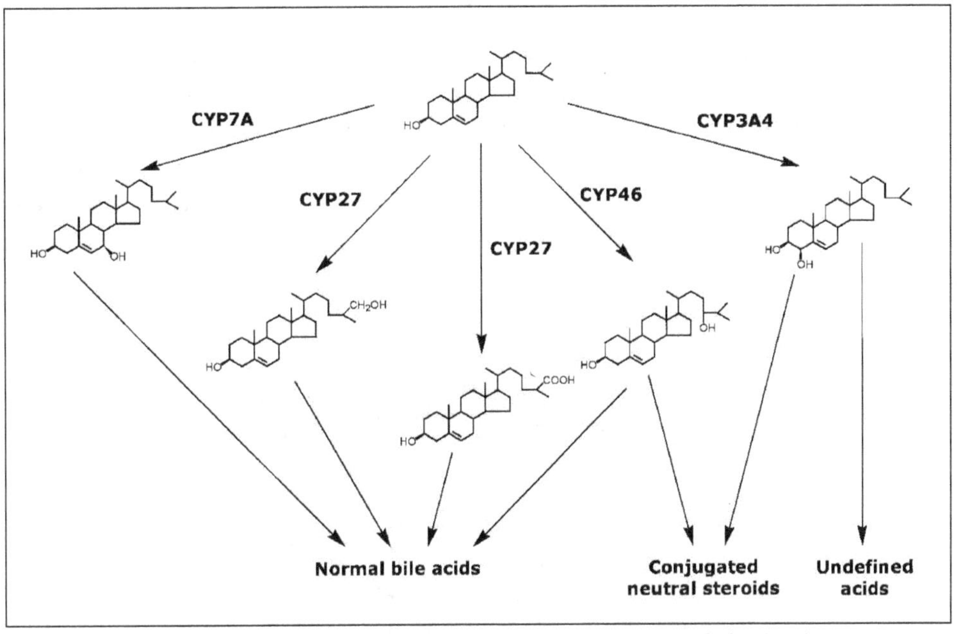

Figure: Major oxysterols in the circulation: formation and elimination.

The major oxysterols in the circulation are transported by lipoproteins, and their distribution between different lipoprotein fractions is similar to that of cholesterol. An exception is cholestenoic acid, which is transported entirely in the lipoprotein-free fraction.

7α-Hydroxycholesterol

The classical and quantitatively most important pathway for bile acid synthesis starts with a 7-hydroxylation of cholesterol.16 This pathway is under strict metabolic control, and there are multiple mechanisms involved in the regulation of the rate-limiting hepatic cytochrome P-450 enzyme, cholesterol 7-hydroxylase (CYP7A1), by hormonal and dietary factors. Under normal conditions, the most important regulation of this enzyme is a negative-feedback suppression caused by bile acids reabsorbed from the intestine and reaching the liver via the portal vein. Details are now known about the transcription factors involved in this regulation, although important, not-yet-defined, speciesrelated differences seem to exist.

The 7α-hydroxycholesterol present in the circulation corresponds to a "leakage" from the liver, and as judged by measurements in the hepatic artery and hepatic vein, the secretion of this oxysterol from the human liver seems to be about the same as the uptake. The concentration of 7α-hydroxycholesterol in the circulation reflects the activity of CYP7A in the liver and can, in fact, be used as an in vivo marker for this activity.

27-Hydroxycholesterol and Cholestenoic Acid

In addition to the 7-hydroxylase pathway, there is an alternative "acidic" pathway, starting with the introduction of a hydroxyl group at the terminal methyl group (C27 position) of the steroid side chain.16 In contrast to CYP7A1, the cytochrome P-450 that is involved in this conversion, CYP27A1, has a broad substrate specificity and is present in most tissues and not only in the liver. The enzyme is capable of oxidizing the C27-methyl group not only into a CH2OH group but also into a carboxylic acid (cholestenoic acid). Although the alternative acidic pathway may be predominantly localized in the liver, there is a continuous flux of 27-oxygenated cholesterol metabolites from extrahepatic sources to the liver. It has been calculated that 5% to 10% of the total conversion of cholesterol into bile acids starts with an extrahepatic 27-hydroxylation. In addition to initiating the pathway starting with 27-hydroxylation, CYP27A1 is an important enzyme in connection with the conversion of the C-27 steroid side chain of cholesterol into the C-24 steroid side chain of bile acids. However, there are alternative mechanisms for degradation of the steroid side chain that are independent on the sterol 27-hydroxylase.

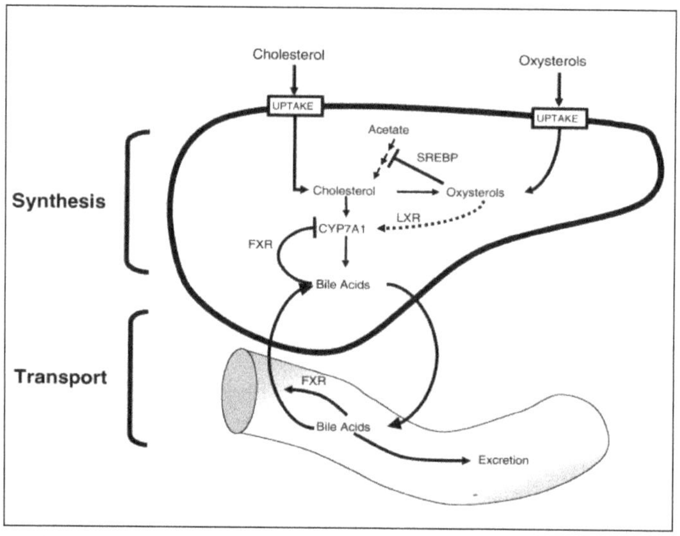

Figure: Suggested mechanism for regulation of cholesterol 7α-hydroxylase and its relation to cholesterol synthesis and bile acid turnover.

According to this suggested mechanism, the LXR is involved in an oxysterol-dependent feedforward regulation of CYP7A1 in rodents. Expansion of the bile acid pool activates the receptor FXR, which leads to a suppressed transcription of the CYP7A1 gene. This effect is not a direct interaction of FXR with a bile acid responsive element in the promoter of the CYP7A1 gene but is mediated by at least 2 other factors. Expansion of the cholesterol pool is assumed to lead to increased levels of oxysterols with subsequent reduced synthesis of cholesterol and increased degradation of cholesterol into bile acids. It should be emphasized that there is no direct experimental evidence for the role of oxysterols in this model.

Numerous regulatory mechanisms involving negative feedback by bile acids have also been described for CYP27A1. In a preliminary study from our laboratory, no clear effect on hepatic messenger RNA levels was seen in human subjects treated with chenodeoxycholic acid or cholestyramine. That CYP27A1 is less sensitive to regulatory factors is also evident from the fact that the plasma levels of the product, 27- hydroxycholesterol, are relatively stable under different conditions.

24S-Hydroxycholesterol

The cytochrome P-450 responsible for formation of 24Shydroxycholesterol, cholesterol 24S-hydroxylase (CYP46), is almost exclusively located in the brain in humans, and most or all of the 24S-hydroxycholesterol present in human circulation is thus derived from the brain. However, the situation is different in the mouse. There is a continuous flux of 24S-hydroxycholesterol across the blood-brain barrier into the circulation, and evidence has been provided that this flux is of importance for the homeostasis of brain cholesterol. Thus, the levels of 24S-hydroxycholesterol in the circulation can be used as a marker for the turnover of cholesterol in the brain. Although the mechanisms regulating the expression and activity of CYP46 have not been extensively studied, in general, the levels of 24S-hydroxycholesterol in the circulation appear to be remarkably stable.

It has recently been shown that only about half of the 24S-hydroxycholesterol formed in humans is converted into bile acids. The other half of it is excreted from the liver in conjugation with sulfuric acid and glucuronic acid. Part of it is -hydroxylated before conjugation and excretion.

4β-Hydroxycholesterol

Very recently, it was shown that this oxysterol is formed by the action of cytochrome P-450 3A4 (CYP3A4) on cholesterol. At the present state of knowledge, the possibility that also other cytochrome P-450 enzymes are involved cannot be excluded. Preliminary experiments in our laboratory have shown that the rate of elimination of 4-hydroxycholesterol is very low, much lower than that for the other circulating oxysterols. This may be the explanation for the relatively high concentration of this oxysterol in the circulation, in spite of a low rate of formation. The metabolic end products of cholesterol 4-hydroxycholesterol have not yet been defined.

Oxysterols Present in Minor Quantities in Circulation and/or Specific Tissues

25-Hydroxycholesterol is present in very low concentrations in the circulation and in tissues. It has been clearly shown that it is enzymatically formed in vivo, and the enzyme responsible for its formation has recently been characterized. Surprisingly, the cholesterol 25-hydroxylase is not a cytochrome P-450 but belongs to a smaller family of nonheme iron–containing proteins. The enzyme is expressed in most tissues, as judged by mRNA blots.

24,25-Epoxycholesterol and 24,25-epoxylanosterol have been suggested to be potential regulatory oxysterols.36–38 According to 1 report, 24,25-epoxycholesterol is present in surprisingly high levels in the human liver (2 > μg/g liver).38 However, this has not been confirmed by other groups, and in an investigation from our group, the level of 24,25-epoxycholesterol in the mouse liver was found to be < 25 ng/g.

Oxysterols as Regulators of Cholesterol Homeostasis

That cholesterol synthesis in isolated cultured cells is suppressed by cholesterol added to the medium is very well documented. Kandutsch et al found that purified cholesterol was only weakly inhibitory, and they suggested that most of the inhibitory actions on the cells are due to oxysterols rather than to cholesterol itself. The inhibitory potency of some oxysterols was found to be orders of magnitude higher than that of cholesterol itself. 25-Hydroxycholesterol was found to be the most efficient inhibitor, and since then, 25- hydroxycholesterol, alone or in combination with cholesterol, has been the most important tool to demonstrate steroidsensitive biochemical effects in cultured cells. In most of these experiments, highly unphysiological concentrations of the free oxysterol have been used.

Under in vitro conditions, oxysterols are able to regulate key enzymes in cholesterol turnover at transcriptional and posttranscriptional levels. The genes affected at the transcriptional level by oxysterols all have a sterol-responsive element (SRE) within the 5-flanking region. The SREcontaining genes are controlled by a sterol-regulated cleavage of the SRE-binding proteins (SREBP1a, SREBP1c, and SREBP2). Sterol-sensitive genes are responsible for the uptake (LDL receptor) and synthesis (3-hydroxy-3-methylglutaryl [HMG]– coenzyme A [CoA] synthase, HMG-CoA reductase, farnesyl diphosphate synthase, and squalene synthase) of cholesterol. Sterol-sensitive genes are also of importance for fatty acid synthesis and desaturation (fatty acid synthase, acetyl CoA carboxylase, and stearyl CoA desaturase) as well as triglyceride synthesis (glycerol-3-phosphate acyltransferase). At the posttranscriptional level, oxysterols accelerate degradation of HMGCoA reductase and activate acyl CoA:cholesterol acyltransferase by a mechanism that is independent of gene transcription and protein synthesis. The gene coding for the rate-limiting enzyme in degradation of cholesterol into bile acids, CYP7A1, is transcriptionally regulated by a mechanism that has been suggested to involve oxysterols, at least in the rat.

Mutant cells of different classes have been shown to resist the regulation of HMG-CoA reductase by LDL and 25-hydroxycholesterol, consistent with a common mechanism for downregulation of cholesterol synthesis by cholesterol itself and oxysterols. This has been suggested to be integrated at the level of the SREBP sterol-sensing domain. In view of this, use of 25-hydroxycholesterol may provide useful information about steroid sensitivity in spite of the fact that the conditions are unphysiological.

Although the results of the above type of experiments are consistent with a role of oxysterols in the normal regulation of cholesterol homeostasis, they do not prove the hypothesis. According to some experiments with cultured cells, a cytochrome P-450 system seems to be involved in the LDL-induced suppression of cholesterol synthesis and uptake.42 However, the specificity of the cytochrome P-450 inhibitors used in these in vitro experiments may be questioned, and it is difficult to draw firm conclusions. In 1 of these experiments, support was obtained for the contention that 24,25-epoxylanosterol is metabolized into an important regulator of cholesterol synthesis by a ketoconazole-sensitive cytochrome P-450. In another study, support was obtained for the contention that LDL-induced downregulation of HMG-CoA reductase in cultured fibroblasts is dependent on the formation of 27-hydroxycholesterol by CYP27A1, and genetically CYP27A1-deficient fibroblasts had a markedly reduced response to LDL cholesterol. In another study, however, CYP27A1-deficient and control fibroblasts responded similarly to suppression by LDL cholesterol.

Saucier et al have shown that several oxysterols accumulate in the mouse liver after cholesterol feeding, and they suggested that these oxysterols, in particular 24S-hydroxycholesterol, are responsible for the downregulation of HMG-CoA reductase. With use of data on the potential of the accumulated oxysterols to suppress HMG-CoA reductase in an in vitro system, it was concluded that the amount of oxysterol that had accumulated in the liver after cholesterol feeding was sufficient to explain the suppression of cholesterol synthesis. However, in view of the unphysiological conditions used for testing the suppressive potential, the conclusion may be questioned. It may be noted that the most important suppressive oxysterol accumulated was 24S-hydroxycholesterol. There is little or no formation of this oxysterol in human liver.

On the basis of recent experiments involving feeding rats an atherogenic diet or the administration of mevalonate, Zhang et al concluded that 25-hydroxycholesterol and 24(S),25- epoxycholesterol, but not 27-hydroxycholesterol, may be of regulatory importance.

Attempts in our laboratory to demonstrate a critical role of the sterol 27-hydroxylase and the cholesterol 24S-hydroxylase in the downregulation of HMG-CoA reductase activity in the mouse liver in vivo have failed. According to another in vivo study using different analogues and deuterium-labeled derivatives of cholesterol, the presence of the 3-hydroxyl group and of the 5,6-double bond in cholesterol is essential for the downregulation. The results also suggest that hydroxylations at positions 7, 24, and 27 are not critical for the downregulation. Even complete elimination of the steroid side chain of cholesterol does not prevent its inhibitory effect on the downregulation of hepatic cholesterol synthesis in mice.

Oxysterol-Binding Proteins

The early observation that several oxygenated cholesterol derivatives are able to regulate HMG-CoA reductase led to a search for proteins that could mediate these effects. Kandutsch and colleagues characterized a protein that had affinities for different oxysterols that were correlated with the potency of the oxysterol to suppress HMG-CoA reductase in fibroblasts. The protein seems to be present in most or all animal cells, suggesting that some vital function may be associated with it. Under normal conditions, the oxysterol-binding protein is mainly present in the cytosol. Addition of 25- hydroxycholesterol leads to a marked change in the distribution of the protein, with the majority of it bound to the Golgi apparatus. The latter finding is consistent with a function of the protein for intracellular transport of the oxysterol or even excretion of it. However, whether or not oxysterols are the natural and most important ligands for the protein is not known with certainty.

Oxysterol-binding protein was later shown to belong to a family with several members occurring in mammals and also in yeast. Some of these proteins seem to be involved in Golgi-derived vesicle transport. Interestingly, at least 1 member of this family does not bind oxysterols but is able to bind phospholipids (in particular, phosphatidic acid) with high affinity.

Recently, orphan nuclear receptors have been shown to be of importance for cholesterol homeostasis. The LXRs (LXR and LXR) and the farnesoid X receptor (FXR) bind and are also activated by oxysterols and bile acids, respectively. These receptors are able to regulate the expression of a number of genes involved in cholesterol metabolism. A prerequisite for the transcriptional activation by LXR and FXR is that the receptor forms a heterodimer with the retinoid X receptor.

LXRα has been shown to mediate the transcriptional induction of cholesterol 7α-hydroxylase, which is the rate-limiting enzyme in the major pathway from cholesterol to bile acids. Mice with a disruption of the gene coding for LXR fail to upregulate the cholesterol 7α-hydroxylase when they are fed a high cholesterol diet, causing accumulation of cholesterol in the liver. This does not occur in LXR-deficient mice. Because the human gene coding for cholesterol 7α-hydroxylase lacks identified LXR-responsive elements in the promoter, this regulatory mechanism probably does not exist in humans. LXRα is also involved in the regulation of the cholesterol transporters ABCA1 and ABCG1, which are involved in the flux of cholesterol from enterocytes and macrophages, respectively. LXR also seems to have a role in the regulation of human cholesterol ester transfer protein, which translocates cholesterol ester between lipoproteins.

According to a recent study, 27-hydroxylation of cholesterol in cholesterol-loaded skin fibroblasts and monocyte-derived macrophages may activate LXR, resulting in increased activity of the cholesterol transporters ABCA1 and ABCG1. According to another study, however, 27-hydroxycholesterol is not an efficient activator of human LXR. The human LXRα gene is itself a target of the LXR-signaling pathway, and autoregulation of LXRα has been suggested to be an important way to amplify the cholesterol catabolic cascade.

Although the importance of the LXRα receptor in the regulation of cholesterol homeostasis has been well established (at least in experimental animals), the physiological ligand of the receptor has not been defined with certainty. Among the oxysterols tested, the receptor was found to have its highest affinity for 24S-hydroxycholesterol and 24,25-epoxycholesterol. The former steroid is present in relatively high concentrations in the brain, whereas the latter has been reported to occur in relatively high levels in the liver, in some but not all studies. At the present state of knowledge, there is no direct experimental evidence that the above 24- and 24,25-oxygenated steroids are the physiological ligands. Interestingly, it has been shown that activation of LXR can be antagonized by some unsaturated fatty acids and geranylgeranyl pyrophosphate. It should be emphasized that the different steroids that are potential ligands to the receptor are present in compartments containing a great excess of cholesterol (10^3 - to 10^6 -fold). It remains to be established that the receptors are able to sense and selectively bind the oxysterol when they are exposed to such mixtures.

The LXR β receptor has a broader distribution than the LXRα receptor, with a particularly high concentration in the brain. In view of the efficient binding of 24S-hydroxycholesterol to this receptor and the high concentration of 24S-hydroxycholesterol in the brain, it is tempting to suggest that there may be an important oxysterol-signaling pathway in the brain involving LXRβ. However, direct evidence for this is still lacking.

Genetically Engineered Animal Models and Metabolic Diseases with Altered Levels of Oxysterols

A number of genetically engineered animal models have provided interesting information concerning the role of some of the oxysterols suggested to be of regulatory importance.

Mice deficient in sterol 27-hydroxylase have a complete lack of 27-hydroxycholesterol and cholestenoic acid. These mice have normal levels of cholesterol in the circulation, suggesting that 27-hydroxycholesterol is not an obligatory factor for the regulation of cholesterol homeostasis. Because of the fact that the sterol 27-hydroxylase is needed for the normal synthesis of bile acids, there are a number of changes that are related to the reduced formation of bile acids, such as reduced cholesterol absorption and compensatory increase in cholesterol synthesis. Thus, the mRNA for the cholesterogenic transcription factor SREBP-2 and mRNAs for SREBP-2–regulated cholesterol biosynthetic genes are elevated in the livers of the transgenic animals. All these changes are reversed by feeding the animals cholic acid. Patients with the human disease cerebrotendinous xanthomatosis (CTX), in which sterol 27-hydroxylase is deficient, are also normolipidemic. In contrast to the situation in mice, there is an accumulation of cholesterol and cholestanol in the human disease, which is probably partially due to the lack of the transport function of the sterol 27-hydroxylase. Also, in humans, most or all of the biochemical changes due to the lack of CYP27A1 can be reversed by feeding with bile acids.

Mice deficient in the oxysterol-catabolizing enzyme oxysterol 7α-hydroxylase (Cyp7b) have markedly elevated levels of 27- hydroxycholesterol and 25-hydroxycholesterol. With the exception of a slight downregulation of cholesterol synthesis in the kidneys, the in vivo sterol biosynthetic rates were unaltered in several tissues in spite of the high levels of the oxysterols.

The accumulated oxysterols were mainly in esterified form, and the possibility must be considered that an esterified oxysterol may be a less potent regulator than an oxysterol in the free form. The other possibility is that the well-documented ability of the side-chain oxidized oxysterols to downregulate genes in vitro is of little or no importance under in vivo conditions.

1 fatal human case with a lack of the oxysterol 7α-hydroxylase has been reported. Also, in this case, there were high levels of 27-hydroxycholesterol and 25- hydroxycholesterol in the circulation. In addition, there were high levels of the bile acid 3-hydroxy-5-cholenic acid. Whether the fatal outcome of the metabolic defect was due to the accumulation of this hepatotoxic bile acid or the hydroxysteroids is not possible to evaluate.

Although the above studies seem to exclude 27-hydroxycholesterol as an obligatory factor in cholesterol homeostasis, the possibility that 25-hydroxycholesterol is obligatory cannot be completely excluded at the present state of knowledge.

Oxysterols as Markers for Oxidative Stress

Oxidative modification of LDL is associated with an oxidation of cholesterol to yield 7- and 5,6-oxygenated species. It has been suggested that the above oxysterols may be used as markers for in vivo lipoprotein oxidation. The great problems in connection with the analyses and the risk for artifactual formation of oxysterols from the parent cholesterol during the workup procedures make it difficult to draw conclusions from the results of several old studies with less accurate techniques. One recent study has demonstrated an association between plasma levels of 7β-hydroxycholesterol and the progression of carotid atherosclerosis. Long-term vitamin E supplementation reduced plasma levels of 7β-hydroxycholesterol. Plasma oxysterols have been reported to be higher in smokers than in nonsmokers. In view of the methodological problems, oxysterols do not appear to have clear merits in relation to other markers for oxidative modification of LDL.

Oxysterols and Atherosclerosis

On the basis of a great number of in vitro studies demonstrating cytotoxic effects on cultured endothelial cells and arterial smooth muscle cells, oxysterols have been suggested to be atherogenic. Highly unphysiological conditions have been used, and, it is not possible to draw valid conclusions from them about the role of the oxysterols under in vivo conditions.

In view of the fact that oxysterols mimic increasing concentrations of cholesterol, one would expect them to be atherogenic when they are fed to experimental animals. A great number of such studies have been carried out; most of these studies used highly unphysiological doses of the oxysterol. Of 13 studies on the effect of dietary oxysterols, 6 indicated a proatherogenic effect, and 4 indicated an antiatherogenic effect, whereas 3 showed no clearcut activity. It has been suggested that the markedly higher incidence of heart disease in Indians living in London compared with non-Indians in the same city may be due to their high consumption of ghee, which is known to contain very high concentrations of oxysterols. However, other explanations are possible, and there is a very low incidence of heart disease in India itself, where ghee is also an important part of the diet in many regions. At the present stage of knowledge, there is no direct evidence that dietary oxysterols contribute to atherogenesis in humans.

Oxidatively modified LDL contains elevated levels of 7- and 5, 6-oxygenated oxysterols. The possibility cannot be excluded that some of these species may be of importance in atherogenesis. The primary 7-oxygenated product of cholesterol, 7-hydroperoxycholesterol, seems to be the most cytotoxic oxygenated lipid present in oxidized LDL. This oxysterol is rapidly decomposed into 7α-hydroxycholesterol, 7-hydroxycholesterol, and 7-oxocholesterol, which may be found in relatively high concentrations in foam cells and fatty streaks. Whether the 7-oxygenated steroids present in foam cells and atheromas are of pathogenetic importance or if they just reflect the uptake of oxidized LDL particles is not known. Interestingly, the accumulation of 7-oxocholesterol in macrophages seems to be prevented by the action of the sterol 27-hydroxylase.

7α-Hydroperoxycholesterol has been detected in low concentrations in atheromas. In view of the short half-life of this compound, the measurements are likely to underestimate the true levels. In view of its cytotoxicity, this oxysterol may be of pathogenetic importance.

The most dominant oxysterol in human atheromas is 27- hydroxycholesterol. There is a continuous flux of 27-hydroxycholesterol and cholestenoic acid from peripheral cells to the liver; thus, the presence of 27- hydroxycholesterol in the atheromas may reflect a defense mechanism to prevent the accumulation of cholesterol. The fact that patients with CTX, lacking the sterol 27-hydroxylase, often develop premature atherosclerosis is in accordance with the hypothesis. The importance of the mechanism is underlined by the fact that CTX patients most often have normal levels of plasma cholesterol. The concentration of 27-hydroxycholesterol seems to be increased with the severity of the lesion, and the ratio between 27-hydroxycholesterol and cholesterol may be 100-fold higher in an atheroma than in the circulation. However, the "trapping" of 27-hydroxycholesterol and other oxysterols in the atheromas may be secondary to a high rate of esterification by the acyl CoA : cholesterol acyltransferase enzyme, and oxysterols are known to be a better substrate for this enzyme than cholesterol. Almost all of the oxysterols in the atheromas are thus esterified.

Information Obtained from Measurements of the Levels of Oxysterols in the Circulation

The levels of 7-hydroxycholesterol seems to reflect the presence of oxidatively modified LDL and "oxidative stress." 7α-Hydroxycholesterol and its metabolic product 7α-hydroxy-4-cholesten-3-one may be used as mark ers for the activity of the rate-limiting enzyme in bile acid synthesis, cholesterol 7α-hydroxylase. The levels of 27- hydroxycholesterol in the circulation seem to be elevated more often in a population of patients with atherosclerosis than in control subjects, possibly reflecting an activated defense mechanism in the patients. However, measurement of this oxysterol does not seem to be of diagnostic value. In some cases, the levels of 27-hydroxycholesterol may be elevated by a reduced metabolism for various reasons.

The levels of 24S-hydroxycholesterol in the circulation reflect the balance between cerebral production and hepatic degradation and may provide some information concerning cholesterol homeostasis in the brain in connection with neurological and neurodegenerative diseases. The levels of 4β-hydroxycholesterol may give information about upregulation of the CYP3A4 gene.

Chlorinated Cholesterol

α-chlorohydrins (6-β-chloro-cholestane-(3β, 5α)-diol) and other chlorinated products are generated by myeloperoxidase–H2O2–Cl system after reaction with double bond of cholesterol second ring. Chlorohydrins are relatively firm, and have been found valuable as a lipoproteins markers oxidatively damaged by triggered phagocytes secreting myeloperoxidase. These stuffs were also produced in LDL and membrane of the cell ensuing exposure to myeloperoxidase system or HOCl. The chlorohydrins formation in cells might be unruly to membranes resulting in the lysis of the cell and death. They might also be used as biomarkers for monocyte / neutrophil activated oxidative damage.

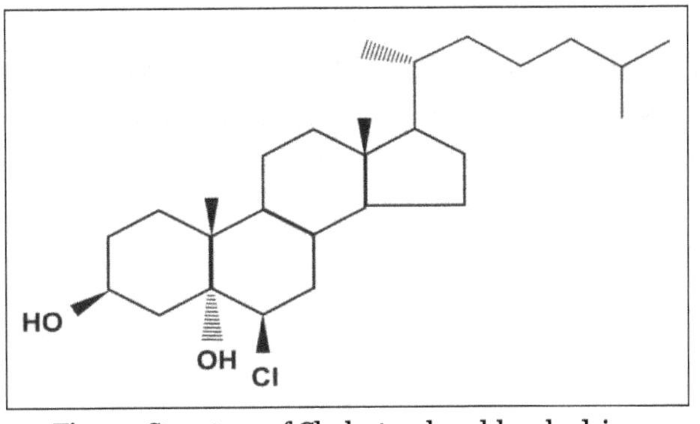

Figure: Structure of Cholesterol α-chlorohydrins.

Phytosterols

Phytosterols and phytostanols, also referred to as plant sterols and stanols, are common plant and vegetable constituents and are therefore normal constituents of the human diet. They are structurally related to cholesterol, but differ from cholesterol in the structure of the side chain.

Phytosterols has been found more competent than cholesterol in expanding the temperature dimension in which membrane-coupled biological action takes place as plants have to counter elevated temperature disparity than animals.

Mostly, phytosterols have 28-30 carbon units with 1-2 C=C double bonds, usually one in the nucleus of the sterol and occasionally a second in the side chain of alkyl residue. In plants, cycloartenol give rise to phytosterols whereas in fungi (including yeasts), vertebrates they derived from lanosterol through squalene cyclization. Facts also specifies the occurrence of biosynthetic pathway for phytosterol in plant cells through lanosterol and was chosen as the 'Lanosterol Pathway'.

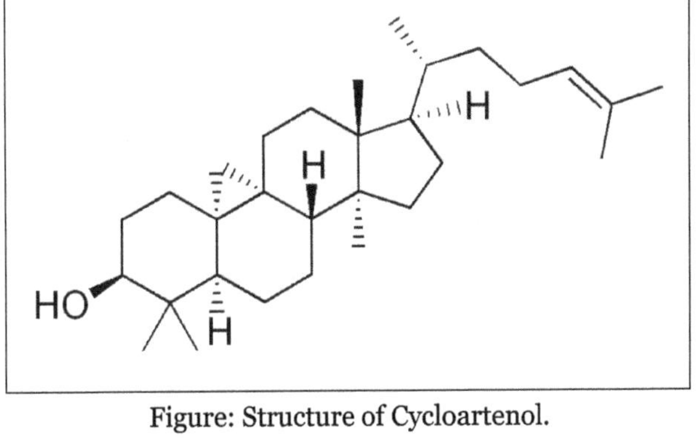

Figure: Structure of Cycloartenol.

In plant species, over and above 250 diverse types of phytosterols have been identified and reported. Some of the envoys are campesterol, β-sitosterol and stigmasterol. β-Sitosterol is used for steroid synthesis; exist in the entire plant lipids.

Stigmasterol, also known as 'Wulzen factor', is needed for the synthesis of Vitamin D3 and progesterone. It is a potential anti-inflammatory compound that inhibits numerous matrix degradation and pro-inflammatorymediators implicated in cartilage degradation induced by osteoarthritis.

In brown algae, the prevailing sterol is fucosterol and cholesterol exists in diminutive amounts. Ergosterol (Mycosterol), is the C28 imperative sterol from ergot and yeast. Ergosterol was discovered by Tanret C followed by Gerard E who observed that fungi also have ergosterol. In yeast, Gerard was the first to identify ergosterol as a cell membrane component limited to fungi, and it could be applicable as an indicator molecule for these microbes in an existing biomass. Upon irradiation, ergosterol gives rise to calciferol (Vitamin D2).

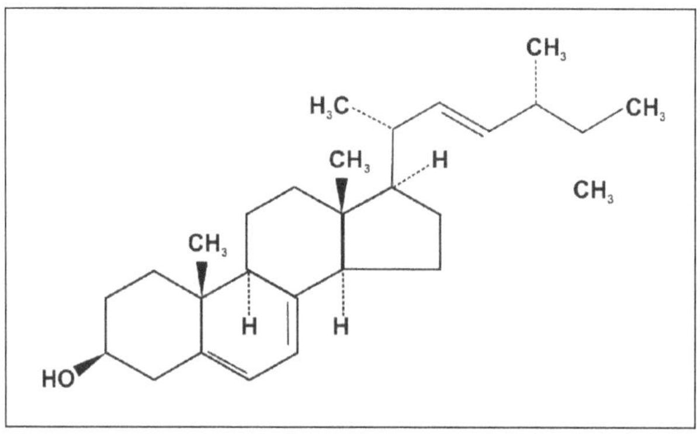

Figure: Structure of Ergosterol.

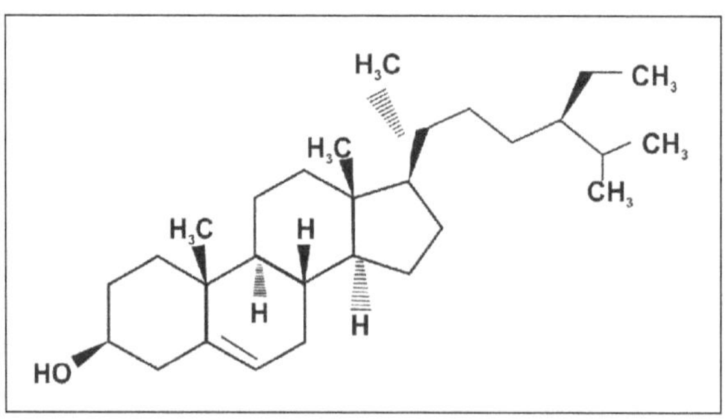

Figure: Structure of β-Sitosterol.

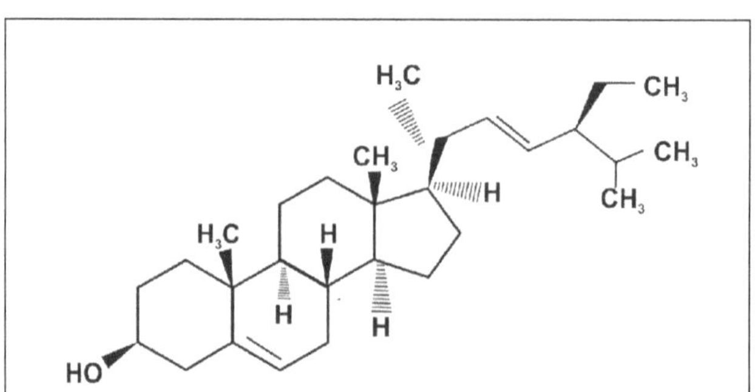

Figure: Structure of Stigmasterol.

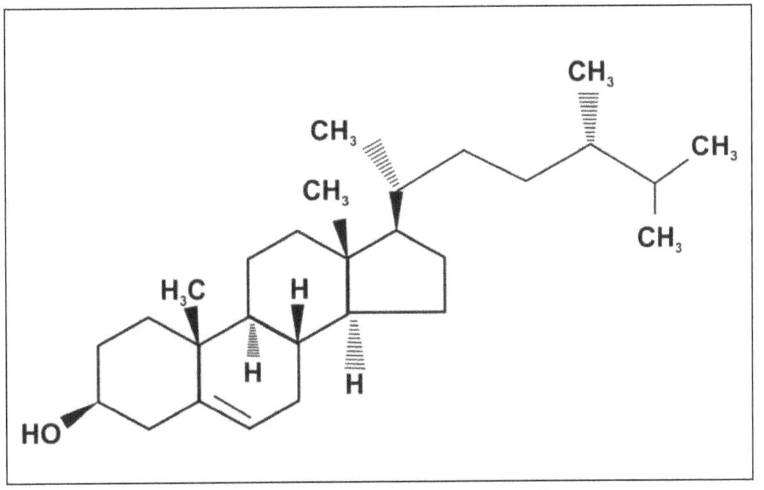

Figure: Structure of Campesterol.

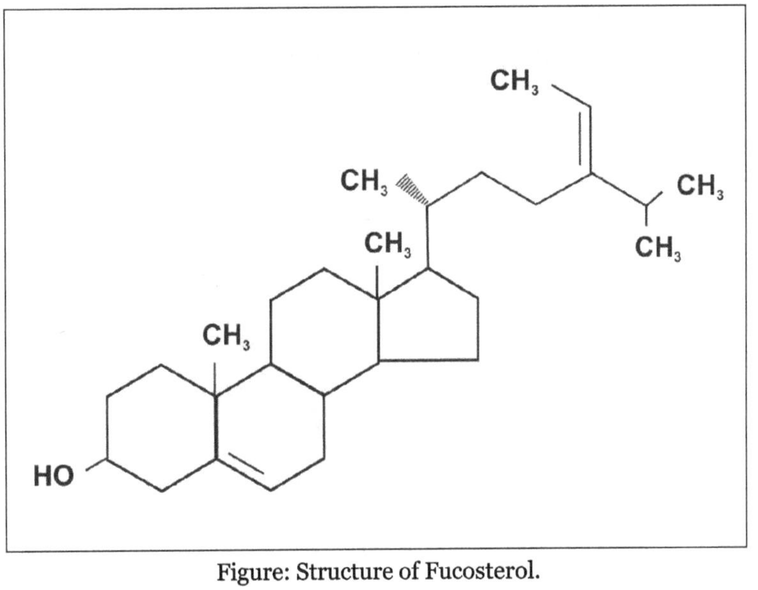

Figure: Structure of Fucosterol.

Antrosterol (ergosta-7, 9, 22-trien-3β-ol), an isomeric form of ergosterol, from Antrodia camphorate (a fungus), is efficient in hepatoprotection via its anti-inflammation ability. This compound is used in Chinese traditional medicine for the treatment of cancer, diarrhoea, drug intoxication and hypertension.

EuroFIR-BASIS is exclusive online database containing information on the content of the selected phytosterols in plant based foods and different plants. Phytosterols detailed for an extensive portion of total dietary sterols in vertebrates and generate a broad range of biological actions in humans and animals. They possess efficient cholesterol- lowering and anti-tumor properties, decreases serum β-carotene concentration, without varying vitamins A, D and E absorption.

In brown algae like Sargassum ringgoldianum and Lessonia nigrescens (Phaeophyta), Saringosterol (a derivative of fucosterol) has been revealed to restrain the Mycobacterium tuberculosis growth comparable to that of rifampicin but without considerable toxicity against mammalian cells.

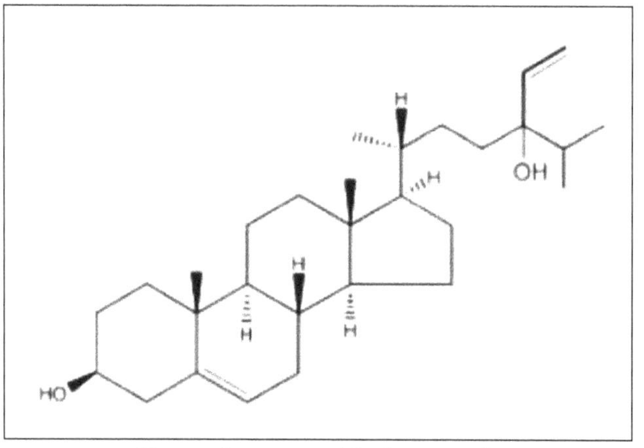

Figure: Structure of Saringosterol.

In 2004, the European Commission endorsed the toting up of phytosterols and phytostanols in food products with conditions of appropriate cataloging. Phytosterols similar to cholesterol undergo oxidative processes and the resultant oxyphytosterols have been revealed to display beneficial biological properties.

Phytostanols

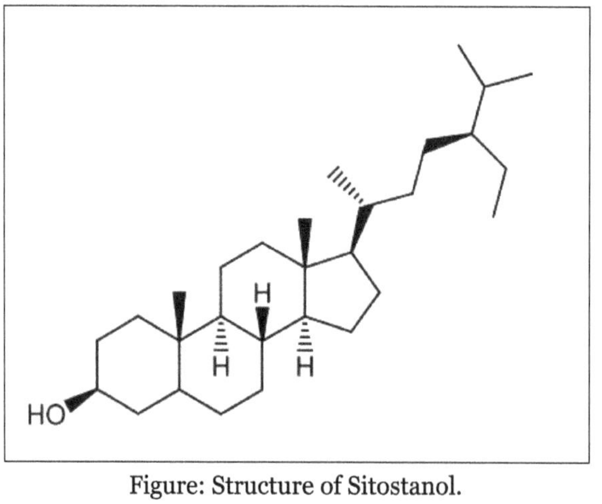

Figure: Structure of Sitostanol.

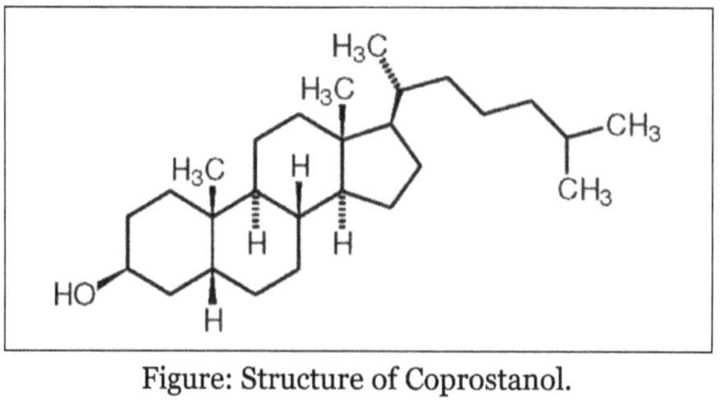

Figure: Structure of Coprostanol.

Are fully saturated sub-group of phytosterols with no double bonds, produced by hydrogenation. They are found in very minute amounts in legumes, nuts and seeds but in high amounts in few cereal species. To advance their solubility, they combine with the esters of fatty acid to generate esters of plant stanol. Sitostanol is the utmost commonly found stanol and stanols frequently crop up in dinoflagellates, the chief source of 5α (H)-stanols in marine sediments. In animals, fully saturated sterols of bacterial origin are conjointly established. 5β (H)-stanol and coprostanol, constitutes more or less 60% of the overall sterols in human faeces.

Avenasterol

It is sequested from the oil of oat and has been revealed to preserve explicitly frying oils from oxidation due to its ethylidene group in the lateral chain.

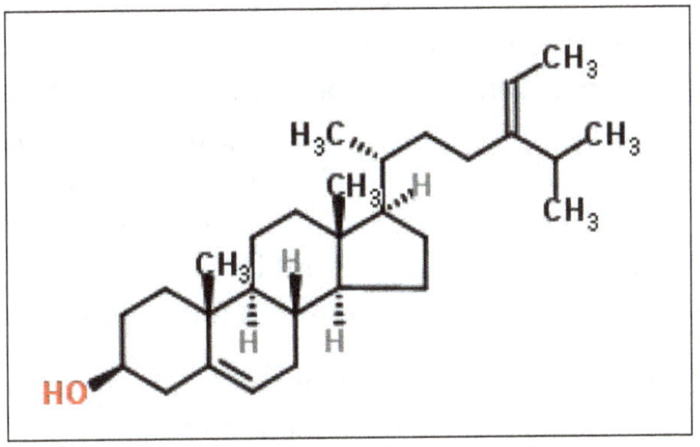

Figure: Structure of Avenasterol.

Phytosterols are of wide occurrence and its persistence in sediment has been reported in aged geological samples. The analysis, diversity and health-promoting uses of phytosterols and phytostanols have already been revealed and reviewed.

Stigmastanol, dihydrocholesterol and coprostanol are recurrently exist in sediments and occasionally utilized as biomarkers primarily for the existence of life in petroleum and sedimentary organic matter. The majority of steroids exist in sediments are C27-C29 steranes which are pivotal intermediates in the exchange of steroids (oxygenated) into steranes (saturated). In addition, a wide variety of sterenes (monounsaturated) have been detected in immature sediments. Marine demosponges created 24-Isopropylcholestane (hydrocarbon remnants of C30 sterols) accounted the charisma of Metazoa presence in the geological documentation during the Neo-proterozoic era (542-1,000 million years ago). This sterol symbolizes a powerful biomarker for early animal diversification and endows with the primeval facts for animals in the fossil record.

Sterol Esters

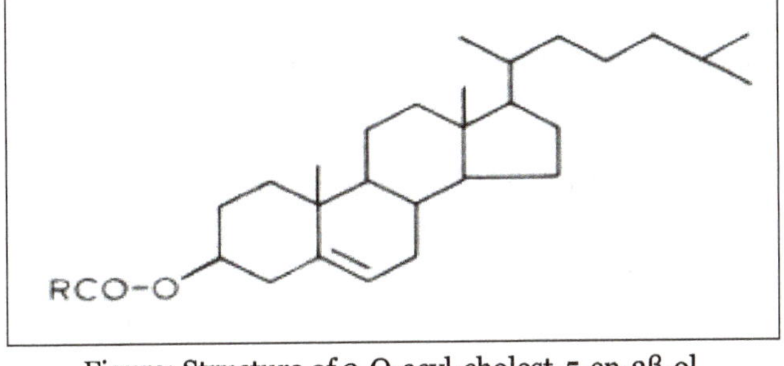

Figure: Structure of 3-O-acyl-cholest-5-en-3β-ol.

Sterols are often endowed esterified to fatty acids. In the tissues of animal (adrenals, liver, plasma), cholesterol is esterified by an array of fatty acids and recurrently by essential fatty acids, accordingly forming esters of cholesterol. Esters of sterol are vital but exceedingly erratic part of yeast cell with values varies from remnants to 50% of the total lipids.

In animals, the cholesterol esterification inside intestinal cells (acyl CoA cholesterol acyltransferase, ACAT) allocates the free cholesterol to be accumulated as a neutral lipid in cytosolic droplets and in the stuffing of cholesterol into lipoprotein particles for the export by means of the liver cells via plasma. The human meibomian glands secretions (sebaceous glands at the edge of eyelids) are predominantly rich in esters of cholesterol (~30% of the lipid puddle), exemplified by a monounsaturated or saturated fatty acid moiety with C18-34 carbon chain. Cholesteryl arachidonate and cholesteryl linoleate present in nasal fluid have been revealed to bequeath to the intrinsic antibacterial activity. Cholesteryl nitrolinoleate, a nitrated lipid perceived in human lipoproteins and blood plasma is a prospective marker of the chain breaking antioxidant role of nitric oxide radical during lipid peroxidation. Nitrated linoleic acid, a potent signaling molecule and a down regulator of inflammatory response revealed to be fundamentally enhanced after macrophage activation, signifying that nitration of lipid happens as part of the retort to inflammatory stimuli.

β-sitosteryl esters, stigmasteryl and ergosteryl are present in plant cell membranes and seed oils. Cycloartenol and esters of stigmasterol have been isolated from Bryophytes (Hepaticeae). In addition, cholesterol also forms ester linkages with secreted polypeptide and signaling molecules preset by the hedgehog gene family during the growth and development of metazoan.

Manufacturing

Production of Sterols from Vegetable Oil Distillates

Edible vegetable oils, extracted from oil seeds, are typically refined to remove minor oil components like phosphatides, free fatty acids, pigments and odours, with the least possible damage to the glycerides and with minimal loss of oil. The conventional or caustic refining procedure comprises degumming, neutralization, bleaching and deodorization. In physical refining, the neutralization step is omitted and the residual free fatty acids are removed in the final deodorization step.

Deodorization is the last step in the edible oil refining process in which volatiles are removed, that can cause deterioration of the oil quality during use in products (flavour, odour, colour and taste stability). This process relies on the large volatility differences between the oil itself (triglycerides) and the volatile compounds to be removed and is carried out under reduced pressure, an elevated temperature in the presence of a stripping gas. The volatiles are recovered in a vapor condenser. With caustic refining the yield of volatiles distillate is approximately 0.3 - 0.4% on the processed oil volume. This distillate mainly contains free fatty acids, but also significant levels of tocopherols (5-15%) and phytosterols (8- 20%).

In a transesterification (methanolysis) step, the glycerides are converted into fatty acid methyl esters and glycerol and the phytosterol-esters into free phytosterols and fatty acid methyl esters. After removal of the methanol/glycerol phase, the methyl esters are removed and the free phytosterols and tocopherols removed by distillation. The phytosterols are separated from the tocopherols by solvent crystallization and filtration using food grade solvent. The phytosterols are further purified by re-crystallisation, mainly to remove wax-esters.

Production of Sterols from Wood Pulp/Tall Oil

Commercially grown coniferous trees (Pinus sp.) are the usual source of wood that is chemically digested in the so-called Kraft pulping process. In this alkaline process the wood chips are digested at pH 14 (hence the term "soap") for about 18 hours at 50 °C to free the wood fibers. The soapy material (black liquor pulping soap) is then separated from the cellulose pulp.

The soapy lipid phase which is obtained contains more than 2% phytosterols. One way of recovering these sterols is via solvent (methanol) extraction directly from the soap, after which the phytosterols are purified by precipitation from the solvent. More commonly the tall oil soap is acidified to produce an oily phase which is a mixture of free rosin and fatty acids and neutral components, most importantly consisting of sterols, fatty alcohols, squalene, waxes and other esters. This mixture is referred to as crude tall oil.

Crude tall oil is refined into different fractions (e.g. rosin acids, fatty acids) by distillation, where the phytosterols are concentrated, mostly as phytosterol esters, in the residue. This is known as tall oil pitch and serves as the raw material for the production of tall oil phytosterols. The concentration of phytosterols in tall oil pitch is in the range of 5-15%.

Pure phytosterols are obtained from the tall oil pitch, mainly containing high boiling fatty acid esters, resin acids and the phytosterols. The tall oil pitch is saponified with food-grade caustic soda to hydrolyze phytosterols esters and saponify the fatty acids. The mixture is then neutralized with a foodgrade mineral acid (such as sulfuric acid, hydrochloric acid or phosphoric acid). Thereafter the aqueous phase is removed and any remaining water is removed by flash evaporation. The residual pitch is distilled in a number of steps to recover the phytosterol fraction. This fraction is finally purified via solvent re-crystallization using food-grade solvents.

Production of Phytostanols from Phytosterols

Starting with the unsaturated phytosterols from any of the processes described above, pure saturated phytostanols can be obtained by hydrogenation. In this process the double-bond in the sterol molecule is saturated by the addition of hydrogen. This reaction is carried out in a suitable solvent under high hydrogen pressure, generally using a noble-metal based catalyst (e.g. Pd or Pt).

Phytostanols thus produced mainly consist of sitostanol and campestanol. Phytostanols produced from tall oil sterols typically contain ~ 90% sitostanol and ~ 10% campestanol, whereas a blend of stanols obtained from vegetable oils, typically from soybean oil, contains 68 – 75% sitostanol and 25 - 32% campestanol. It should be noted that stanols are also naturally-occurring. Especially in tall oil phytosterols, the level of phytostanols can be as high as 15%.

Production of Phytosterol and Phytostanol Esters

Phytostanol and phytosterol esters are produced via esterification of plant stanols or sterols with fatty acids from common vegetable oils. Thus, the fatty acid composition of the esters is similar to the parent vegetable oil used as a source of the fatty acids.

Esterification of phytosterols or phytostanols modifies the physical properties from high-melting crystalline powders with low oil solubility into liquid or semi-liquid substances that can easily be incorporated into a variety of (fat containing) foods. The proportion of the phytosterol backbone is approximately 60 % by weight of the ester and that of the fatty acid tail approximately 40 % by weight.

The phytosterols and phytostanols can be esterified with fatty acids from vegetable oils by two different routes:

- Direct esterification using free fatty acids.
- Trans-esterification using fatty acid methylesters.

Free Fatty Acid Route

This process consists of two consecutive steps:

- Preparation of free fatty acids.
- Esterification of free fatty acids and phytosterols/phytostanols.

The first step comprises hydrolysis of edible vegetable oil, e.g. sunflower oil triacylglycerides to form free fatty acids. After separation of the glycerol formed, the free fatty acids are purified from the unsaponifiable fraction and residual partial glycerides by distillation. In the second step the free fatty acids and phytosterols/phytostanols are esterified to form phytosterol or phytostanol esters in a process similar to the conventional manufacture of mono-acylglycerides. This reaction is carried out at elevated temperature (>180 °C) using a food grade catalyst. The reaction is carefully controlled with respect to reaction temperature and time. After the esterification reaction the excess of free fatty acids is removed by distillation.

Methylester Route

In this process two similar steps are involved:

- Preparation of fatty acid methyl esters.
- Inter-esterification of the fatty acid methyl esters and the phytosterols/phytostanols.

The first step comprises the methanolysis of edible vegetable oils to fatty acid methylesters and glycerol. In the second step these fatty methylesters are interesterified with the phytosterols and/or phytostanols by a similar process as used for the conventional chemical interesterification of fat blends. Also here the final purification involves deodorization to remove the excess methylesters and produce bland tasting and stable esters.

Commercial Suppliers

Depending on the manufacturer, the commercial product may be a mixture of the extracted sterols, a mixture of free sterols and stanols, sterol and stanol esters or stanol esters. A list of commercial suppliers is given in table.

Table: Commercial suppliers of phytosterols, phytostanols and/or their esters; 1) TO: Tall oil; VO: Vegetable oil.

Manufacturer	Brand name	Source[1]
Raisio Plc.	Benecol ®	TO & VO stanol esters
Cognis	Vegapure ®	TO & VO sterol esters
Archer Daniels Midland Co.	CardioAid TM	VO sterols & esters
Cargill Inc.	Corowise TM	VO sterols
Triple Crown	Prolocol TM	TO sterols
Pharmaconsult Oy Ltd	Multibene ®	TO & VO sterols/esters
Teriaka Ltd.	Diminicol TM	TO & VO sterols
Forbes Medi-Tech Inc.	Reducol TM	TO sterols & stanols
Arboris	AS-2 TM	TO sterols
PrimaPharm B.V.	Beta sitosterol	TO sterols
Fenchem Enterprises Ltd.	Cholevel TM	VO sterols
DRT	Phytopin ®	TO sterols
DDO Processing LLC	Nutraphyl TM	TO sterols
Degussa Food Ingr. GmbH	Cholestatin TM	VO sterols
Phyto-Source LP	Phyto-S-Sterol TM	TO sterols
Lipofoods	Lipophytol TM	VO sterols
Enzymotech Ltd.	CardiaBeat TM	TO & VO sterols

Chemical Characterization

Composition and Properties

The physical characteristics and composition of different commercial phytosterols, phytostanols and their esters are summarized in table.

Phytostanol and phytosterol esters are chemically stable, fat-type materials, having comparable chemical and physical properties to edible fats and oils. The product is insoluble in water, but soluble in non-polar solvents, such as hexane, iso-octane and 2-propanol. The esters are also soluble in vegetable fats and oils.

Heat stability of the esters is comparable to or even better than that of the parent vegetable oil or oil blend from which the fatty acids were derived. During shelf-life studies (long term storage), as pure material or in a product, phytostanol- and phytosterol esters produce similar decomposition products to those of edible oils and fats as oxidation of the fatty acid moiety is the major cause of the quality deterioration and formation of off-flavors in oils and fats. The phytosterol and phytostanol moieties are very stable at ambient temperatures. At higher temperatures some oxidation may occur.

Table: Physical characteristics and composition of different commercial phytosterols, phytostanols and their esters; 1) from TO sterols; 2) from VO sterols; 3) mainly sitosterol and campesterol.

	Tall oil sterols / stanols	Vegetable oil Sterols	Sterol- and stanol esters
Colour	White to off-white.		Whitish solid, pale yellow liquid.
Appearance	Crystalline waxy powder or prills.		Solid or liquid.
Solubility	<0.01 g in 100 g water; 2.5% in fat at ambient temperature; soluble in acetone, ethyl acetate, isopropanol.		Insoluble in water; soluble in oils and fats; soluble in non-polar solvents.
Melting point	138 - 158 °C		−25.7 - 38.8 °C
Sitosterol (%)	36 - 80	43 - 52	
Sitostanol (%)	6 - 34	0 - 3	90[1] / 68 – 75[2]
Campesterol (%)	4 - 25	24-28	
Campestanol (%)	0 – 14	0 - 2	10[1] / 25 - 32[2]
Stigmasterol (%)	< 1	14 - 24	
Brassicasterol (%)	< 1	0 - 9	
Minor sterols (%)	0 - 2	< 3	<5[3]
Solvents	MeOH <0.2%	MeOH 40 mg/kg; Ethanol 1.0 g/kg; Hexane 15 mg/kg	
Water (%)		<1	<0.1
Ash, residue on ignition (%)	<0.1		
Mercury (mg/kg)		<0.1	
Lead (mg/kg)	<0.1	<0.1	<0.1
Cadmium (mg/kg)		<0.1	
Arsenic (mg/kg)		<0.1	<0.1
Total heavy metals (mg/kg)		<20	
PAHs (ppb)	< 2		
Dioxins and Dioxin-like PCB's (ng-TEQ/kg)	< 1.5		
Pesticides (ppb)	absent		

Quality of Phytosterols, Phytostanols and their Esters

Phytosterols, phytostanols and their fatty acid esters should not contain contaminants or other impurities in concentrations that may prevent or limit their use in food products. Major contaminants to analyze or monitor in the final ingredients are:

- Heavy metals (Cd, Pb, Hg, As).
- Pesticides.
- Dioxins/furans/PCBs.
- Polycyclic aromatic hydrocarbons (PAHs).

Analysis of total phytostanol/phytosterol and ester content is generally measured on both the pure ingredients as well as the final food products. If required the level of free phytosterols and phytostanols can be separately analyzed. The product specification of the esters may set a minimum value for esterification and limit the free stanol/sterol content.

The conventional fats and oils analyses of water content, free fatty acids and peroxide value are generally performed to ensure good stability of the phytosterol, phytostanols and their esters during storage in pure form as well as in the supplemented foods.

Analytical Methods

Quantitative determination of phytosterols and phytostanols as their trimethylsilyl (TMS) derivatives in the presence of an internal standard is carried out by capillary gas chromatography (GC) equipped with a flame ionization detector (FID). In addition to quantitative amounts of stanols/sterols in the sample, the GC analysis provides information also on the distribution of individual sterol components and possible sterol degradation products. So far, there are no official reference methods developed particularly for the analysis of phytosterols or phytostanols present in sterol-enriched foodstuffs or ingredients. Some international reference methods exist for the analysis of sterols as natural minor food components (sterol content 1 % or less), whereas the sterol/stanol concentrations in phytostanol ester ingredients or enriched functional food products may be as high as 8%.

A number of methods are available for the analysis of naturally occurring or added phytosterols/ phytostanols in foods. Most methods are based on hot saponification (ISO 12228 or similar standard methods). In principle, the sample is saponified in the presence of an internal standard with potassium hydroxide in ethanol to break the ester bonds. The unsaponifiable material containing phytosterols/phytostanols is extracted with an organic solvent and evaporated to dryness under a stream of nitrogen gas. Sterols/stanols are silylated and analyzed by GC-FID. A GC/FID method commonly used is based on the AOAC Official Method 994.10 for “Cholesterol in Foods”. This method has been developed and validated for determining sitosterol, campesterol and stigmasterol in saw palmetto and dietary supplements, and it may be suitable for analyzing the main sterols/stanols in different types of food. However, different sample preparation procedures may be required, and saponification of the sample is required if esterified plant sterols are present. Alternatively, a GC/FID method for the analysis of phytosterols in margarine may be used.

For mixtures of phytosterols, phytostanols and their esters the separation of esterified and unesterified sterols and stanols is necessary. This may be achieved by using thin layer chromatography (TLC) or solid phase extraction. The fraction containing free phytosterols and phytostanols can be quantified by separating them from the esters and other fat and oil components by TLC.

The purity of sterol and stanol raw materials may be monitored by GC-FID. The sample is silylated in the presence of an internal standard and analyzed by GC. No saponification step is necessary, as the sterols and stanols are unesterified. TLC or LC may be used as a preparative step, if necessary. Quantification relies on relating peak area to the internal standard concentration. GS-MS provides data that can be used for sterol structure elucidation and to confirm peak identification. However, identification of trace level sterols is more difficult. Only some of the peaks can be identified as sterols, whereas other peaks remain unidentified. Because many compounds are not commercially available at sufficient purity, a single response factor is often used. Other properties of fats and oils, e.g. free fatty acids, peroxide value and moisture content, are determined using standard methods, e.g. AOCS methods.

Functional Use

Function in Products

Phytosterols and phytostanols, in free or esterified form, are added to foods for their properties to reduce absorption of cholesterol in the gut and thereby lower blood cholesterol levels. It is now generally accepted that sterols and stanols have the same cholesterol lowering efficacy.

The daily doses, considered optimal for the purpose of lowering blood cholesterol levels, are 2-3 g of phytostanols and/or phytosterols, which translates to 3.4-5.2 g in esterified form. This recommended daily dose is typically divided in 1-3 portions of food providing 1.7-5.2 g ester, which equals 1-3 g phytostanol and/or phytosterol equivalents.

In some cases phytostanol- and phytosterol esters can be used as a fat replacer because the phytostanol/ sterol moiety of the ester molecule does not provide any energy to the body. Moreover, phytostanol esters may be used to modify the fatty acid composition of a fat blend and replace part of the hard fat in margarines and spreads. Furthermore these esters can provide a crispy texture (prevents sogginess) to cereal products by coating the product surface. Both phytosterol and phytostanol esters give an enhanced creamy texture to low fat dairy products (yoghurt/ drinking yoghurt). They may also improve the taste of food products by masking bitterness and hence reduce the amount of sugar or other sweetener required to obtain a pleasant taste and mouth feel (e.g. in soy drinks).

Food Categories and Use Levels

Phytosterols, phytostanols and their esters are incorporated into a variety of foods and beverages and supplements, produced by a growing number of food- and beverage manufacturers.

Reactions and Fate in Foods

Stability at High Temperatures

Phytosterols and their fatty acid esters are quite stable compounds and undergo only limited degradation during oil processing. Only under harsh conditions, such as high temperatures (>100°C) in the presence of oxygen, oxidation of the phytosterol moiety may occur, in the same way as for cholesterol. Phytosterols are mono-unsaturated compounds (double bond in the B-ring), which are much more stable than the mono-unsaturated fatty acids (e.g. oleic acid), because of steric hindrance by the ring structure. Therefore, even under severe conditions, such as during shallow frying, sterol oxidation products form only slowly. Under conditions of use for shallow frying by consumers (temperatures 160-200 °C, 5-10 minutes of frying) the level of oxidation of sitosterol esters remains below 1.3% when the matrix consists of liquid oil of liquid margarine. Using free sterols these levels are somewhat higher at 2.5% and 5.1%, respectively.

Factors affecting phytosterol oxidation include, as would be expected, temperature and heating time, but also the composition of the lipid matrix. Phytosterol esters were found to be more susceptible to oxidation at elevated temperatures than free phytosterols.

Phytostanols are generally heat stable and phytostanol esters also show an oxidative stability. It should be noted that most spreads on the market are actually low fat spreads (< 40% fat). These are intended for spreading, not for shallow frying. Higher fat spreads (>60% fat) spreads however are suitable for shallow frying but are certainly not intended for deep fat frying.

Stability during Product Manufacturing and Storage

Phytosterols and phytostanols are microbiologically largely inert as shown by the absence of an effect during the fermentation process used to produce yoghurt. Furthermore, the ester added to various food products show excellent stability at different pH values during long term storage (up to at least a year). Phytostanol and phytosterol esters are also stable in milk and fermented milk and products with viable bacteria like yoghurts and yoghurt drinks.

References

- Biochemistry-genetics-and-molecular-biology, simple-lipid: sciencedirect.com, Retrieved 21, June 2020
- Chemistry-for-the-Life-Sciences-(Cessna), fats-and-oils: chem.libretexts.org, Retrieved 30, March 2020
- Waxes, lipid: britannica.com, Retrieved 13, May 2020
- Phospholipid: biologydictionary.net, Retrieved 29, January 2020
- Membrane-glycolipids-functional-heterogeneity-a-review-2161-1009: longdom.org, Retrieved 20, April 2020
- Phytosterols: fao.org, Retrieved 08, August 2020

3

Lipid Biosynthesis

When the supply of dietary carbohydrates is high, the excess carbohydrate is converted into triglycerides. It involves two processes: production of fatty acids from acetyl-CoA and esterification of fatty acids during the synthesis of triglycerides. The topics elaborated in this chapter will help in gaining a better perspective about the lipid biosynthesis.

Lipids not only play an array of cellular roles including energy storage, membrane constituents but also serve as anchors for membrane proteins (prenyl groups, phosphatidylinositol), cofactors (vitamin K), detergents (bile salts), extracellular and intracellular messengers (eicosanoids, phosphatidylinositol derivatives), hormones (vitamin D derivatives, sex hormones), pigments (Retinal, carotene) and transporters (dolichols). Essentially, all organisms has the ability to synthesize a variety of lipids and in this module we study the biosynthetic pathways for some of the most common cellular lipids. The reaction sequences in the biosynthetic pathways of lipids are endergonic and reductive. They usually utilize NADPH (reduced electron carrier) as a reductant and ATP as a source of metabolic energy.

In animals, an excess of carbohydrate is transformed to triacylglycerol. The process involves synthesis of fatty acids from acetyl-CoA followed by esterification of fatty acids in triacylglycerol production by a mechanism called lipogenesis. FAs are prepared by FAS that polymerizes and subsequently reduce acetyl-CoA units. The FAs acyl chains extends by a series of reactions which add the acetyl group, ebb it to an alcohol, desiccate it to an alkene group and after that condense it again to an alkane group. The enzymes involved in the biosynthesis of fatty acid are divided into two groups: (1) In animals and fungi, FAS reactions are carried out by a solitary multifunctional protein and (2) In plant plastids and bacteria, separate enzymes execute each step in the pathway. The synthesis of fatty acids might be later converted to TAGs through packaging in lipoproteins and finally secreted from the liver. TAGs synthesis takes place in the endoplasmic reticulum where acyl groups in fatty acyl-CoAs are shifted to the hydroxyl groups of glycerol-3-phosphate and diacylglycerol.

However, the synthesis of unsaturated FAs engrosses a desaturation reaction, where a double bond is introduced into the fatty acyl chain. In humans, for example, stearoyl-CoA desaturase-1 produces oleic acid by the desaturation of stearic acid. The linoleic acid (doubly unsaturated) as well as α-linolenic acid (triply unsaturated) cannot be amalgamated in mammalian tissues, and are therefore termed as essential fatty acids and must be taken through the diet.

Isoprenoids, Carotenoids and Terpenoids are primed by the assemblage and amendment of isoprene units bestowed from the reactive precursor's isopentenyl pyrophosphate (IPP) and dimethylallyl pyrophosphate (DMPP). In animals and archaea, the mevalonate (MVA) pathway produces these compounds from acetyl-CoA units, while in bacteria and plants, the non-mevalonate pathway uses glyceraldehyde 3-phosphate and pyruvate as substrates. These activated isoprene donors is also important in steroid biosynthesis where the isoprene units are tied together to make squalene followed by folding into a set of rings to make lanosterol which further be converted into cholesterol and ergosterol.

Biosynthesis of Fatty Acids

The biosynthesis of Fatty Acids (Fas) is the conception from acetyl CoA and malonyl CoA precursors via the deed of enzymes complex called FAS (fatty acid synthases). It is an imperative part of the lipogenesis process, which collectively with glycolysis, function to make fats from blood sugar.

Synthesis of Straight Chain Fas

Straight chain Fas synthesis crop up in two categories: Saturated and Unsaturated.

In E. coli, synthesis of saturated Fas by FAS II occurs by means of the six recurring reactions until the 16- carbon palmitic acid is produced (as shown in the Table below). FAS II complex restrain multiple enzymes that act as one unit and exists in prokaryotes, fungi, parasites and plants.

In animals and some yeast/fungi system, the synthesis of saturated Fas occurs via similar reactions using FAS I, a hefty dimeric protein demeanouring all enzymatic activities necessary to create a fatty acid. FAS I is however, less efficient than FAS II; nevertheless, it permit formation of medium chain Fas via premature chain termination. As soon as 16:0 carbon Fas has been created, it endures a number of modifications, resulting in elongation and/or desaturation. Elongation starts with stearate (18:0) primarily in the ER by some membrane bound enzymes. The enzymatic steps implicated in the elongation procedure are predominantly the same as those demeanoured by FAS, but the four major consecutive steps of the elongation are executed by physically associated individual proteins.

Type of Reaction Catalyzed	Enzyme Involved	Mechanism
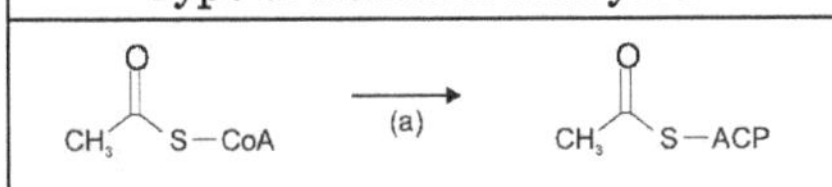	Acetyl CoA: ACP transacylase	Trigger acetyl CoA for reaction with malonyl ACP
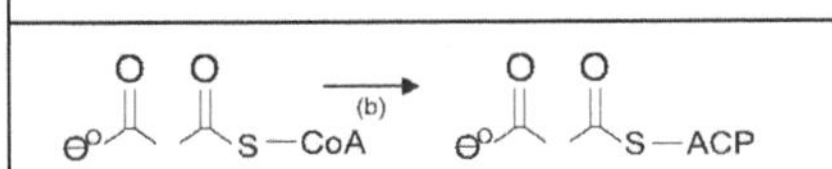	Malonyl CoA: ACP transacylase	Trigger malonyl CoA for reaction with acetyl ACP
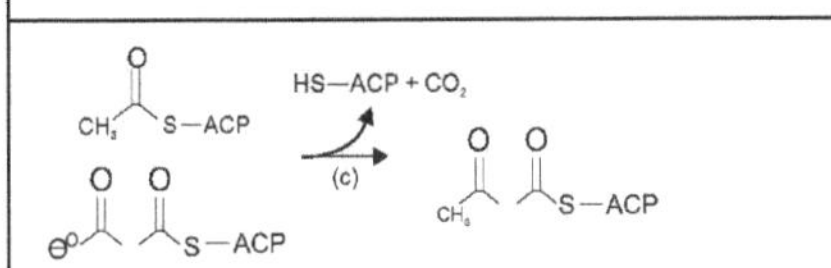	3-ketoacy -ACP synthase	Priming of acetyl ACP with chain extending malonylACP.
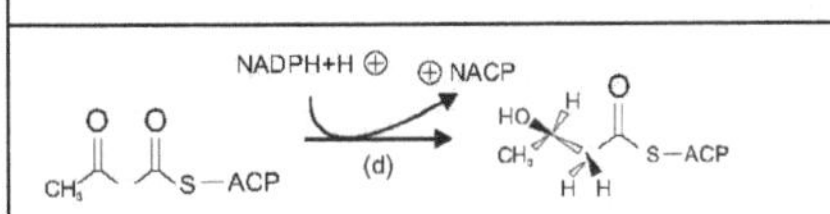	3-ketoacyl ACP reductase	Trim down the carbon 3 ketone to a hydroxyl group
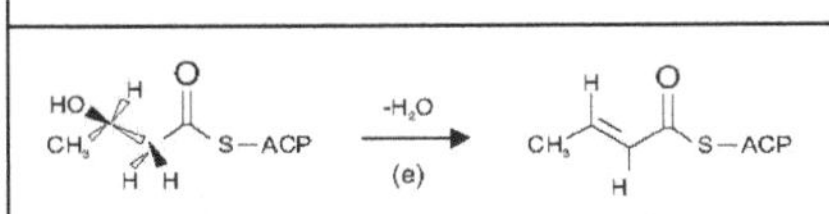	3-Hydroxyacyl ACP dehydrase	Eliminate water
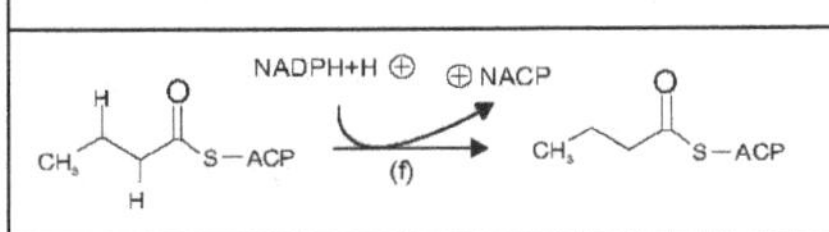	Enoyl ACP reductase	Lessen the C2-C3 double bond

Abbreviations: Acyl Carrier Protein (ACP), Coenzyme A (CoA), Nicotinamide Adenine Dinucleotide Phosphate (NADP).

FAs Biosynthesis Regulation

The Acetyl-CoA gets transformed to malonyl CoA by acetyl CoA carboxylase and ordained to nosh into the FA synthesis pathway. The enzyme Acetyl-CoA carboxylase is the point of regulation in saturated straight chain FA synthesis focusing both allosteric regulation and phosphorylation. Allosteric regulation takes place in most organisms while regulation by phosphorylation occurs typically in mammals. Allosteric control occurs via activation by citrate and feedback inhibition by palmitoyl-CoA. During the high levels of palmitoyl-CoA, the final product of saturated FA synthesis allosterically inactivates acetyl-CoA carboxylase to avert a build-up of FAs in cells. Activation of acetyl-CoA carboxylase by citrate occurs under high levels as it indicates sufficient acetyl-CoA to equip into the Krebs cycle and generate energy.

Fatty Acids are Synthesized and Degraded Pathway

Fatty acid synthesis is not simply a reversal of the degradative pathway. Rather, it consists of a new set of reactions, again exemplifying the principle that synthetic and degradative pathways are almost always distinct. Some important differences between the pathways are:

- Synthesis takes place in the cytosol, in contrast with degradation, which takes place primarily in the mitochondrial matrix.
- Intermediates in fatty acid synthesis are covalently linked to the sulfhydryl groups of an acyl carrier protein (ACP), whereas intermediates in fatty acid breakdown are covalently attached to the sulfhydryl group of coenzyme A.
- The enzymes of fatty acid synthesis in higher organisms are joined in a single polypeptide chain called fatty acid synthase. In contrast, the degradative enzymes do not seem to be associated.
- The growing fatty acid chain is elongated by the sequential addition of two-carbon units derived from acetyl CoA. The activated donor of twocarbon units in the elongation step is malonyl ACP. The elongation reaction is driven by the release of CO_2.
- The reductant in fatty acid synthesis is NADPH, whereas the oxidants in fatty acid degradation are NAD^+ and FAD.
- Elongation by the fatty acid synthase complex stops on formation of palmitate (C_{16}). Further elongation and the insertion of double bonds are carried out by other enzyme systems.

The Formation of Malonyl Coenzyme A is the Committed Step in Fatty Acid Synthesis

Fatty acid synthesis starts with the carboxylation of acetyl CoA to malonyl CoA. This irreversible reaction is the committed step in fatty acid synthesis.

The synthesis of malonyl CoA is catalyzed by *acetyl CoA carboxylase*, which contains a biotin prosthetic group. The carboxyl group of biotin is covalently attached to the ϵ amino group of a lysine residue, as in pyruvate carboxylase and propionyl CoA carboxylase. As with these other enzymes, a carboxybiotin intermediate is formed at the expense of the hydrolysis a molecule of ATP. The activated CO_2 group in this intermediate is then transferred to acetyl CoA to form malonyl CoA.

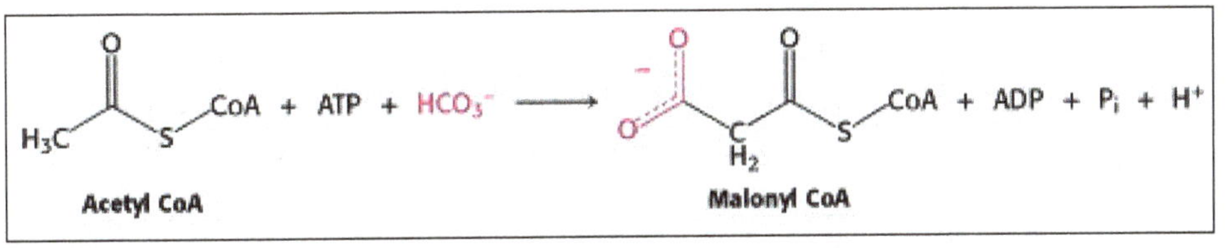

$$\text{Biotin} - \text{enzyme} + \text{ATP} + \text{HCO}_3 \rightleftharpoons \text{CO}_2 - \text{biotin} - \text{enzyme} + \text{ADP} + \text{P}_i$$

$$\text{CO}_2 - \text{biotin} - \text{enzyme} + \text{acetyl CoA} \rightarrow \text{malonyl CoA} + \text{biotin} - \text{enzyme}$$

This enzyme is also the essential regulatory enzyme for fatty acid metabolism.

Intermediates in Fatty Acid Synthesis are Attached to an Acyl Carrier Protein

The intermediates in fatty acid synthesis are linked to an acyl carrier protein. Specifically, they are linked to the sulfhydryl terminus of a phosphopantetheine group, which is, in turn, attached to a serine residue of the acyl carrier protein. Recall that, in the degradation of fatty acids, a phosphopantetheine group is present as part of CoA instead. ACP, a single polypeptide chain of 77 residues, can be regarded as a giant prosthetic group, a “macro CoA.”

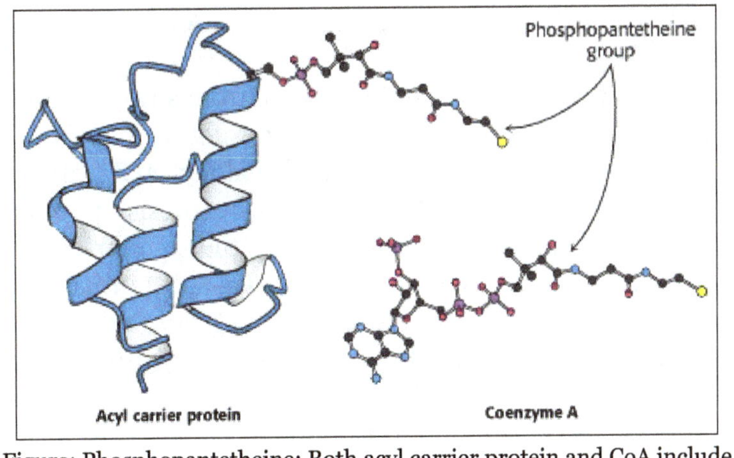

Figure: Phosphopantetheine: Both acyl carrier protein and CoA include phosphopantetheine as their reactive units.

The Elongation Cycle in Fatty Acid Synthesis

The enzyme system that catalyzes the synthesis of saturated long-chain fatty acids from acetyl CoA, malonyl CoA, and NADPH is called the fatty acid synthase. The constituent enzymes of bacterial fatty acid synthases are dissociated when the cells are broken apart. The availability of these isolated enzymes has facilitated the elucidation of the steps in fatty acid synthesis. In fact, the reactions leading to fatty acid synthesis in higher organisms are very much like those of bacteria.

Table: Principal reactions in fatty acid synthesis in bacteria.

Step	Reaction	Enzyme
1	Acetyl CoA + HCO_3^- + ATP → malonyl CoA + ADP + P_i + H^+	Acetyl CoA carboxylase
2	Acetyl CoA + ACP ⇌ acetyl ACP + CoA	Acetyl transacylase
3	Malonyl CoA + ACP ⇌ malonyl ACP + CoA	Malonyl transacylase
4	Acetyl ACP + malonyl ACP → acetoacetyl ACP + ACP + CO_2	Acyl-malonyl ACP condensing enzyme
5	Acetoacetyl ACP + NADPH + H^+ ⇌ D-3-hydroxybutyr-yl ACP + $NADP^+$	β-Ketoacyl ACP reductase
6	D-3-Hydroxybutyryl ACP ⇌ crotonyl ACP + H_2O	3-Hydroxyacyl ACP dehydratase
7	Crotonyl ACP + NADPH + H^+ → butyryl ACP + $NADP^+$	Enoyl ACP reductase

The elongation phase of fatty acid synthesis starts with the formation of acetyl ACP and malonyl ACP. Acetyl transacylase and malonyl transacylase catalyze these reactions:

$$\text{Acetyl CoA} + \text{ACP} \rightleftharpoons \text{acetyl ACP} + \text{CoA}$$
$$\text{Malonyl CoA} + \text{ACP} \rightleftharpoons \text{malonyl ACP} + \text{CoA}$$

Malonyl transacylase is highly specific, whereas acetyl transacylase can transfer acyl groups other than the acetyl unit, though at a much slower rate. Fatty acids with an odd number of carbon atoms are synthesized starting with propionyl ACP, which is formed from propionyl CoA by acetyl transacylase.

Acetyl ACP and malonyl ACP react to form acetoacetyl ACP. The acyl-malonyl ACP condensing enzyme catalyzes this condensation reaction:

$$\text{Acetyl ACP} + \text{malonyl ACP} \rightarrow \text{acetoacetyl ACP} + \text{ACP} + CO_2$$

In the condensation reaction, a four-carbon unit is formed from a two carbon unit and a three-carbon unit, and CO_2 is released. Why is the four-carbon unit not formed from 2 two-carbon units? In other words, why are the reactants acetyl ACP and malonyl ACP rather than two molecules of acetyl ACP? The answer is that the equilibrium for the synthesis of acetoacetyl ACP from two molecules of acetyl ACP is highly unfavorable. In contrast, the equilibrium is favorable if malonyl ACP is a reactant because its decarboxylation contributes a substantial decrease in free energy. In effect, ATP drives the condensation reaction, though ATP does not directly participate in the condensation reaction. Rather, ATP is used to carboxylate acetyl CoA to malonyl CoA. The free energy thus stored in malonyl CoA is released in the decarboxylation accompanying the formation of acetoacetyl ACP. Although HCO_3^- is required for fatty acid synthesis, its carbon atom does not appear in the product. Rather, all the carbon atoms of fatty acids containing an even number of carbon atoms are derived from acetyl CoA.

The next three steps in fatty acid synthesis reduce the keto group at C-3 to a methylene group. First, acetoacetyl ACP is reduced to D-3-hydroxybutyryl ACP. This reaction differs from the corresponding one in fatty acid degradation in two respects: (1) the D rather than the L isomer is formed; and (2) NADPH is the reducing agent, whereas NAD^+ is the oxidizing agent in β oxidation. This difference exemplifies the general principle that NADPH is consumed in biosynthetic reactions, whereas NADH *is generated in energy-yielding reactions*. Then D-3-hydroxybutyryl ACP is dehydrated to form crotonyl ACP, which is a trans-Δ^2-enoyl ACP. The final step in the cycle reduces crotonyl ACP to butyryl ACP. NADPH is again the reductant, whereas FAD is the oxidant in the corresponding reaction in β-oxidation. The enzyme that catalyzes this step, enoyl ACP reductase, is inhibited by triclosan, a broad-spectrum antibacterial agent. Triclosan is used in a variety of products such as toothpaste, soaps, and skin creams. These last three reactions—a reduction, a dehydration, and a second reduction—convert acetoacetyl ACP into butyryl ACP, which completes the first elongation cycle.

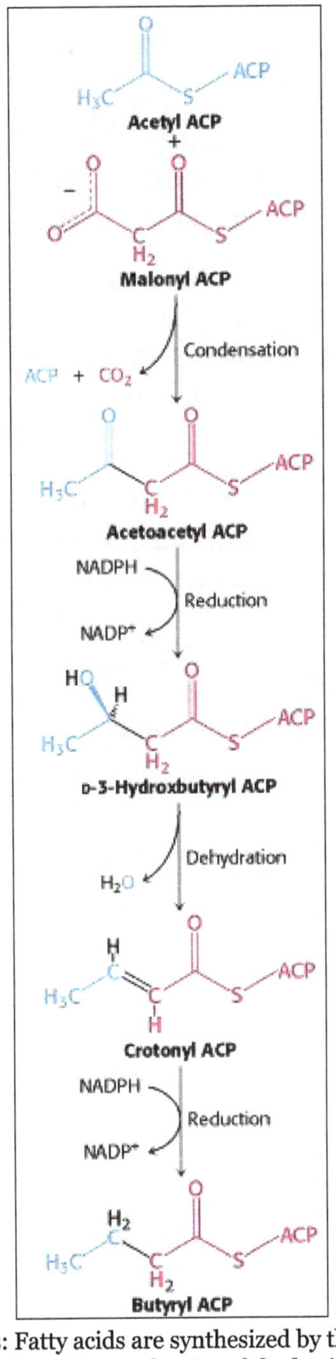

Figure: Fatty Acid Synthesis: Fatty acids are synthesized by the repetition of the following reaction sequence: condensation, reduction, dehydration, and reduction. The intermediates shown here are produced in the first round of synthesis.

In the second round of fatty acid synthesis, butyryl ACP condenses with malonyl ACP to form a C_6^--β-ketoacyl ACP. This reaction is like the one in the first round, in which acetyl ACP condenses with malonyl ACP to form a C_4^--β-ketoacyl ACP. Reduction, dehydration, and a second reduction convert the C_6^--β-ketoacyl ACP into a C_6^--acyl ACP, which is ready for a third round of elongation. The elongation cycles continue until C_{16}^--acyl ACP is formed. This intermediate is a good substrate for a thioesterase that hydrolyzes C_{16}^--acyl ACP to yield palmitate and ACP. The thioesterase acts as a ruler to determine fatty acid chain length.

Fatty Acids are Synthesized by a Multifunctional Enzyme Complex in Eukaryotes

Although the basic biochemical reactions in fatty acid synthesis are very similar in E. coli and eukaryotes, the structure of the synthase varies considerably. The fatty acid synthases of eukaryotes, in contrast with those of E. coli, have the component enzymes linked in a large polypeptide chain.

Mammalian fatty acid synthase is a dimer of identical 260-kd subunits. Each chain is folded into three domains joined by flexible regions. Domain 1, the substrate entry and condensation unit, contains acetyl transferase, malonyl transferase, and β-ketoacyl synthase (condensing enzyme). Domain 2, the reduction unit, contains the acyl carrier protein, β-ketoacyl reductase, dehydratase, and enoyl reductase. Domain 3, the palmitate release unit, contains the thioesterase. Thus, seven different catalytic sites are present on a single polypeptide chain. It is noteworthy that many eukaryotic multienzyme complexes are multifunctional proteins in which different enzymes are linked covalently. An advantage of this arrangement is that the synthetic activity of different enzymes is coordinated. In addition, a multienzyme complex consisting of covalently joined enzymes is more stable than one formed by noncovalent attractions. Furthermore, intermediates can be efficiently handed from one active site to another without leaving the assembly. It seems likely that multifunctional enzymes such as fatty acid synthase arose in eukaryotic evolution by exon shuffling, because each of the component enzymes is recognizably homologous to its bacterial counterpart.

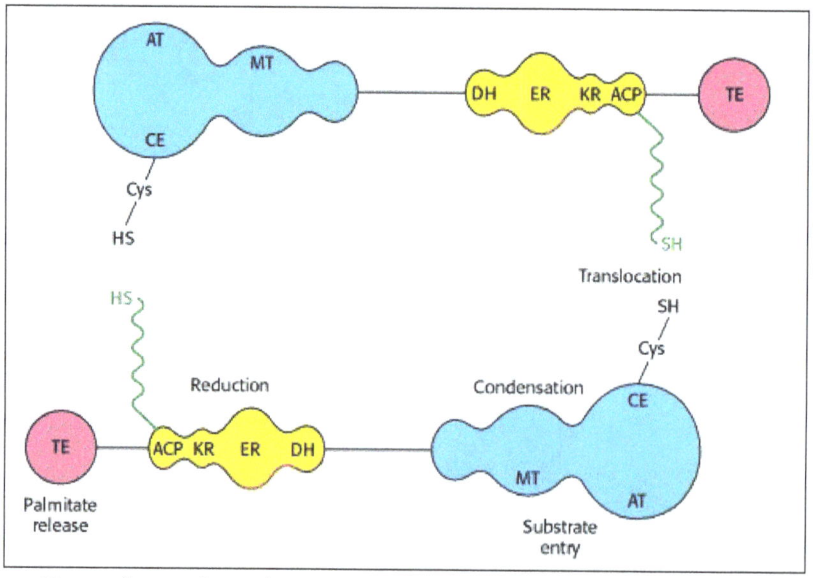

Figure shows schematic Representation of Animal Fatty Acid Synthase.

Each of the identical chains in the dimer contains three domains. Domain 1 (blue) contains acetyl transferase (AT), malonyl transferase (MT), and condensing enzyme (CE). Domain 2 (yellow) contains acyl carrier protein (ACP), β-ketoacyl reductase (KR), dehydratase (DH), and enoyl reductase (ER). Domain 3 (red) contains thioesterase (TE). The flexible phosphopantetheinyl group (green) carries the fatty acyl chain from one catalytic site on a chain to another, as well as between chains in the dimer.

The Flexible Phosphopantetheinyl Unit of ACP Carries Substrate from One Active Site to Another

Fatty acid synthesis begins with the transfer of the acetyl group of acetyl CoA first to a serine residue in the active site of acetyl transferase and then to the sulfur atom of a cysteine residue in the active site of the condensing enzyme on one chain of the dimeric enzyme. Similarly, the malonyl group is transferred from malonyl CoA first to a serine residue in the active site of malonyl transferase and then to the sulfur atom of the phosphopantetheinyl group of the acyl carrier protein on the other chain in the dimer. Domain 1 of each chain of this dimer interacts with domains 2 and 3 of the other chain. Thus, each of the two functional units of the synthase consists of domains formed by different chains. Indeed, the arenas of catalytic action are the interfaces between domains on opposite chains.

Elongation begins with the joining of the acetyl unit on the condensing enzyme (CE) to a two-carbon part of the malonyl unit on ACP.

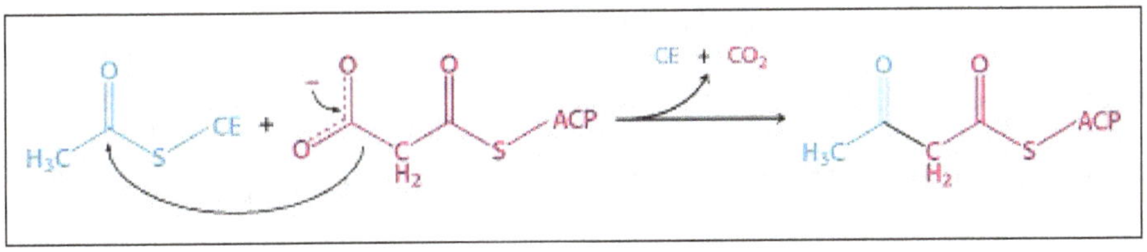

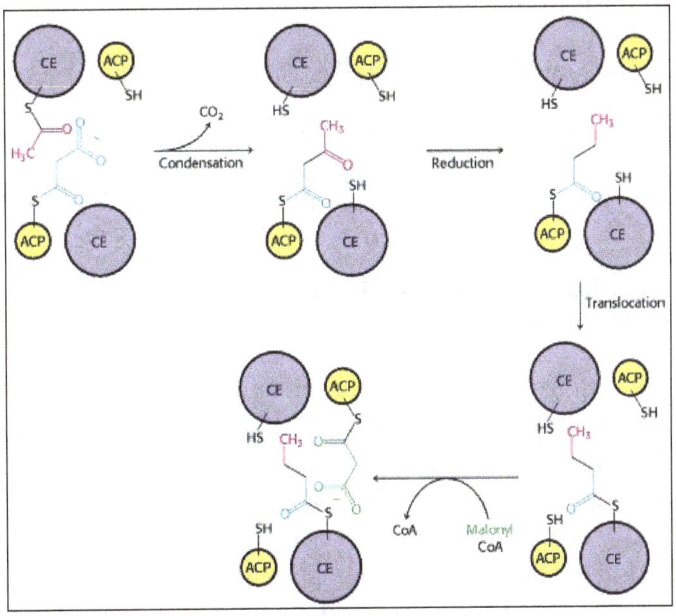

Figure: Reactions of Fatty Acid Synthase: Translocations of the elongating fatty acyl chain between the cysteine sulfhydryl group of the condensing enzyme (CE, blue) and the phosphopantetheine sulfhydryl group of the acyl carrier protein (ACP, yellow) lead to the growth of the fatty acid chain.

The reactions are repeated until the palmitoyl product is synthesized. CO_2 is released and an aceto-acetyl-S-phosphopantetheinyl unit is formed on ACP. The active-site sulfhydryl group on the condensing enzyme is restored. The acetoacetyl group is then delivered to three active sites in domain 2 of the opposite chain to reduce it to a butyryl unit. This saturated C_4 unit then migrates from the phosphopantetheinyl sulfur atom on ACP to the cysteine sulfur atom on the condensing enzyme. The synthase is now ready for another round of elongation. The butyryl unit on the condensing enzyme becomes linked to a two-carbon part of the malonyl unit on ACP to form a six-carbon unit on ACP, which undergoes reduction. Five more rounds of condensation and reduction produce a palmitoyl (C_{16}) chain on the condensing enzyme, which is hydrolyzed to palmitate by the thioesterase on domain 3 of the opposite chain. The migration of the growing fatty acyl chain back and forth between ACP and the condensing enzyme in each round of elongation is analogous to the translocations of growing peptide chains that take place in protein synthesis.

The flexibility and 20-Å maximal length of the phosphopantetheinyl moiety are critical for the function of this multienzyme complex. The enzyme subunits need not undergo large structural rearrangements to interact with the substrate. Instead, the substrate is on a long, flexible arm that can reach each of the numerous active sites. Recall that biotin and lipoamide also are on long, flexible arms in their multienzyme complexes. The organization of the fatty acid synthases of higher organisms enhances the efficiency of the overall process because intermediates are directly transferred from one active site to the next.

The Stoichiometry of Fatty Acid Synthesis

The stoichiometry of the synthesis of palmitate is:

$$\text{Acetyl CoA} + 7\,\text{malonyl CoA} + 14\,\text{NADPH} + 20\,\text{H}^+ \rightarrow \text{palmitate} + 7\,\text{CO}_2 + 14\,\text{NADP}^+ + 8\,\text{CoA} + 6\,\text{H}_2\text{O}$$

The equation for the synthesis of the malonyl CoA used in the preceding reaction is:

$$7\,\text{Acetyl CoA} + 7\,\text{CO}_2 + 7\,\text{ATP} \rightarrow 7\,\text{malonyl CoA} + 7\,\text{ADP} + 7\,\text{P}_i + 14\,\text{H}^+$$

Hence, the overall stoichiometry for the synthesis of palmitate is:

$$8\,\text{Acetyl CoA} + 7\,\text{ATP} + 14\,\text{NADPH} + 6\,\text{H}^+ \rightarrow \text{palmitate} + 14\,\text{NADP}^+ + 8\,\text{CoA} + 6\,\text{H}_2\text{O} + 7\,\text{ADP} + 7\,\text{P}_i$$

Citrate Carries Acetyl Groups from Mitochondria to the Cytosol for Fatty Acid Synthesis

The synthesis of palmitate requires the input of 8 molecules of acetyl CoA, 14 molecules of NADPH, and 7 molecules of ATP. Fatty acids are synthesized in the cytosol, whereas acetyl CoA is formed from pyruvate in mitochondria. Hence, acetyl CoA must be transferred from mitochondria to the cytosol. Mitochondria, however, are not readily permeable to acetyl CoA. Recall that carnitine carries only long-chain fatty acids. The barrier to acetyl CoA is bypassed by citrate, which carries acetyl groups across the inner mitochondrial membrane. Citrate is formed in the mitochondrial matrix by the condensation of acetyl CoA with oxaloacetate. When present at high levels, citrate is transported to the cytosol, where it is cleaved by ATP-citrate lyase.

$$\text{Citrate} + \text{ATP} + \text{CoA} + \text{H}_2\text{O} \rightleftharpoons \text{acetyl CoA} + \text{ADP} + \text{P}_i + \text{oxaloacetate}$$

Lyases

Enzymes catalyzing the cleavage of C-C, C-O, or C-N bonds by elimination. A double bond is formed in these reactions.

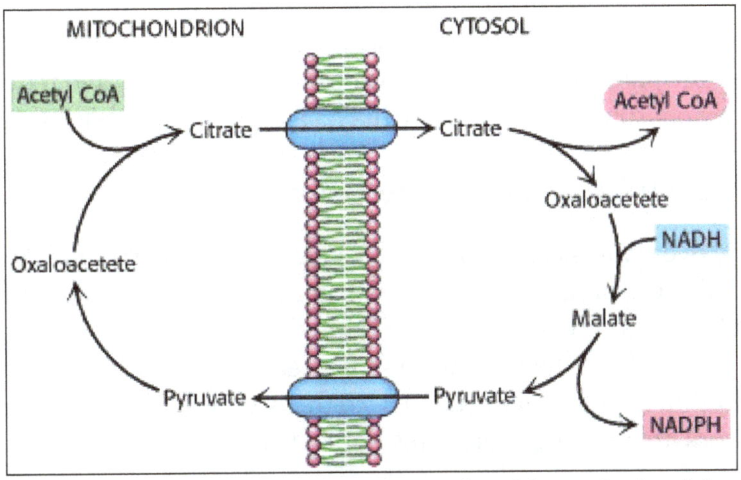

Figure: Transfer of Acetyl CoA to the Cytosol: Acetyl CoA is transferred from mitochondria to the cytosol, and the reducing potential NADH is concomitantly converted into that of NADPH by this series of reactions.

Thus, acetyl CoA and oxaloacetate are transferred from mitochondria to the cytosol at the expense of the hydrolysis of a molecule of ATP.

Sources of NADPH for Fatty Acid Synthesis

Oxaloacetate formed in the transfer of acetyl groups to the cytosol must now be returned to the mitochondria. The inner mitochondrial membrane is impermeable to oxaloacetate. Hence, a series of bypass reactions are needed. Most important, these reactions generate much of the NADPH needed for fatty acid synthesis. First, oxaloacetate is reduced to malate by NADH. This reaction is catalyzed by a malate dehydrogenase in the cytosol.

$$\text{Oxaloacetate} + \text{NADH} + \text{H}^+ \rightleftharpoons \text{malate} + \text{NAD}^+$$

Second, malate is oxidatively decarboxylated by an $NADP^+$-linked malate enzyme (also called malic enzyme).

$$\text{Malate} + \text{NADP}^+ \rightleftharpoons \text{pyruvate} + \text{CO}_2 + \text{NADPH}$$

The pyruvate formed in this reaction readily enters mitochondria, where it is carboxylated to oxaloacetate by pyruvate carboxylase.

$$\text{Pyruate} + \text{CO}_2 + \text{ATP} + \text{H}_2\text{O} \rightleftharpoons \text{oxaloacetate} + \text{ADP} + \text{P}_i + 2\text{H}^+$$

The sum of these three reactions is:

$$\text{NADP}^+ + \text{NADH} + \text{ATP} + \text{H}_2\text{O} \rightleftharpoons \text{NADPH} + \text{NAD}^+ + \text{ADP} + \text{P}_i + \text{H}^+$$

Thus, one molecule of NADPH is generated for each molecule of acetyl CoA that is transferred from mitochondria to the cytosol. Hence, eight molecules of NADPH are formed when eight molecules of acetyl CoA are transferred to the cytosol for the synthesis of palmitate. The additional six molecules of NADPH required for this process come from the pentose phosphate pathway.

The accumulation of the precursors for fatty acid synthesis is a wonderful example of the coordinated use of multiple processes to fulfill a biochemical need. The citric acid cycle, subcellular compartmentalization, and the pentose phosphate pathway provide the carbon atoms and reducing power, whereas glycolysis and oxidative phosphorylation provide the ATP to meet the needs for fatty acid synthesis.

Fatty Acid Synthase Inhibitors may be Useful Drugs

Fatty acid synthase is overexpressed in some breast cancers. Researchers intrigued by this observation have tested inhibitors of fatty acid synthase on mice to see how the inhibitors affect tumor growth. A startling observation was made: mice treated with inhibitors of the condensing enzyme showed remarkable weight loss due to inhibition of feeding. The results of additional studies revealed that this inhibition is due, at least in part, to the accumulation of malonyl CoA. Thus, fatty acid synthase inhibitors are exciting candidates both as antitumor and as antiobesity drugs.

Polyketide and Nonribosomal Peptide Synthetases Resemble Fatty Acid Synthase

The mammalian multifunctional fatty acid synthase is a member of a large family of complex enzymes termed megasynthases that participate in step-by-step synthetic pathways. Two important classes of compounds that are synthesized by such enzymes are the polyketides and the nonribosomal peptides. The antibiotic erythromycin is an example of a polyketide, whereas penicillin is a non-ribosomal peptide.

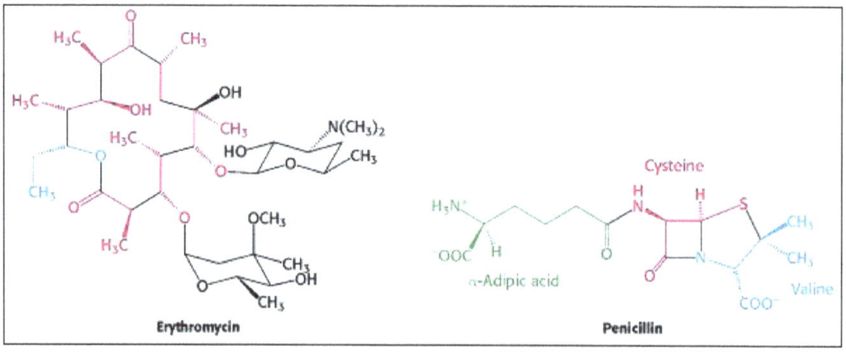

The core of erythromycin (deoxyerythronolide B, or Deb) is synthesized by the following reaction:

$$\text{Propionyl CoA} + 6\,\text{methylmalonyl CoA} + 6\,\text{NADPH} + 6\text{H}^{+} \rightarrow \text{Deb} + 7\,\text{CoA} + 6\,\text{NADP}^{+} + 2\,\text{H}_2\text{O}$$

This reaction is accomplished by three megasynthases consisting of 3491, 3567, and 3172 amino acids. The synthesis of deoxyerythronolide B begins with propionyl CoA linked to a phosphopantetheine chain connected to an acyl carrier protein domain. Similarly, the precursor of penicillin [Δ-(l-aminoadipyl)-l-cysteinyl-D-valine, or ACV] is generated by the following reaction:

$$\text{L-Aminoadipic acid} + \text{L-Cys} + \text{L-Val} + 3\,\text{ATP} + \text{H}_2\text{O} \rightarrow \text{ACV} + 3\,\text{ADP} + 3\,\text{P}_\text{i}$$

Which is catalyzed by a megasynthase consisting of 3791 amino acids. Each amino acid is activated by a specific adenylation domain within the enzyme—a domain that is homologous to acyl CoA synthase. Additional domains are responsible for peptide-bond formation and for the epimerization of the valine residue. Again, during synthesis, the components are linked to phosphopantetheine chains. Members of this remarkably modular enzyme family generate many of the natural products that have proved to be useful as drugs.

De Novo FA Synthesis in Humans

In humans, FAs are produced predominantly in liver, lactating mammary glands and to a slighter extent, the adipose tissue. The majority of acetyl-CoA is produced from pyruvate by pyruvate dehydrogenase in the mitochondria followed by condensation with oxaloacetate by citrate synthase to form citrate, which is subsequently elated into the cytosol and busted to yield acetyl-CoA and oxaloacetate by ATP citrate lyase. Oxaloacetate in reduced in the cytosol to malate by cytoplasmic malate dehydrogenase and malate is again transported backside into the mitochondria to play a part in the Citric acid cycle.

Fatty Acid Desaturases

The enzymes that insert unsaturated bonds to FA molecules are referred to as fatty acid desaturases. These enzymes catalyze the biosynthesis of monounsaturated fatty acids (MUFAs) and polyunsaturated fatty acids (PUFAs) by conversion of single bonds (C-C) into double bonds (C = C) in the acyl chain. Conversion of saturated FAs into MUFAs is catalyzed by stearoyl-CoA desaturase-1 (SCD1). PUFAs are formed in the reactions catalyzed by another group of desaturases, unsaturated fatty acid desaturases. The type, expression level, and activity of unsaturated fatty acid desaturases determine the qualitative and quantitative composition of PUFA profile. These enzymes are present in all living organisms, but their role in bacteria, fungi, plants, and animals differ considerably.

PUFAs, formed in the reactions catalyzed by unsaturated fatty acid desaturases, are important bioactive compounds that regulate many physiological processes. The most important PUFAs in this context seem to be omega-3 and omega-6 FAs, the names of which derive from the position of the first double bond from the methyl end. The group of omega-3 FAs includes alpha-linolenic acid (ALA, 18:3), eicosapentaenoic acid (EPA, 20:5), and docosahexaenoic acid (DHA, 22:6), whereas linoleic acid (LA, 18:2), gamma-linolenic acid (GLA, 18:3), dihomo-gamma-linolenic acid (DGLA, 20:3), and arachidonic acid (ARA, 20:4) are the examples of omega-6 FAs. PUFAs can produce various biological effects due to their ability to change the composition of plasma membranes, to regulate gene transcription, and to modulate cell signaling. PUFAs are precursors of various lipid mediators, including pro- and anti-inflammatory eicosanoids and docosanoids. Prostaglandins, prostacyclins, and leukotrienes derived from PUFAs are involved in inflammatory reactions and immune response. ARA is a precursor of pro-inflammatory eicosanoids, whereas EPA and DGLA are substrates for the synthesis of anti-inflammatory eicosanoids. DHA is also a precursor of anti-inflammatory and immunomodulatory docosanoids, such as resolvins and protectins. Both omega-3 and omega-6 FAs are considered key factors in the prevention of some undesirable bodily reactions, such as autoimmune response. Moreover, PUFAs play a significant role in chronic diseases, including cardiovascular disorders, cancers, and diabetes mellitus.

In addition to controlling the PUFA content, recent studies have shown that unsaturated fatty acid desaturases play an important role in glycolytic nicotinamide adenine dinucleotide (NAD+) recycling in the cell and as a consequence, they can be an alternative route for the flow of reducing equivalents generated during glycolysis. Under anaerobic conditions, when mitochondrial respiration is impaired, not only anaerobic glycolysis but also unsaturated fatty acid desaturases can oxidize the reduced form of nicotinamide adenine dinucleotide (NADH) to regenerate the NAD+ pool, which can be used by glyceraldehyde 3-phosphate dehydrogenase during glycolysis. Considering the complex function of PUFAs and unsaturated fatty acid desaturases itself (regardless from their products), changes in the expression or activity of fatty acid desaturases may have important consequences for the regulation of many processes taking place in the body, either those associated with basic metabolism or those involved in the development of various pathological conditions.

Activity of Unsaturated Fatty Acid Desaturases

Desaturase genes are widely spread in plant and animal kingdoms. Fatty acid desaturases convert MUFAs to PUFAs or insert additional unsaturated bonds into already existing PUFAs, but in some organisms not all of these activities are present. A plethora of various fatty acid desaturases have been identified, each of them inserting the double bond into a unique position. The regional selectivity of desaturases allows them to distinguish between the two ends of the FA acyl chain. A given desaturase always inserts the new double bond between the already existing unsaturated bond and the methyl end (methyl-end desaturase, e.g., delta-12 or delta-15 desaturase) or carboxyl end (front-end desaturase, e.g., delta-4, delta-5, delta-6 desaturase). The name of the given enzyme reflects the position of the acyl chain in which the double bond is inserted. According to a traditional naming convention, an enzyme from this group is referred to as 'delta X desaturase', where X is the number corresponding to the position from the molecule's end at which the double bond is inserted, whereas delta (Δ) means that the positions are counted from the carboxyl end.

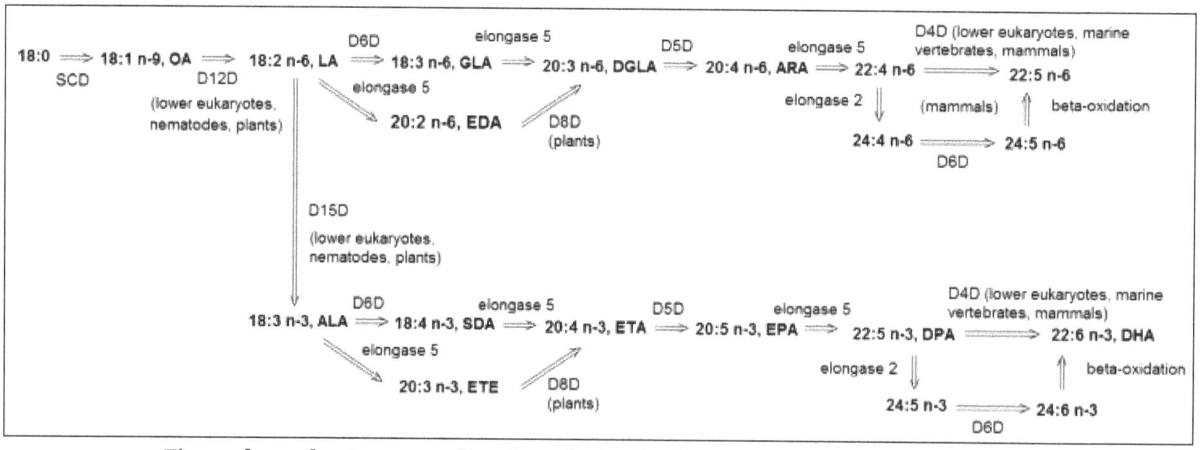

Figure shows desaturase-mediated synthesis of polyunsaturated fatty acids (PUFAs).

ALA—alpha-linolenic acid, 18:3 n-3, ARA—arachidonic acid, 20:4 n-6, DGLA—dihomo-gamma-linolenic acid, 20:3 n-6, DHA—docosahexaenoic acid, 22:6 n-3, DPA—docosapentaenoic acid, 22:5 n-3, D4D—delta-4 desaturase, D5D—delta-5 desaturase, D6D—delta-6 desaturase, D8D—delta-8 desaturase, D12D—delta-12 desaturase, D15D—delta-15 desaturase, EDA—eicosadienoic acid, 20:2 n-6, ETA—eicosatetraenoic acid, 20:4 n-3, ETE—eicosatrienoic acid, 20:3 n-3, EPA—eicosapentaenoic acid, 20:5 n-3, GLA—gamma-linolenic acid, 18:3 n-6, LA—linoleic acid, 18:2 n-6, OA—oleic acid, 18:1 n-9, SCD—stearoyl-CoA desaturase (delta-9 desaturase), SDA—stearidonic acid, 18:4 n-3.

The reaction catalyzed by desaturase is an aerobic process. Aside from the presence of oxygen (as a hydrogen acceptor), reaction requires cytochrome b5 (electron carrier) and NADH-dependent cytochrome b5 reductase. During the first stage of the desaturation process, reduced cytochrome b5 interacts with the non-heme iron active site of the desaturase, which enables the enzyme to react with O_2 and a substrate. This results in the activation of hydrogen at the carbon of incipient double bond in substrate. The product of this relatively slow stage is a very short-lived carbon-centered radical. During the second stage, this transient molecule loses the second hydrogen via rapid disproportionation, which leads to the formation of the final product with a new double bond, along with two molecules of water.

Eukaryotic desaturases share similar structural motifs. These enzymes have an N-terminal cytochrome b5 domain, which contains the heme-binding motif, crucial for catalytic activity of desaturases. The C-terminal domain contains three histidine motifs with seven histidines (the so-called His box) which are also vital for the catalytic activity. Substitution of histidine with glutamine in the third motif from the C-end is a specific feature of all desaturases that insert a double bond between the already existing bond and carboxyl end. Separation of the first and second histidine motif by a long histidine fragment, aside from fatty acid desaturases, is also observed in aldehyde hydroxylases and carboxylases.

The most well-known examples of fatty acid desaturases are delta-6 desaturase (D6D), delta-5 desaturase (D5D), delta-12 desaturase (D12D), and delta-15 desaturase (D15D). D6D catalyzes the insertion of the double bond between sixth and seventh carbon from the carboxyl end. Reaction of D6D is the rate-limiting step in the formation of GLA from LA and stearidonic acid (SDA, 18:4 n-3) from ALA. D6D can use at least two other substrates, tetracosatetraenoic acid (24:4 n-6) and tetracosapentaenoic acid (24:5 n-3), converting them into tetracosapentaenoic acid (24:5 n-6) and tetracosahexaenoic acid (24:6 n-3), respectively. Aside from the activity of delta-6 desaturase, this enzyme can also show other desaturase activities, e.g., synthesize DGLA and eicosatetraenoic acid (ETA, 20:4 n-3) from eicosadienoic acid (EDA, 20:2 n-6) and eicosatetraenoic acid (ETE, 20:3 n-3), respectively. Moreover, D6D is responsible for the synthesis of sapienic acid, the most abundant FA of human sebum, catalyzing desaturation of 16:0 to 16:1 n-10. The latter reaction is a unique example of this enzyme activity because it leads to the synthesis of MUFA, and this reaction occurs exclusively in sebaceous glands. D5D inserts double bond at the fifth position from the carboxyl end. This enzyme is primarily involved in the synthesis of ARA and EPA, using DGLA and ETA, respectively, as the substrates. Other acids, such as 20:3 n-3 and 20:2 n-6, can also serve as the substrates for D5D. Moreover, some isoforms of D5D can catalyze the synthesis of atypical products, such as non-methylene-interrupted fatty acids (FA where double bound are interrupted with group other than methylene), e.g., 18:3 Δ5, 9, 12 and 20:4 Δ5, 11, 14, 17. D12D and D15D are the key enzymes involved in the synthesis of omega-3 and omega-6 FAs. D12D is responsible for the conversion of oleic acid (OA, 18:1 n-9) to LA, and D15D desaturates LA to ALA. These enzymes are not expressed in all mammals, including humans; as a result, they cannot synthesize LA and ALA de novo and need to obtain these essential FAs from the diet. Delta-12 and/or delta-15 desaturases are expressed in lower eukaryotes and most plants, as well as in some animals, such as cockroaches, house crickets, and parasitic wasps.

Regulation of the expression and activity of unsaturated fatty acid desaturases is a very complex biochemical phenomenon. Dietary PUFA intake and transcriptional regulation are well-recognized determinates of unsaturated fatty acid desaturases activity. Many studies proved that dietary FA can influence gene expression of unsaturated fatty acid desaturases. Both D5D and D6D expression were reduced by PUFAs in multiple experimental models, ranging from cell lines cultured in vitro and treated with different PUFAs to animals fed with PUFA-enriched diets (including grouper larvae, European sea bass, mice, piglets, and baboons). Unsaturated fatty acid desaturases gene expression is also regulated by insulin and gonadal hormones. Several transcription factor binding site motifs were identified in promoters regions of unsaturated fatty acid desaturases, including nuclear factor Y (NF-Y), CCAAT enhancer binding protein (C/EBP), sterol regulatory element (SRE), nuclear factor 1 (NF-1), stimulatory protein 1 (Sp1), activated protein 1 (AP1), hepatocyte nuclear factor 4α (HNF4α), and peroxisome proliferator activated receptor γ (PPARγ). Epigenetic alterations (i.e., methylation of DNA) may also contribute to changes in desaturase activity. Furthermore, recent findings suggest that D5D and D6D activity can change in response to the cytosolic NAD+/NADH ratio. The decreased cytosolic NAD+/NADH ratio (for example, as a result of inhibition of aerobic cellular respiration) increases the activity of unsaturated fatty acid desaturases, without change in mRNA or protein levels.

Role of Unsaturated Fatty Acid Desaturases

The activity of desaturases determines, together with dietary FA intake, the qualitative and quantitative composition of unsaturated FAs. Products of these enzymes are essential components of all biological systems. The exact role of desaturases depends on the organism in which they are expressed.

Role of Unsaturated Fatty Acid Desaturases in Microorganisms

Firstly, the desaturases genes were discovered, and their functions were recognized in microorganisms. For example, the first D6D gene was cloned from cyanobacterium Synechocystis, the first D5D gene was cloned from fungus Mortierella alpina, and the first D8D gene was cloned from the protist Euglena gracilis. Up to now, microorganisms remain a subject of interest in many researches on the structure, function, and activity of unsaturated fatty acid desaturases.

In bacteria, changes in the PUFA composition in plasma membranes can be helpful in the adjustment to unfavorable environmental conditions. These organisms can change the activity of desaturases to adapt the PUFA content to changing temperature, pH, and atmospheric pressure to maintain appropriate fluidity of plasma membranes. Long-chain PUFAs are detected in the largest amount in bacteria inhabiting cold marine environments, such as Polar Regions or deep seas. Transfer of such bacteria into a warmer environment has resulted in changes in expression of desaturases genes and in a decrease in the PUFA content in membrane phospholipids.

PUFAs in microorganisms usually have 16 or 18 carbons, nevertheless, some microorganisms were shown to synthesize ARA, EPA, and DHA. Nowadays primary sources of EPA and DHA in our diet are marine water fish species. However, their availability is limited due to catch limits implemented to prevent resource depletion. Hence, the use of microorganisms expressing specific homologous or heterologous unsaturated fatty acid desaturases capable of synthesizing EPA and DHA seems to be a promising alternative source of omega-3 FAs. Current commercially used microorganisms to produce n-3 PUFA, such as Crypthecodinium cohnii, Schizochytrium sp., or Nannochloropsis sp., are the result of intensive screening and selective breeding procedures. Genetic engineering of unsaturated fatty acid desaturases in microorganisms opens up new possibilities to maximize the efficiency of upstream and downstream processing.

Role of Unsaturated Fatty Acid Desaturases in Plants

FA composition is crucial for the growth and vegetation of plants. FA desaturation is a key factor determining the tolerance of plants to various environmental stressors. Enhanced accumulation of PUFAs, due to changes in desaturases gene expression level, is postulated to facilitate cold adaptation, maintaining normal fluidity and integrity of plasma membranes. For example, in many plants the expression of FAD8 gene (genes encoding fatty acid desaturases in plants belong to the FAD family) is strongly induced by low temperatures (product of this gene catalyzed by the conversion of diene fatty acids to triene fatty acids). However, excessive accumulation of PUFAs may also exacerbate a thermal injury. Transgenic tobacco transformed with FAD7 gene showed greater resistance to cold, whereas the plants with silenced FAD7 contained less trienoic fatty acids and showed better tolerance to high temperatures than the wildtype plants. Furthermore, FAD2 and FAD6 genes in seedlings were shown to be activated by salt and osmotic stress. Mutants with FAD2 deficiency accumulated more sodium ions $(Na+)^+$ in the cytoplasm of root cells and were highly sensitive to salt stress during seed germination and early seedling growth.

The activity of various fatty acid desaturases is also crucial in the context of nutritional properties of edible oils, as these enzymes determine the contents of individual PUFAs in the final product. Progress in genetic engineering opened a perspective for modification of fatty acid desaturases to enrich cultivated plants in particularly desirable FAs. Many research groups obtained transgenic plants that can synthesize and accumulate selected FAs, such as OA, EPA, DHA, and SDA. For example, expression of borage (Borago officinalis) D6D in oilseed crops contributed to increase in SDA levels up to 20% of all FAs present in these plants. High levels of EPA were reported in Ethiopian mustard (Brassica carinata) after its transformation with the fungal D5D gene. Few research groups were able to produce DHA in plants by using the D4D gene from microorganisms. A higher specific PUFA consumption is recommended by doctors and nutritionists in several diseases, such as cardiovascular diseases, diabetes, and inflammatory disorders. However, the efficacy, quality, and safety of PUFA supplements available on the market are questionable because they are beyond any pharmaceutical control. Several studies conducted in geographically diverse locations have shown that vegans and vegetarians have lower levels of DHA and EPA in the serum compared to omnivores. Therefore, the use of plants with modified fatty acid desaturases seems to be a promising alternative for dietary n-3 PUFA consumption. This can be especially important for vegan and vegetarian mothers during breastfeeding.

Role of Unsaturated Fatty Acid Desaturases in Humans and other Mammals

PUFA profile is a consequence of diet and genetic makeup. Even one single nucleotide polymorphisms (SNPs) in FADS genes (mammalian desaturase genes that belong to the FADS family) were shown to be associated with altered activity of fatty acid desaturases and significant changes in plasma lipid profile. People with high dietary intake of ALA who carried one or two altered FADS1 alleles (the so-called rs174546 variant), presented with lower concentrations of total cholesterol and non-HDL cholesterol than those with other allelic variants and the same intake of this FA (FADS1 encodes D5D). Only children with certain FADS1 SNPs benefit from fish oil supplementation. Recent studies of the mouse FADS2 gene (encodes D6D) demonstrated that even one non-synonymous SNP can alter properties of the enzyme. SNP A→G was associated with the change of amino acid in the region responsible for iron binding. The presence of G allele seems to be associated with lower activity of the desaturase since mice carrying this allele had higher levels of LA and ALA and lower levels of ARA, EPA, and DHA. Moreover, the G allele was shown to be more common in mice with higher basal metabolic rate. The protein encoded by the third mammalian fatty acid desaturase gene, FADS3, has yet not been ascribed any specific function. Expression of FADS3 was found in many human organs. Surprisingly, recent in vitro studies demonstrated that the product of FADS3 in rat encodes the enzyme which can catalyze delta-13 desaturation of trans-vaccenic acid (VA, trans-11 18:1) to trans-11, cis-13 conjugated linoleic acid (CLA; trans-11, cis-13 18:2). Hence, FADS3 might be the first gene encoding methyl-end desaturase in mammals, but this concept still has not been confirmed in vivo. What is interesting, some studies have shown that unsaturated fatty acid desaturases activity resulting from genetic variants of FADS genes can be masked. Alterations with the gene-related PUFA profile was only observed in normal weight women, whereas overweight/obese women (with a BMI ≥ 25) are less affected by FADS genetic variants in this regard. Exploration of gene-BMI interactions in the general population or male subpopulation is still needed.

Fatty acid desaturases seem to exert a variable effect on human health. This functional link results from a relationship between specific polymorphisms of desaturase genes and various metabolic phenotypes. Many studies demonstrated a relationship between the activity of these enzymes and various complex diseases, such as coronary artery disease (CAD). Among patients with CAD, the activity of D6D was higher in comparison to healthy subjects. Moreover, specific SNP can alter the probability of CAD occurrence. For example, SNP in position rs174537G/T is associated with a higher risk of CAD. Variant rs174537T is associated with a lower risk of CAD, while variant rs174537G is more common in CAD patients. Fortunately, there is evidence that appropriate modification of PUFA intake from the diet can prevent the unfavorable effects of FADS polymorphism.

Since n-3 PUFA are considered as anti-inflammatory substances, altered PUFA metabolism facilitates the chronic inflammation observed in Crohn's disease (CD). Data from the genome-wide association study point to a few SNP in FADS genes associated with CD risk. Analysis of D6D in mesenteric adipose tissue biopsies (abnormal mesenteric adipose tissue is the hallmark of CD) shows a lower level of D6D in CD specimens than in normal tissue at both mRNA and protein level. However, the activity of D6D, calculated as enzyme activity indexes based on the fatty acid concentration in plasma, was significantly higher in the CD patients than in the healthy volunteers. Moreover, studies suggest that patients with CD benefit from varying degrees of n-3 PUFA supplementations. Lack of consistency across studies may be explained by the presence of gene-diet interactions. For example, patients with specific FADS variants associated with higher endogenous production of n-6 PUFA have a greater risk of developing CD even with lower dietary intake of n-6 PUFA, so they can benefit from n-3 PUFA supplementation, whereas patients with FADS variants associated with a lower endogenous production of n-3 PUFA, even with high dietary PUFA intake, still tends to have a lower n-3 PUFA level, so they benefit less from n-3 PUFA supplementations than patients with other FADS variants. This suggests that in the future, a more personalized approach is needed.

Many studies conducted in humans and mice showed correlations between the activities of delta-5 and delta-6 desaturases and metabolic syndrome, insulin resistance, and obesity. They reported that there is a strong positive correlation between D6D activity and the risk of these conditions, whereas the activity of D5D is inversely related to metabolic syndrome, insulin resistance, and obesity (no genetic backgrounds were investigated in these studies). Comparative genetic analysis of persons with insulin resistance and normal sensitivity to insulin demonstrated down regulation of D5D expression in adipose tissue and muscles of the former group (no activity measurements were made in this study). The problem is even more complicated if we take into consideration the fact that certain SNP can modulate desaturases activity. Some studies from Japanese populations show that some variant of FADS1 (SNP rs174550) was significantly associated with an increased risk of type 2 diabetes. Interestingly, the same SNP in the European population was associated with the lower fasting plasma glucose level in a normal physiological range, but not with pathological glucose levels. Data from multiple experiments on mice suggest that the development of selective desaturases inhibitors can be beneficial in the treatment of human diabetes, obesity, and atherosclerotic cardiovascular disease. Mice with D6D knockdown developed resistance against high-fat diet induced obesity, so maybe some metabolic benefits for obese people can be gained with D6D inhibitors.

Since obesity is considered as a chronic low-grade inflammation state, D5D inhibitors can be also beneficial. Potential D5D inhibitor can increase DGLA level (precursor of anti-inflammatory eicosanoids) and decrease AA level (precursor to pro-inflammatory eicosanoids). Insulin resistance is a complicated condition and several factors including genetic makeup, lifestyle, and diet contribute to its development. Considering them separately may lead to inconsistencies across different studies, such as in case of studies on the effectiveness of PUFA supplementation. Studies have shown, that n-3 PUFA supplementation can be effective in prevention or reversion of insulin resistance, but only together with reducing the intake of saturated FA, trans-FA, and n-6 PUFA. Supplementation of n-6 PUFA may also decrease insulin resistance but only if n-6 PUFA replaces SFA, trans-FA, or sucrose. Replacing n-3 PUFA with n-6 PUFA in the diet can worsen insulin resistance. n-3 PUFA may prevent insulin resistance by a number of different mechanisms, including regulation of inflammation, modulation of adiponectin and leptin secretion, and influencing the expression of several genes involved in carbohydrate and lipid metabolism. It seems that associations between desaturases and complex metabolic conditions such as metabolic syndrome, insulin resistance, and obesity are still not fully understood and some mechanisms still remain elusive.

Patients with non-alcoholic fatty liver disease (NAFLD) present a higher level of free FA in the serum and altered profile of FA in comparison to healthy subjects, including higher total SFA, higher total MUFA, lower total PUFA. Available evidence shows that altered activity of fatty acid desaturases is one of the risk factors of NAFLD. Patients with NAFLD were shown to present with a higher activity of D6D and lower activity of D5D than healthy volunteers. A similar observation was made in the case of pediatric non-alcoholic liver disease. Differences in activity of desaturases are often connected to naturally existing genetic variations of those genes.

Better understanding of the connection between locus polymorphism and level of functional enzyme can result in the personalized treatment strategy for NAFLD in the near future. A pilot study from 2018 has shown that NAFLD patients with alleles connected to a low D5D activity benefit more from n-3 PUFA supplementation (significant reduction in steatosis, fibrosis, ballooning, and NAFLD activity score) than patients with others SNP variants. So far, SNPs in patatin-like phospholipase domain-containing protein 3 (PNPLA3) and transmembrane 6 superfamily member 2 (TM6SF2) regions have been identified and validated in large cohorts of patients as a biomarker for NAFLD risk. Maybe in the near future the SNPs in FADS1 will be the third biomarker for NAFLD but some large-scale research is still needed. However, the situation is complicated by the fact that in NAFLD, as well as, in CAD disturbances in unsaturated fatty acid desaturases are observed and NAFLD is often accompanied by CAD. What is more, CAD is one of the leading causes of death in this group of patients. Nevertheless, there is almost no research on the associations among NAFLD,

CAD, and FADS polymorphism. A deeper understanding of FADS SNP in the context of NAFLD with CAD is essential if we want to use a genetic background for early diagnosis and prevention.

Unsaturated fatty acid desaturases also play a significant role in carcinogenesis, including cancer cell survival, metastasis, and drug resistance. Genome-wide association study in East Asians population identifies a loci map to FADS1 and FADS2 genes as associated with colorectal cancer and laryngeal squamous cell carcinoma risk. Unfortunately, there is no information on this matter from populations from other continents, so it is not clear if this can be treated as a universal prediction marker. Moreover, in patients with malignant melanoma, breast, brain and lung malignancies, expression and activity of D6D in cancer tissues turned out to be higher than in adjacent normal tissues. It cannot be excluded that this is a common feature of all malignant neoplasms. In animal studies, inhibition of delta-6 desaturase (both through knockdown and via RNAi) effectively prevented cancer growth. Since D6D is an enzyme catalyzing the rate-limiting step of AA synthesis, its activity is crucial for the production of pro-inflammatory eicosanoids, downstream metabolites of AA, such as prostaglandins, leukotrienes, and epoxy-eicosatrienes. Hence, it seems logical that inhibition of the principal metabolite (AA) formation might be a more effective way to reduce the synthesis of pro-inflammatory eicosanoids than blocking each specific pathway (i.e., cyclooxygenase, lipoxygenase, and cytochrome P450 epoxygenase pathway), separately. Reduced unsaturated fatty acid desaturases activity can have an anti-proliferative effect not only because of reduced eicosanoid availability for cell signaling. Another possibility is the altered cytosolic NAD+/NADH ratio, which can modulate glycolysis and lactate fermentation—two important sources of respiratory fuel in cancer cells. Studies have shown that transient knockdown of D5D and D6D in cells cultured in vitro resulted in changes in the reactive oxygen species (ROS) level, pyruvate consumption, and cell proliferation. However, further studies are required to resolve whether unsaturated fatty acid desaturases-mediated NAD+/NADH ratio play an important role in the Warburg effect in cancer cells. Taking all information together, fatty acid desaturation can be a new metabolic marker and therapeutic target in certain types of cancer.

Fatty acid desaturases also play an important role during pregnancy and breastfeeding. Maternal levels of ARA, EPA, and DHA influence the amount of these acids delivered to the fetus directly or provided with maternal milk after birth. Hence, desaturase activity has an impact on the qualitative and quantitative composition of n-3 and n-6 FAs delivered to the child. FAs, among them ARA, EPA, and DHA, play a key role in growth, development of neurons, and functioning of the immune system in infants. Moreover, the development of cognitive function and intellectual development of the child may be determined by a unique composition of FAs contained in maternal milk. Interestingly, few studies also mention a relationship between breast milk fatty acids composition and postnatal HIV-1 transmission. ALA can be elongated to create ETE or desaturated to create SDA. When D6D is less active than elongase 5 the amount of ETE in breast milk increases. A higher concentration of ETE was associated with a lower risk of HIV-1 transmission. Unfortunately, a large prospective cohort study is still lacking in this matter.

FAs, especially n-3, can modulate behavior and cognitive function. Animals maintained on a diet excluding or restricting substrates for fatty acid desaturases showed greater locomotor hyperactivity and worse cognitive skills. It was observed that children with specific SNPs in FADS genes (so called DHA-increasing cluster) score more in personal and social skills questionnaire. Moreover, studies demonstrated a link between the variability of nucleotide sequences within desaturase encoding genes and the incidence of attention-deficit hyperactivity disorder (ADHD).

The type of FAs contained in membrane phospholipids may modulate a broad range of biological mechanisms and pathways in the brain, including neurons, glial cells, and endothelial cells integrity and survival, neurotransmission (dopaminergic, serotonergic, glutamatergic, and cholinergic), and neuroinflammation. PUFAs, especially ARA and DHA, are the major constituents of the neuronal membranes. However, the cause and effect relationship between mental disorders and unsaturated fatty acid desaturases is still unclear. Patients with neuropsychiatric disorders, such as bipolar affective disorders and schizophrenia often present overexpression and increased activity of D6D and at the same time, the levels of n-3 PUFA (DPA, DHA) and n-6 PUFA (EDA, ARA, 22:4 n-6) in neuronal membrane phospholipids. Evidence from in vitro animal and clinical studies suggest that long-term treatment with risperidone (an antipsychotic drug) upregulate the expression and activity of multiple lipogenic genes, including D6D. Interestingly, some studies in animal models suggest that prenatal dietary PUFA deficiency (especially ARA and DHA) may affect early neurodevelopment of the offspring and thereby predispose to the development of schizophrenia. Mice that had experienced gestational and early postnatal dietary ARA and DHA deprivation have shown schizophrenia-like phenotypes at adulthood (lowered levels of motivation, depressive symptoms, and impaired cognitive functions). Since the PUFA profile can be determined both by dietary PUFA intake and desaturase activity, it seems reasonable to assume that maternal activity of desaturases may also play a role in the risk of schizophrenia in offspring.

A few studies on rats from the late 1970s and early 1980s mentioned changes in maternal, fetal and pups D5D and D6D activities in response to mothers' low essential FA diet, nevertheless, until this day there are no studies on the relationship between mothers' unsaturated fatty acid desaturases (activity and/or genetic variants) and schizophrenia risk in the offspring. Furthermore, a relatively small number of studies have examined the therapeutic utility of PUFA supplementation in schizophrenia and none of them mentioned the potential use of FADS inhibitors. Moreover, results of these studies are inconsistent. Observed effects of PUFA supplementation vary from a significant increase in severity of schizophrenia symptoms, through no major effect on symptoms severity to decrease the intensity of symptoms and improve the level of patients functioning.

Synthesis of Branched Chain FAs

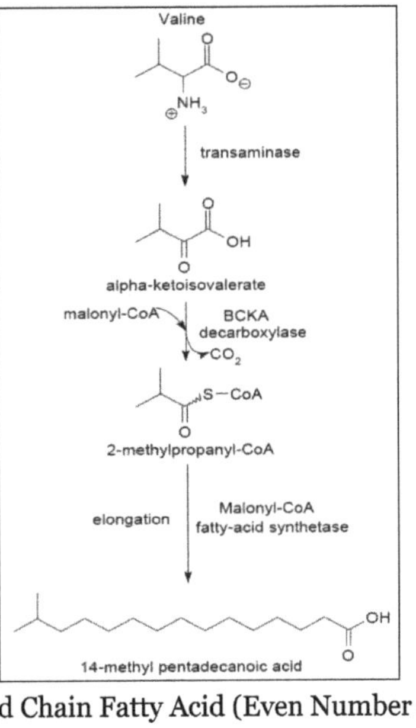

Figure: Iso-branched Chain Fatty Acid (Even Number of Carbon Atoms).

Branched chain FAs are typically saturated and are originated in two divergent families: (i) the iso series and (ii) the ante-iso series. It has been established that Actinomycetales include unique branch chain FA synthesis mechanisms together with tuberculosteric acid.

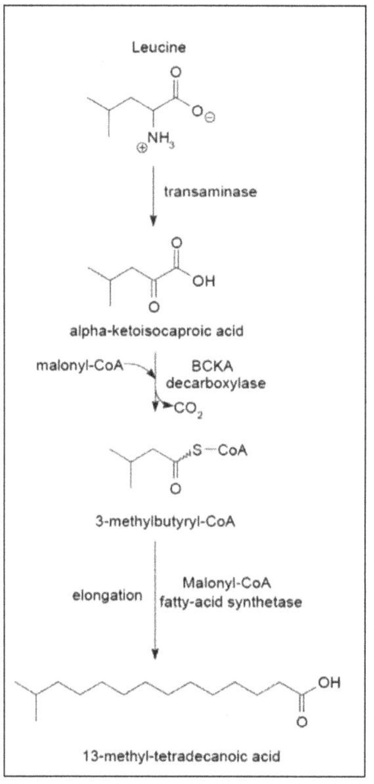

Figure: Iso-branched Chain Fatty Acid (Odd Number of Carbon Atoms).

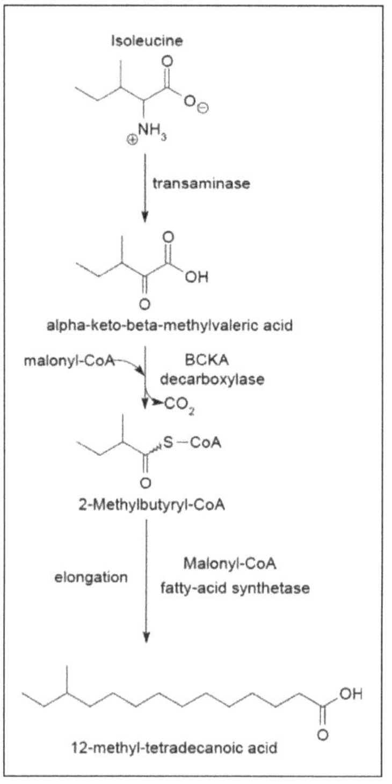

Figure: Anteiso-branched Chain Fatty Acid.

Branched-Chain FA Synthesis Pathway with Differing Primers

The branched chain FAs synthesizing system make use of α keto acids (leucine, Isoleucine and Valine) as primers in contrast to the branched chain FAS that exploits short-chain acyl CoA esters as primers. α Keto acid primers are derived from the decarboxylation and transamination of leucine, isoleucine and valine to form 3 methylbutyryl CoA, 2-Methylbutyryl-CoA and 2 methylpropanyl CoA respectively. 2 Methylpropanyl CoA primers resulting from valine are extended to make even-numbered iso-series fatty acids like 14 methyl pentadecanoic (isopalmitic) acid, 3 methylbutyryl CoA primers from leucine is being used to form oddnumbered iso series FAs (e.g. 13-methyl-tetradecanoic acid). 2 Methylbutyryl CoA primers from isoleucine are lengthened to outline ante-iso series FAs enclosing an odd number of carbon atoms such as 12-Methyl tetradecanoic acid. The decarboxylation of the primer precursors occurs via the branched chain α keto acid decarboxylase (BCKA) enzyme. In E. coli, elongation of the fatty acid pursues the identical biosynthetic pathway to produce straight chain FAs where malonyl CoA is used as a chain extender. The key end products are branched chain FAs (12-17 carbon) and their concerto apt to be consistent and trait for several bacterial species.

BCKA Decarboxylase and Relative Activities of α-Keto Acid Substrates

The enzyme BCKA decarboxylase is poised of two sub-units in a tetrameric structure (A_2B_2) and is indispensable for the synthesis of branchedchain FAs. It is accountable for the decarboxylation of α keto acids produced by the transamination of leucine, isoleucine and valine resulting in the formation of primers used for branched chain FA synthesis. The enzyme activity is exceedingly higher with branched chain α keto acid substrates as compared to straight chain substrates. In Bacillus species its specificity is maximum for the isoleucine derived α keto β methylvaleric acid, trailing by α ketoisocaproate and α ketoisovalerate. The high affinity of the enzyme for branched-chain α keto acids assign it to function as the primer donating system for branched chain FAS.

Activity of BCKA (%)	Substrate Used	Product (CO_2, nM/min/mg)	V_{max} (nM/min/mg)	Km (μM)
25	Pyruvate	4.90	15.20	51.1
38	α-Ketoisocaproate	7.40	5.60	< 1.0
63	α-Ketoisovalerate	12.40	13.30	< 1.0
100	L-α-keto-β-methyl-valerate	19.7	17.80	< 1.0

Factors Affecting Branched FAs Chain Length and Pattern Distribution

The α Keto acid primers (12-17 carbon) are utilized to make branched chain FAs and the size tend to be unswerving and uniform among a meticulous bacterial species which may be changed due to malonyl-CoA concentration, heat-stable factors (HSF) present or temperature. Since these factors affect chain length, HSFs modify the specificity of BCKA decarboxylase for a specific α-keto acid substrate, accordingly varying the ratio of branched chain FAs formed. It has been observed that an increase in malonyl CoA concentration result affect a bigger proportion of C17 FAs formed till an optimal concentration of malonyl CoA (~20μM) is accomplished. In Bacillus species, decrease in temperature also apt to move the FA distribution vaguely toward C17 FAs.

Branch Chain FAS

Bacteria that do not have the aptitude to execute the branch chain FA system via α keto primer apply this procedure using short chain carboxylic acids as primers. Distinctive short-chain primers consist of isobutyrate, 2 methyl butyrate and isovalerate. Commonly, the acids required for these primers are taken up from the environment which is frequently seen in ruminal bacteria. The disparity amid straight chain FAS and branch chain FAS is substrate specificity of the enzyme converting acyl CoA to acyl ACP. In general, the overall reaction is as under:

Isobutyryl CoA + 6 malonyl CoA + 12 NADPH + $12H^+$

$\downarrow$

Isopalmitic acid + 6 CO_2 + 12 NADP + 5 H_2O + 7 CoA

Omega Alicyclic FAs

In numerous species of bacteria, omega alicyclic FAs (structure shown below) are the foremost membrane FAs, classically enclosing an omega terminal propyl or butyryl cyclic group. The FAS employed to build omegaalicyclic FAs is also used to make membrane branched chain FAs. The regale of cyclic carboxylic acid CoA esters is a lot superior to that of branched-chain primers in bacteria with omega alicyclic FAs. The cyclic primers synthesis is not well-known but it has been recommended that procedure engrosses the conversion of sugars to shikimic acid followed by cyclohexylcarboxylic acid CoA esters that dole out as primers for omega alicyclic FA synthesis.

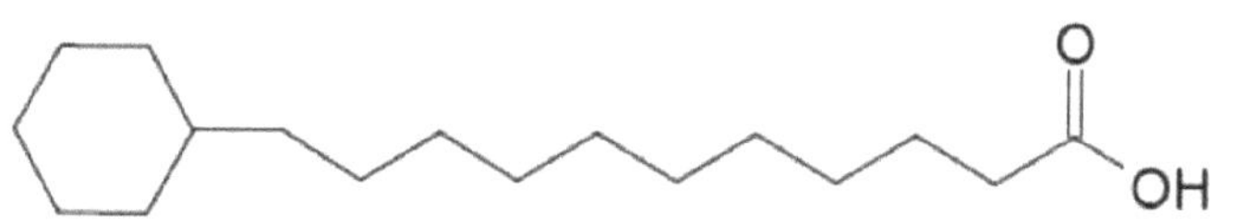

11-cyclohexylundecanoic acid

Omega-alicyclic fatty acid

Synthesis of Tuberculostearic Acid

Tuberculostearic acid chemically known as D-10 Methylstearic acid is a saturated FA well-known to be produced by Mycobacterium spp. and two species of Streptomyces. Oleic acid (a monosaturated FA) is the precursor that forms the Tuberculostearic acid by esterification to a phospholipid (S-adenosyl-methionine) that donates a methyl group to the double bond of oleic acid. The methylation results in the formation of intermediate 10 methylene octadecanoyal which on successive reduction of the residue with NADPH as a cofactor results in the formation of 10-methylstearic acid.

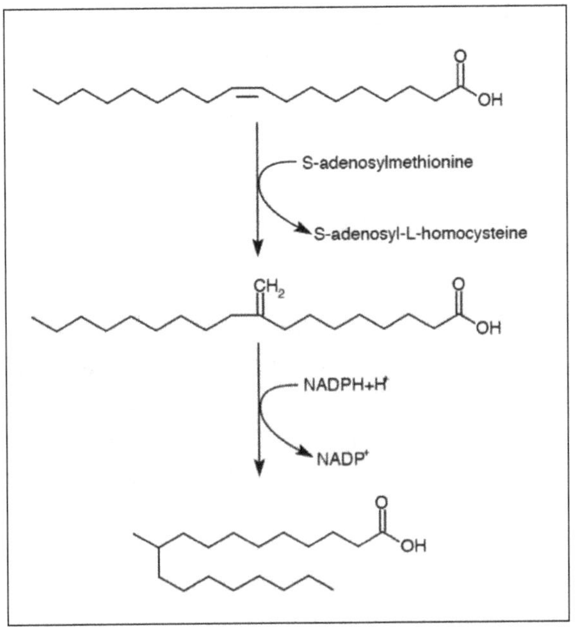

Fatty Acid Synthase (FAS): The Charismatic Enzyme Complex

In humans, FAS is encoded by the FASN gene. It is a multi-enzyme protein that catalyzes FA synthesis. FAS is not a solitary enzyme though a whole enzymatic system poised of two identical 272 kDa multifunctional polypeptides where substrates are handed from one functional domain to the subsequent domain catalyzing the synthesis of palmitate (C16:0, a long-chain saturated FA) from acetyl-CoA and malonyl-CoA, in the charisma of cofactor NADPH.

Metabolic Function and Classes

FAs are aliphatic acids essential to energy formation and storage, cellular structure and as intermediates in hormone biosynthesis and other biologically significant molecules. They are synthesized from acetyl CoA and malonyl CoA by a series of decarboxylative Claisen condensation reactions. Subsequent to every round of elongation the β-keto group is abridged to the effusive saturated carbon chain by the chronological action of a ketoreductase (KR), dehydratase (DH) and enol reductase (ER). The budding FA chain is carried among these active sites while fasten covalently to the phosphopantetheine prosthetic group of an acyl carrier protein (ACP) followed by its release through the accomplishment of a thioesterase (TE) abreast of reaching a 16C chain length (palmitic acid).

There are two major classes of FAS namely Type I and Type II:

- Type I FAS systems make the most of a single large, multifunctional polypeptide and are widespread to both mammals and fungi (even if the structural arrangement of mammalian and fungus FAS vary). This FAS system is also found in the CMN group of bacteria (corynebacteria, mycobacteria and nocardia) where it produces palmitic acid and collaborates with FAS II system to produce a superior range of lipid products.
- Type II FAS system is present in archaea and bacteria and is exemplified by the use of

distinct monofunctional enzymes for FAS. Investigators are searching for the inhibitors of this pathway (FAS II) as probable antibiotics.

The mechanism of elongation and reduction in both FAS I and FAS II is similar because the domains of the FAS II enzymes are mostly homologous to their domain counterparts in FAS I multienzyme polypeptides. Conversely, the disparity in the organization of the enzymes i.e. coordinated in FAS I and distinct in FAS II gives rise to a lot of essential biochemical differences.

The evolutionary history of FAS is very much disheveled with that of polyketide synthases (PKS). PKS make use of an analogous mechanism and homologous domains to produce lipids (secondary metabolite). Besides, PKS also reveal a Type I and Type II organization. In animals, the FAS system I is thought to have arisen through alteration of PKS I in fungi, whereas in fungi and the CMN group of bacteria, FAS I appear to have arisen discretely through the fusion of FAS II genes.

Structure of FAS

The mammalian FAS embodies a homodimer of two identical protein subunits, in which three catalytic domains in the N-terminal section namely dehydrase (DH), ketoacyl synthase (KS) and malonyl/acetyltransferase (MAT) are alienated by a core region of 600 residues from four C-terminal domains i.e. acyl carrier protein (ACP), enoyl reductase (ER), ketoacyl reductase (KR) and thioesterase (TE). The traditional model for organization of FAS (the head-to-tail model on the right) is basically based on the observations that the bifunctional reagent 1,3-dibromopropanone (DBP) is competent to crosslink the cysteine thiol (active site) of the KS domain in one FAS monomer with the phosphopantetheine (prosthetic group) of the ACP domain in the other monomer.

Complementation analysis of FAS dimers haulage diverse mutations on each monomer has established that the KS and MAT domains can work together with the ACP of either monomer and a reinvestigation of the DBP cross linking experiments discovered that the KS Cys161 thiol (active site) could be cross linked to the ACP 4'- phosphopantetheine thiol (prosthetic group) of either monomer. It has also been revealed that a heterodimeric FAS having only one competent monomer is proficient of palmitate synthesis.

Substrate Shuttling Mechanism

The solved structures of FAS (both yeast and mammalian) illustrate two distinct organizations of extremely conserved catalytic domains/enzymes in the multienzyme cellular mechanism. Mammalian FAS has an open lithe structure with only two reaction chambers while yeast FAS has an exceedingly well-organized rigid barrel like structure with 6 reaction chambers that synthesize FA independently. Nevertheless, the conserved ACP in both the cases acts as the mobile domain liable for shuttling the intermediate FA substrates to different catalytic sites. Using cryo EM analysis, a first direct structural insight into the substrate shuttling mechanism was acquired where ACP is experiential bound to the different catalytic domains in the barrel-shaped yeast FAS. The cryo EM results imply that the binding of ACP to diverse sites is asymmetric and stochastic, as also signified by computer-simulation studies.

Regulation of FAS

The homeostasis and metabolism of FAS is transcriptionally regulated by Upstream Stimulatory Factors (USF1 and USF2) and sterol regulatory element binding protein-1c (SREBP-1c) in retort to feeding insulin in living animals. Even though, liver X receptor (LXRs) alter the expression of SREBP-1c in feeding, regulation of FAS by SREBP-1c is USF-dependent. Acylphloroglucinols isolated from the fern (Dryopteris crassirhizoma) demonstrated a FAS inhibitory activity.

Clinical Significance of FAS

The gene that codes for FAS has been examined as a possible oncogene, up regulated in breast cancers as well as being criteria of poor prognosis might also be valuable as a chemotherapeutic target. It could also be implicated in the assembly of an endogenous ligand for the nuclear receptor PPARα, the target of the fibrate drugs for hyperlipidemia. It is also being investigated as a possible drug target for treating the metabolic syndrome. In several cancer cell lines, this protein has been found to be complexed with ERα (estrogen receptor alpha) where the N-terminus of FAS is fused in frame with the C-terminus of ERα. The association of FAS with uterine leiomyomata has also been reported.

Biosynthesis of Steroid

It is an anabolic metabolic alleyway which utilizes simple precursors to make steroids. In animals and several other living organisms, this conduit is being carried out in discrete ways making it a frequent target for anti-infective agents and other antibiotics. Furthermore, in humans, statins, the cholesterol lowering drugs targets steroid metabolism. It starts with the Acetyl CoA in the MVA pathway as building blocks that configure IPP and DMAPP followed by the formation of lanosterol, the first steroid. Additional modifications give rise to the ensuing steroidogenesis.

Steroidogenesis is the natural course of action by which steroids are enkindled from cholesterol and gets altered into several other steroids. The biosynthetic pathways of steroidogenesis vary amid distinct species, but in humans, the products of the biosynthetic pathways of steroidogenesis include: aldosterone, androgens, corticoids, cortisol, estrogens, progesterone and testosterone.

Cholesterol Biosynthesis

Cholesterol is synthesized from acetyl-CoA via the isoprenoid pathway, and at least four enzymes in the biosynthetic pathway are regulated by cellular cholesterol levels. Essential non-steroidal isoprenoids, such as dolichol, prenylated proteins, heine A and isopentenyl adenosinecontaining tRNAs are also synthesized by this pathway. In extrahepatic tissues, most cellular cholesterol is derived from de novo synthesis, whereas hepatocytes obtain most of their cholesterol via the receptor-mediated uptake of plasma lipoproteins, such as low-density lipoprotein (LDL). LDL is bound and internalized by the LDL receptor and delivered to the acidic late endosomes and lysosomes, where hydrolysis of the core cholesteryl esters occurs. The cholesterol that is released is transported throughout the cell. Normal mammalian cells tightly regulate cholesterol synthesis and LDL uptake to maintain cellular cholesterol levels within narrow limits and supply sufficient isoprenoids to satisfy metabolic requirements of the cell. Regulation of cholesterol biosynthetic enzymes takes place at the level of gene transcription, mRNA stability, translation, enzyme phosphorylation and enzyme degradation. Cellular cholesterol levels are also modulated by a cycle of cholesterol esterification by acylCoA : cholesterol acyltransferase (ACAT) and hydrolysis of the cholesteryl esters, and by cholesterol metabolism to bile acids and oxysterols.

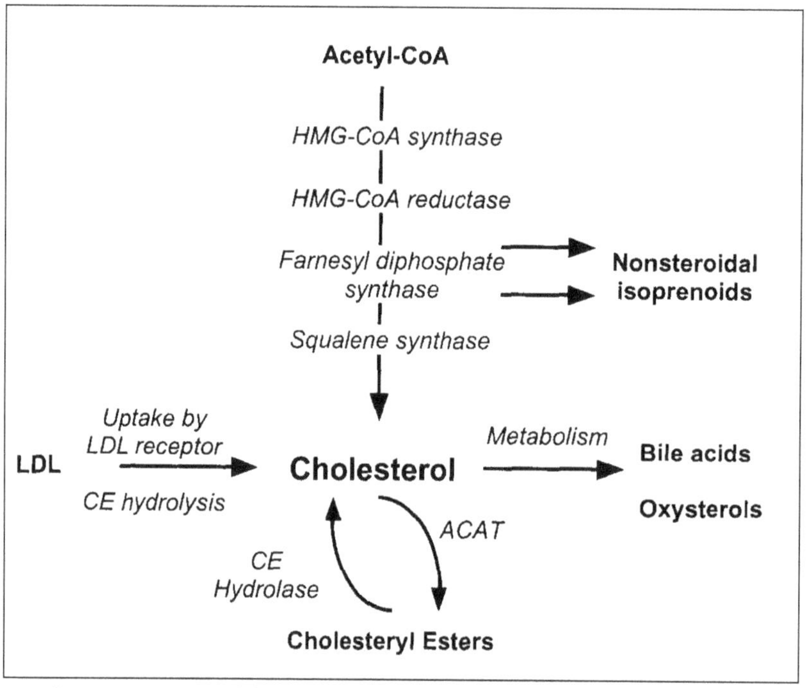

Figure: Overview of the metabolic and transport pathways that control cholesterol levels in mammalian cells.

Cholesterol is synthesized from acetyl-CoA and the four key enzymes that regulate cholesterol synthesis are indicated. Cells also obtain cholesterol by uptake and hydrolysis of LDL's cholesteryl esters (CE). End products derived from cholesterol or intermediates in the pathway include bile acids, oxysterols, cholesteryl esters and non-steroidal isoprenoids. ACAT, acyl-CoA: cholesterol acyltransferase.

The Cholesterol Biosynthetic Pathway

Figure takes a closer look at the cholesterol biosynthetic pathway, focusing on the enzymes that are regulated, sterol intermediates and the location of enzymes in the cell. Sterols are synthesized from the two-carbon building block, acetyl-CoA. The soluble enzyme acetoacetyl-CoA thiolase interconverts acetyl-CoA and acetoacetylCoA, which are then condensed by 3-hydroxy-3-methylglutaryl (HMG)-CoA synthase to form HMG-CoA. There are two forms of HMG-CoA synthase. A mitochondrial form, involved in ketogenesis, predominates in the liver. In extrahepatic tissues, the most abundant form is a soluble enzyme of 53 kDa that is highly regulated by supply of cholesterol. Like acetoacetyl-CoA thiolase, HMG-CoA synthase has classically been described as a cytosolic enzyme because it is found in the 100,000 x g supernatant of homogenized cells and tissues. However, both enzymes contain peroxisomal targeting sequences and may reside in multiple cellular compartments.

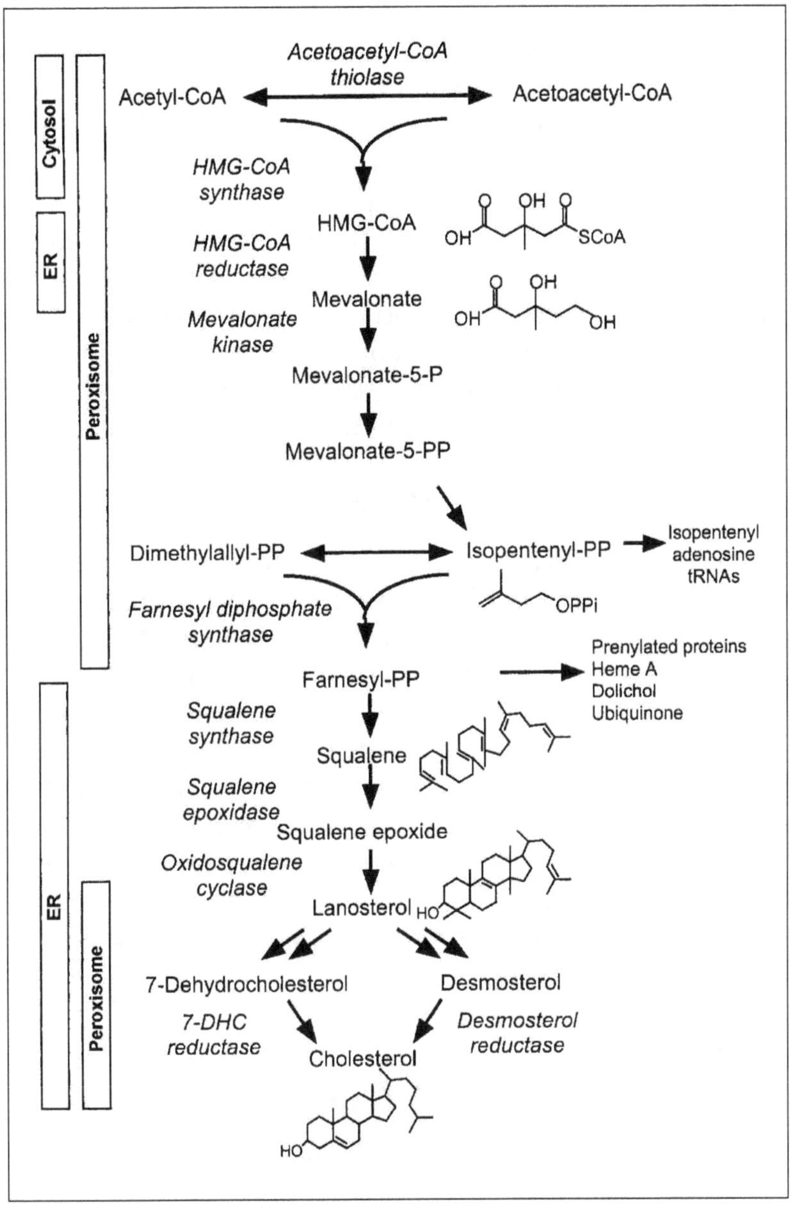

Figure: The cholesterol biosynthetic pathway. Some of the major intermediates and end-products are indicated. Enzymes in the pathway are found in cytosol, endoplasmic reticulum (ER) and peroxisomes, as noted. HMG, 3-hydroxy-3-methylglutaryl; DHC, dehydrocholesterol.

HMG-CoA reductase catalyzes the reduction of HMG-CoA to mevalonate, utilizing two molecules of NADPH. HMG-CoA reductase is a 97-kDa glycoprotein of the endoplasmic reticulum and peroxisomes. Analysis of the endoplasmic reticulum enzyme's domain structure revealed an N-terminal membrane domain with eight transmembrane spans, a short linker, and a C-terminal catalytic domain facing the cytosol. Transmembrane spans 2-5 shares a high degree of sequence similarity with several other key proteins in cholesterol metabolism; this region is termed the sterol-sensing domain. Elucidation of the crystal structure of the HMG-CoA reductase catalytic domain indicated that the active protein is a tetramer, which is consistent with biochemical analysis. The monomers appear to be arranged in two dimers, with the active sites at the monomer-monomer interface. The dimer-dimer interface is predominantly hydrophobic.

HMG-CoA reductase is the rate-determining enzyme of the cholesterol biosynthetic pathway and, like HMG-CoA synthase, is highly regulated by supply of cholesterol. Thus, the enzyme has received intense scrutiny as a therapeutic target for treatment of hypercholesterolemia. The enzyme is inhibited by a class of pharmacological agents, generally called statins, which have an HMG-like moiety and a bulky hydrophobic group. Statins occupy the HMG-binding portion of the active site, preventing HMG-CoA from binding. Also, the bulky hydrophobic group causes disordering of several catalytic residues. Thus, statins are potent, reversible competitive inhibitors of HMG-CoA reductase with K i values in the nanomolar range. Elevated plasma cholesterol levels are a primary risk factor for coronary artery disease, and statin inhibition of HMG-CoA reductase effectively reduces cholesterol levels and decreases overall mortality. However, complete inhibition of HMG-CoA reductase by statins will kill cells, even if exogenous cholesterol is supplied. That is because complete inhibition deprives cells of all mevalonate-derived products, including essential non-steroidal isoprenoids. To survive, cells must produce a small amount of mevalonate that, when limiting, is used preferentially by higher affinity pathways for non-steroidal isoprenoid production.

Mevalonate is metabolized to farnesyl-diphosphate (-PP) by a series of enzymes localized in peroxisomes. First, mevalonate kinase phosphorylates the 5-hydroxy group of mevalonic acid. The enzyme is a homodimer of 40 kDa that is subject to feedback inhibition by several isoprenoid intermediates. Mutations in the mevalonate kinase gene lead to the human genetic disease mevalonic aciduria. The product of mevalonate kinase, mevalonate-5-R is then phosphorylated to form mevalonic acid-5-PP, which is decarboxylated and dehydrated by mevalonatePP decarboxylase to form isopentenyl-PE Isopentenyl-PP is in equilibrium with its isomer, dimethylallyl-PR Farnesyl-PP synthase catalyzes the head to tail condensations of two molecules of isopentenyl-PP with dimethylallyl-PP to form famesyl-PR The enzyme is part of a large family of prenyltransferases that synthesize the backbones for all isoprenoids, including cholesterol, steroids, prenylated proteins, heine A, dolichol, ubiquinone, carotenoids, retinoids, chlorophyll and natural rubber.

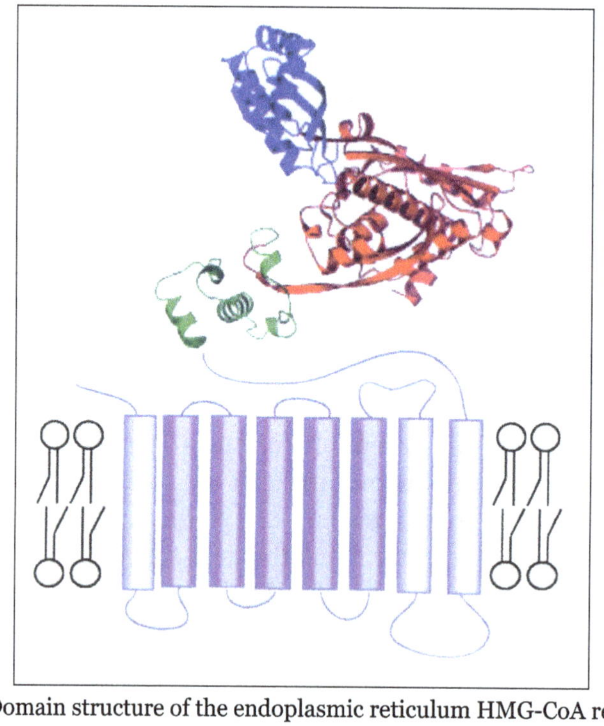

Figure: Domain structure of the endoplasmic reticulum HMG-CoA reductase.

The crystal structure of the catalytic domain has been determined and is depicted as a ribbon diagram. The catalytic domain consists of a small helical domain (green), a large central element resembling a prism (red), which contains the HMG-CoA-binding site, and a small domain to which NADPH binds (blue). The structure of the membrane domain has not been solved; however, it is known that eight transmembrane spans embed the protein into the endoplasmic reticulum membrane. Spans 2-5 (darker cylinders) are termed the sterol-sensing domain and mediate the regulated degradation of the enzyme.

Squalene synthase is a 47-kDa protein of the endoplasmic reticulum and catalyzes the first committed step in cholesterol synthesis. The enzyme condenses two molecules of farnesyl-PP and then reduces the presqualene-PP intermediate to form squalene. A large N-terminal catalytic domain faces the cytosol, anchored to the membrane by a C-terminal domain. This orientation may allow the enzyme to receive the hydrophilic substrates from the cytosol and release the hydrophobic product into the endoplasmic reticulum membrane for further metabolism. Squalene synthase is highly regulated by the cholesterol content of the cell. Thus, it plays an important role in directing the flow of farnesyl-PP into the sterol or non-sterol branches of the pathway.

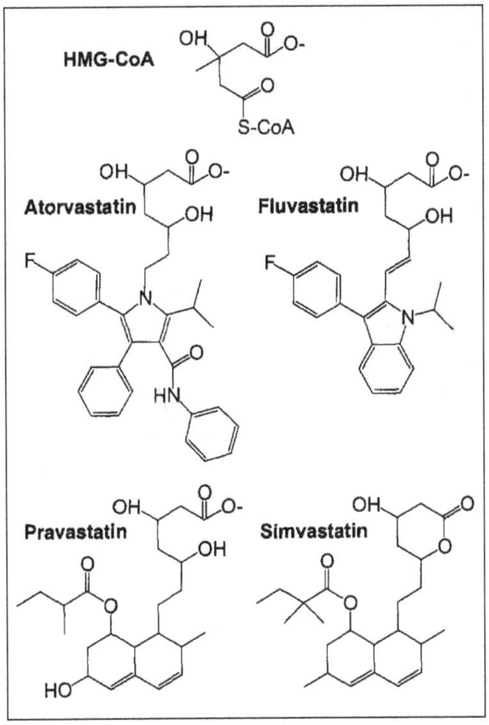

Figure: Chemical structures of HMG-CoA and several statin inhibitors of HMG-CoA reductase. Atorvastatin (Lipitor), fluvastatin (Lescol), pravastatin (Pravachol) and simvastatin (Zocor) are widely prescribed cholesterol-lowering drugs.

Squalene is converted into the first sterol, lanosterol, by the action of squalene epoxidase and oxidosqualene cyclase. Lanosterol is then converted to cholesterol by a series of oxidations, reductions, and demethylations. The required enzyme reactions have been defined and metabolic intermediates identified; however, the precise sequence of reactions between lanosterol and cholesterol remains to be established. There is evidence for two alternative pathways that differ in when the D24 double bond is reduced. Both 7-dehydrocholesterol and desmosterol have been postulated to be the immediate precursor of cholesterol. One of the key enzymes in the latter part of the pathway is 7-dehydrocholesterol Δ7-reductase, a 55-kDa integral membrane protein. Mutations in the gene for 7-dehydrocholesterol Δ7-reductase cause the human genetic disease Smith-Lemli-Opitz syndrome.

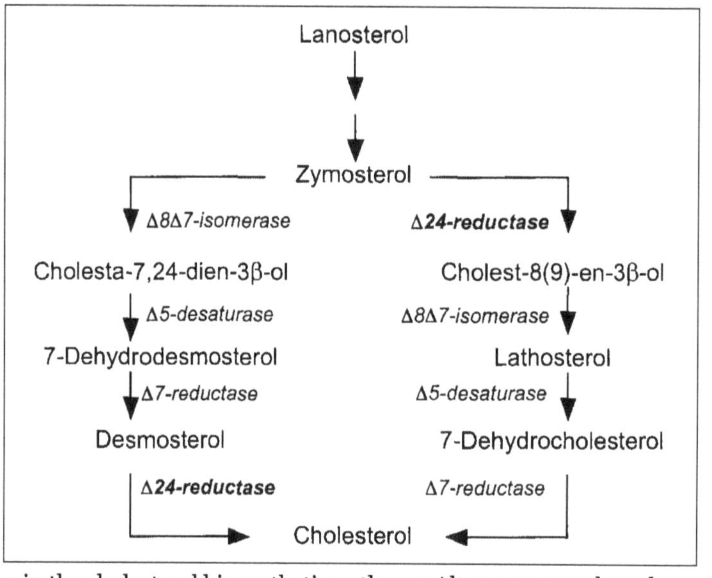

Figure: Final steps in the cholesterol biosynthetic pathway. Alternate steps have been proposed for the conversion of zymosterol to cholesterol, which differ in when the A24-reductase reaction occurs.

Enzyme Compartmentalization

Where does cholesterol synthesis take place? All of the enzymes that convert acetyl-CoA to farnesyl-PP have classically been thought of as cytosolic enzymes, with the exception of HMG-CoA reductase, which is typically depicted as an endoplasmic reticulum enzyme with the catalytic site facing the cytosol. Enzymes that convert farnesyl-PP to cholesterol are classically described as microsomal. However, there is now strong evidence that all but one of these enzymes is also, or exclusively, peroxisomal. The molecular cloning of cDNAs encoding many of these enzymes has revealed peroxisomal targeting sequences. The availability of antibodies has allowed immunocytochemical localization to peroxisomes. Together these data suggest that peroxisomes may play an active role in all steps in the cholesterol biosynthetic pathway except the conversion of farnesyl-PP to squalene, which is catalyzed by squalene synthase found solely in the endoplasmic reticulum.

HMG-CoA reductase is the one exception to the rule. Immunocytochemistry and immunoblotting have localized HMG-CoA reductase to both the endoplasmic reticulum and peroxisomes; however, no peroxisomal targeting motif has been found in the HMGCoA reductase protein sequence. Furthermore, the peroxisomal HMG-CoA reductase has an apparent molecular weight of 90 kDa whereas the endoplasmic reticulum enzyme is 97 kDa. The peroxisomal enzyme exhibits other distinct properties: it is resistant to statin inhibition, the enzyme's activity is not regulated by phosphorylation, and the protein's turnover is not regulated by mevalonate. Altogether, this evidence suggests that the endoplasmic reticulum and peroxisome enzymes are functionally and structurally distinct.

Additional evidence for the involvement of peroxisomes in cholesterol biosynthesis comes from analysis of diseases of peroxisomal deficiency. Zellweger syndrome, neonatal adrenoleukodystrophy, and infantile Refsum's disease are all diseases of peroxisome biogenesis. In most of these peroxisomal disorders, the peroxisomal matrix proteins are synthesized in the cytosol as normal, but they cannot be assembled into nascent peroxisomes due to mutations in one of at least 12 different genes encoding proteins necessary for peroxisomal protein targeting and import. Fibroblasts from individuals with peroxisome biogenesis disorders show reduced enzymatic activities of cholesterol biosynthetic enzymes, reduced levels of cholesterol synthesis and lower cholesterol content. These data support the hypothesis that part of the cholesterol synthesis pathway is peroxisomal.

It is not clear why cholesterol synthesis is compartmentalized and requires intermediates to cycle between peroxisomes and the cytosol. It is also unclear why some of the enzymes are found in multiple compartments and others are solely in endoplasmic reticulum or peroxisomes. As noted, cholesterol synthesis is a very complex process and compartmentalization may represent another level of regulation.

Mevalonic Aciduria

Cholesterol synthesis is essential for normal development and maintenance of tissues that cannot obtain cholesterol from plasma lipoproteins, such as brain. Furthermore, the biosynthetic pathway supplies non-steroidal isoprenoids that are required by all cells. Thus, it is not surprising that metabolic defects in the cholesterol biosynthetic pathway have devastating consequences.

The first recognized human metabolic defect in the biosynthesis of cholesterol and isoprenoids was mevalonic aciduria. Mevalonic aciduria is an autosomal recessive disorder that is quite rare, with only 19 known patients. In normal individuals, a small amount of mevalonic acid diffuses into the plasma at levels proportional to the rate of cellular cholesterol formation. Patients with mild mevalonic aciduria excrete 3000-6000 times the normal amount of mevalonic acid and patients with the severe form of the disease excrete 10,000-200,000 times the normal amount. Enzyme assays using cell lysates showed that mevalonate kinase activity was markedly deficient in patient samples and genetic analysis has revealed nucleotide changes in the mevalonate kinase gene that lead to amino acid substitutions. Because of this enzyme deficiency, there is little to no feedback inhibition of HMG-CoA reductase and, thus, mevalonate is overproduced.

Clinical features of mevalonic aciduria include failure to thrive, anemia, gastroenteropathy, hepatosplenomegaly, psychomotor retardation, hypotonia, ataxia, cataracts, and dysmorphic features. Surprisingly, patients with severe deficiencies in mevalonate kinase show normal plasma cholesterol levels and cultured mevalonic aciduria fibroblasts show rates of cholesterol synthesis half that of normal cells. Close examination of cholesterogenic enzymes in mevalonic aciduria fibroblasts has revealed a 6-fold increase in HMG-CoA reductase activity, which is postulated to compensate for the low mevalonate kinase activity.

Smith-Lemli-Opitz Syndrome

A second metabolic defect in cholesterol synthesis leads to Smith-Lemli-Opitz syndrome (SLOS). SLOS is a relatively common autosomal recessive disorder, with estimates of incidence ranging from 1 in 10,000 to 1 in 60,000. Four lines of evidence pointed to the metabolic defect in SLOS patients. (1) Individuals with SLOS were found to have markedly elevated levels of plasma 7-dehydrocholesterol and low plasma cholesterol levels. (2) 7-Dehydrocholesterol Δ7-reductase activity was deficient in SLOS patient samples and the amount of residual activity could be correlated with severity of the disease. (3) Rodents treated with AY-9944, an inhibitor of 7-dehydrocholesterol Δ7-reductase, developed SLOS-like malformations. (4) Cloning of the gene for 7-dehydrocholesterol Δ7-reductase led to identification of a splice-site mutation and amino acid substitutions in SLOS patients.

Severely reduced cholesterol synthesis is predicted to have severe consequences on development of the foetus because cholesterol is only obtained from the maternal circulation during the first trimester. In addition, the brain is predicted to be severely affected because plasma lipoproteins cannot cross the blood-brain barrier and most, if not all, cholesterol needed for brain growth is synthesized locally. Indeed, severely affected SLOS infants who died soon after birth were found to have functionally null 7-dehydrocholesterol Δ7-reductase alleles, whereas typical affected individuals likely have some residual 7-dehydrocholesterol Δ7-reductase catalytic activity.

Patients with SLOS have mental retardation and microcephaly, which is consistent with cholesterol synthesis being required for normal brain development. Clinical features also include failure to thrive, and characteristic craniofacial, skeletal and genital anomalies. The clinical phenotype appears to be due to a lack of cholesterol rather than the cellular accumulation of 7-dehydrocholesterol. A recent multicenter clinical trial has shown that SLOS children fed a diet supplemented with cholesterol show improved growth and neurodevelopment (i.e. language and cognitive skills). It is likely that the diet fulfilled the daily requirement for cholesterol and down-regulated endogenous 7-dehydrocholesterol synthesis.

What are the final steps in the cholesterol biosynthetic pathway? SLOS may provide an answer to that question. There is evidence for two alternative pathways, which differ in when the D24 double bond is reduced. In both pathways, lanosterol is demethylated to form zymosterol. Then, zymosterol can be metabolized sequentially by a D24-reductase, Δ8, Δ7-isomerase, and Δ5- desaturase to form 7-dehydrocholesterol, which is reduced at the Δ7 position to form cholesterol. Alternatively, zymosterol can be metabolized by the Δ8, Δ7-isomerase and Δ5-desaturase, to form 7-dehydrodesmosterol. 7-Dehydrodesmosterol is metabolized by the Δ7-reductase to form desmosterol and then by the D24-reductase to form cholesterol. The fact that the SLOS deficiency in Δ7-reductase leads to a buildup of 7-dehydrocholesterol rather than 7-dehydrodesmosterol is interpreted to mean that the former pathway is the principal one. However, the latter pathway must also be used because desmosterol is an abundant cholesterol precursor in certain tissues. It has been suggested that the final steps in the biosynthetic pathway may be tissue specific.

Table: Inborn errors of sterol biosynthesis.

Syndrome	Metabolic defect
Mevalonic aciduria	Mevalonate kinase
Smith-Lemli-Opitz	Sterol Δ7-reductase
Desmosterolosis	Sterol D24-reductase
Rhizomelic chondrodysplasia punctata (CDP)	Pex7 peroxisomal enzyme import
CDP X-linked dominant (CDPX2)	Sterol Δ8, Δ7-isomerase
CHILD syndrome (congenital hemidysplasia with ichthyosis and limb defects)	Sterol Δ8, Δ7-isomerase Sterol C-4 demethylase
Greenberg skeletal dysplasia	Sterol Δ 14-reductase

Perhaps, in SLOS cells, any 7-dehydrodesmosterol that accumulates is metabolized by the available D24-reductase to form 7-dehydrocholesterol.

Other Enzyme Deficiencies

Rhizomelic chondrodysplasia punctata, like Zellweger syndrome, exhibits defective sterol synthesis due to the lack of key peroxisomal enzymes of cholesterol biosynthesis. CDPX2, also known as Conradi-Hfinermann syndrome, and most cases of CHILD syndrome are due to mutations in the sterol Δ8, Δ7-isomerase gene, which is located on the X chromosome. Mutations in a single gene may lead to different syndromes with similar, but distinct, pathologies due to the mosaicism of X-chromosome inactivation. A few cases of CHILD syndrome may be due to mutations in the sterol C-4 demethylase gene, also located on the X chromosome.

Regulation of Cholesterol Synthesis

Isoprenoid synthesis is regulated by the sterol end-product of the biosynthetic pathway, by non-sterol intermediates, and also by physiological factors. The cholesterol content of the cell controls several enzymes in the biosynthetic pathway, but the focus has been on the rate-limiting enzyme, HMG-CoA reductase. Different regulators have different mechanisms of action. For example, sterols have been shown to regulate at the level of HMG-CoA reductase transcription whereas non-sterols regulate HMG-CoA reductase mRNA translation. Both sterols and non-sterols are needed for regulation of HMGCoA reductase protein degradation. Physiological factors that influence cholesterol synthesis include diurnal rhythm, insulin and glucagon, thyroid hormone, glucocorticoids, estrogen and bile acids. These factors regulate HMG-CoA reductase by transcriptional, translational and post-translational mechanisms.

Transcriptional Regulation

HMG-CoA reductase is the rate-limiting enzyme in the cholesterol biosynthetic pathway and combined regulation of HMG-CoA reductase synthesis and turnover can alter steady state levels of the enzyme 200-fold. HMG-CoA reductase is regulated in parallel with at least three other enzymes in the cholesterol biosynthetic pathway, HMG-CoA synthase, farnesyl-PP synthase and squalene synthase, as well as the LDL receptor. This coordinate regulation is due to the fact that each gene has a similar sequence (cis-acting element) within the promoter that recognizes a common trans-acting transcription factor. Availability of the transcription factor to bind to the promoter sequence is influenced by the cellular cholesterol content.

Figure illustrates the current model of cholesterol-mediated transcriptional regulation. The 5' flanking regions of cholesterol-regulated genes have one to three copies of a 10-bp non-palindromic nucleotide sequence termed the sterol regulatory element (SRE). SREs are conditional positive elements that are required for gene transcription in cholesterol-depleted cells. The SRE sequence found in the LDL receptor gene is 5'- ATCACCCCAC-3'. SREs have been identified in the HMG-CoA synthase, HMG-CoA reductase, farnesyl-PP synthase and squalene synthase genes, as well as genes of fatty acid synthesis. However, there is not a strict SRE consensus sequence and identifying functional SREs has been difficult.

The transcription factor that binds the SRE is termed the SRE-binding protein (SREBP). The first SREBP to be identified was the protein that bound to the LDL receptor promoter. Cloning of SREBP cDNAs revealed that there are two SREBP genes that produce three distinct proteins. SREBP-la and -lc are derived from one gene that contains two promoters and differ in the length of the N-terminal transactivation domain. SREBP-2 is derived from a second gene and is 45% identical to SREBP-la. SREBP-lc is the predominant isoform in liver and adipocytes. It was isolated independently and called adipocyte determination and differentiation-dependent factor 1. Here when the term SREBP is used, the information is relevant for all three isoforms.

SREBPs are cytosolic 68-kDa proteins with a canonical basic helix-loop-helix leucine zipper (bHLH-Zip) motif that is present in other transcription factors. Unlike other bHLH-Zip transcription factors, SREBPs have a tyrosine in place of a conserved arginine, which allows them to bind to the inverted E-box motif (5'-CANNTG-3') in addition to SREs. By binding to SREs, SREBPs coordinately regulate multiple enzymes involved in fatty acid synthesis and lipogenesis.

An additional feature that distinguishes SREBPs from other bHLH-Zip transcription factors is that SREBP genes encode 125-kDa membrane proteins that are inserted into the endoplasmic reticulum and serve as precursors for the active transcription factors. SREBPs have three functional domains: an N-terminal 68-kDa fragment containing the bHLH-Zip transcription factor, two membrane-spanning segments, and a C-terminal regulatory domain. It is the sequential two-step cleavage of the full-length precursor SREBP and release of the 68-kDa N-terminal bHLH-Zip domain that is influenced by the cellular cholesterol content.

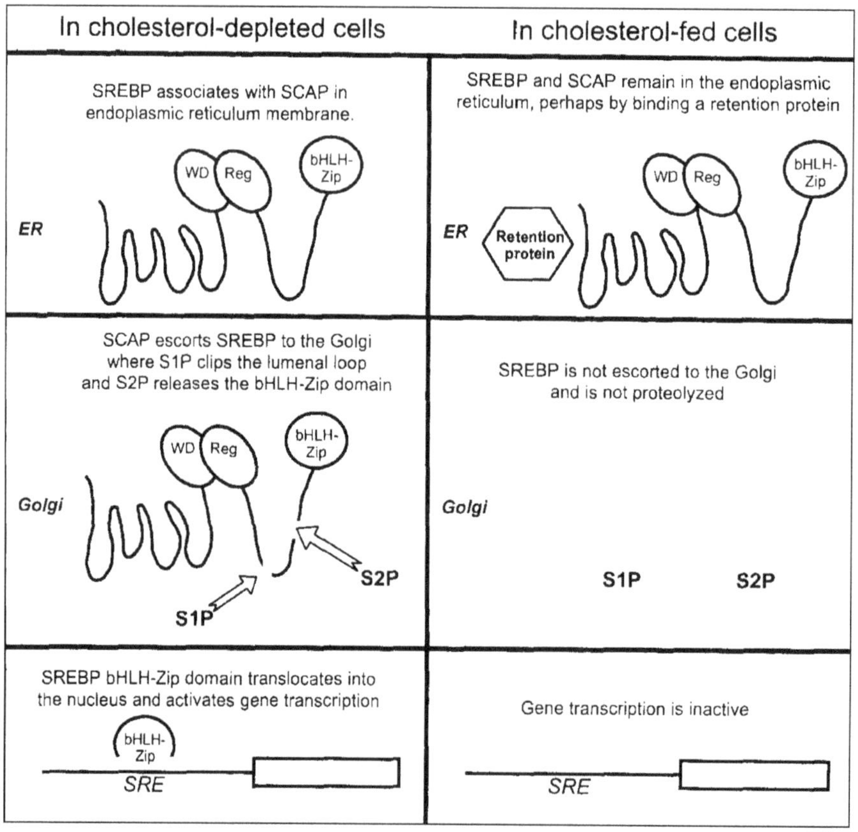

Figure: Current model of cholesterol regulation of SREBP proteolysis.

The sterol regulatory element binding protein (SREBP) precursor is inserted into the endoplasmic reticulum (ER) membrane. The SREBP regulatory domain (Reg) interacts with the SREBP cleavage-activating protein (SCAP), likely through SCAP's WD repeats. When cholesterol levels are low, SCAP escorts SREBP to the Golgi where the bHLHZIP domain is released by site-1 protease (SIP) cleavage of a luminal loop followed by site-2 protease (S2P) cleavage within a transmembrane span. The mature SREBP translocates into the nucleus and activates gene transcription. In cholesterol replete cells, the SREBP precursor and SCAP remain in the endoplasmic reticulum and the SREBP precursor is not proteolyzed to release the bHLH-Zip transcription factor.

Identification of the proteins required for regulated SREBP cleavage was accomplished using somatic cell genetic approaches. Mutant Chinese hamster ovary cells with abnormal regulation of cholesterol and fatty acid metabolism, which were selected over the past 20 years, proved invaluable for this goal. Using expression cloning strategies, genes were isolated that restored SREBP-mediated transcription in each mutant. This work led to identification of two proteases and an escort protein required for SREBP precursor cleavage. Cleavage of tile SREBP precursor at site I requires a subtilisin-like serine protease, whereas cleavage at site 2 requires a zinc metalloprotease. Transcription factor release is controlled by a chaperone, SREBP cleavage-activating protein (SCAP), which escorts the SREBP precursor to the Golgi where the proteases reside.

How is SREBP proteolysis controlled by cholesterol? A hint that this event may involve vesicle trafficking came from the finding that the SREBP precursor's N linked carbohydrates are endoglycosidase-H-resistant (trimmed by Golgi mannosidases) when cellular cholesterol levels are low, and endoglycosidase-H-sensitive when cellular cholesterol levels are high. Thus, release of the mature SREBP transcription factor appeared to coincide with transport to the Golgi. The current model is as follows. When cellular cholesterol levels are low, the SREBP precursor is synthesized and inserted into the endoplasmic reticulum membrane. The C-terminal regulatory domain of the SREBP precursor interacts with the C-terminus of SCAP, likely through SCAP's four WD repeats. SCAP and the SREBP precursor are then transported to the Golgi where the lumenal loop of the SREBP precursor is clipped by the site-1 protease; however, the two halves of the protein remain membrane-anchored. Then the site-2 protease clips the N-terminal SREBP intermediate within the first membrane-spanning segment, releasing the soluble transcription factor, mature SREBR Upon translocation to the nucleus, the mature SREBP binds SRE sequences within the promoters of target genes and enhances their transcription.

When cellular cholesterol levels rise to a threshold level, SCAP and the SREBP precursor no longer travel to the Golgi and the SREBP precursor is not proteolyzed to produce the mature SREBR As a result, transcription of target genes declines to basal levels. Evidence for SCAP-SREBP transport is two-fold. The sterol-dependent movement of SCAP has been directly visualized in cultured cells transfected with a green fluorescent protein-SCAP fusion protein. Furthermore, in vitro vesicle-budding assays have demonstrated that oxysterols suppress SREBP-SCAP complexes from entering into vesicles budding from the endoplasmic reticulum.

The mechanism by which SCAP senses cellular cholesterol is not known. Sensing is postulated to involve a five-transmembrane segment of SCAP with sequence similarity to a five-transmembrane segment of HMG-CoA reductase, called the sterol-sensing domain. Consistent with this hypothesis, several constitutively active SCAP mutants have been isolated that have point mutations in the sterol-sensing domain. Also, there is recent evidence that SCAP interacts with a protein that is retained in the endoplasmic reticulum. Therefore, control of SCAP and SREBP transit to the Golgi could depend upon sterol-dependent binding of SCAP's sterol-sensing domain to a retention protein.

SREBPs control not only cholesterol synthesis, but also the synthesis of fatty acids (via fatty acid synthase, acetyl-CoA carboxylase, and stearoyl-CoA desaturase 2), triacylglycerols and phospholipids (glycerol-3-phosphate acyltransferase). This was illustrated most dramatically in transgenic mice overexpressing the nuclear forms of SREBP-1 a, -lc and -2 in liver. Overexpression of each isoform resulted in activation of a full spectrum of cholesterol and fatty acid biosynthetic enzymes; however, absolute levels of induction of each enzyme and the subsequent liver phenotype varied according to the isoform expressed. From these data, the following conclusions can be drawn. SREBP-la is a strong activator of cholesterol and fatty acid synthesis and is likely to be important in rapidly dividing cells that require lipid for membrane production. SREBP-lc predominates in the liver, where it primarily activates genes of fatty acid synthesis. SREBP-lc appears to be important for maintaining basal transcription levels of fatty acid and cholesterol biosynthetic enzymes during periods of fasting. It also plays a role in the insulin response. SREBP-2 selectively activates cholesterol biosynthetic genes and the LDL receptor, and primarily responds when the liver's demand for cholesterol rises.

mRNA Translation

HMG-CoA reductase is also subject to translational control by a mevalonate-derived non-sterol regulator. This component of the regulatory mechanism can only be observed when cultured cells are acutely incubated with statins, which block mevalonate formation. Under those conditions, sterols have no effect on HMG-CoA reductase mRNA translation; however, mevalonate reduces the HMG-CoA mRNA translation by 80% with no change in mRNA levels. Translational control of hepatic HMG-CoA reductase by dietary cholesterol was shown in an animal model in which polysome-associated HMG-CoA reductase mRNA was analyzed in cholesterol-fed rats. It was found that cholesterol feeding increased the portion of mRNA associated with translationally inactive monosomes and decreased the portion of mRNA associated with translationally active polysomes. The mechanism of HMG-CoA reductase translational control has not been elucidated.

Phosphorylation

Many key metabolic enzymes are modulated by phosphorylation-dephosphorylation and it has long been known that HMG-CoA reductase catalytic activity is inhibited by phosphorylation. Rodent HMG-CoA reductase is phosphorylated on Set 871 by an AMP-activated protein kinase that uses ATP as a phosphate donor. However, examination of HMG-CoA reductase activity in rat liver showed that phosphorylation-dephosphorylation could not account for the long-term regulation that occurred with diurnal light cycling, fasting, or cholesterol-supplemented diet. Approximately 75-90% of HMG-CoA reductase enzyme was found to be phosphorylated (inactive) under all physiological conditions. This reservoir of inactive enzyme may allow cells to respond transiently to short-term cholesterol needs.

The AMP-activated kinase that phosphorylates and inactivates HMG-CoA reductase also phosphorylates and inactivates acetyl-CoA carboxylase. It has been suggested that, when cellular ATP levels are depleted causing AMP levels to increase, the resultant activation of the kinase would inhibit cholesterol and fatty acid biosynthetic pathways, thus conserving energy. Consistent with this hypothesis, cholesterol synthesis was reduced when ATP levels were depleted by incubation with 2- deoxy-o-glucose. However, cholesterol synthesis was not reduced when ATP levels declined in cells expressing a Set 871 to Ala mutant form of HMG-CoA reductase, which is not phosphorylated. Therefore, HMG-CoA reductase phosphorylation appears to be important for preserving cellular energy stores rather than end-product feedback regulation.

Raising the cellular cholesterol content not only stops transcription of the genes encoding the cholesterol biosynthetic enzymes, but it also leads to accelerated degradation of the rate-limiting enzyme, HMG-CoA reductase. In cholesterol-depleted cells, HMGCoA reductase is a stable protein that is degraded slowly ($t_{1/2}$ = 13 h). If cholesterol repletion simply stopped transcription of the HMG-CoA reductase gene, that would lead to a slow decline in HMG-CoA reductase enzyme activity owing to stability of the protein; however, in the presence of excess sterols or mevalonate there is rapid ($t_{1/2}$ = 3.6 h) and selective degradation of the enzyme, which results in more precise control of cellular sterol synthesis.

The HMG-CoA reductase membrane domain is necessary and sufficient for regulated degradation. Expression of the cytosolic catalytic domain results in a stable protein that is not subject to regulated degradation, whereas expression of the N-terminal membrane domain linked to a reporter protein results in regulated degradation of the reporter. Despite the complex topology of HMG-CoA reductase, no proteolytic intermediates have ever been detected. How is HMGCoA reductase proteolyzed? Possibilities include vesicular translocation to lysosomes or autophagy of HMG-CoA reductase-containing endoplasmic reticulum membranes; however, the degradation appears to be rapid and selective for HMG-CoA reductase. It is also unaffected by inhibitors of protein transport through the Golgi. Another possibility is HMG-CoA reductase digestion by resident endoplasmic reticulum proteases. Subcellular fractionation and use of specific protease inhibitors has revealed that HMG-CoA reductase degradation occurs in purified endoplasmic reticulum membranes and is inhibited by lactacystin, an inhibitor of the proteasome. Additional evidence that the proteosome is involved comes from study of the yeast ortholog, Hmg2p, which is subject to regulated degradation like the mammalian enzyme. A yeast genetic approach has led to the identification of three proteins that are involved in the reverse translocation of endoplasmic reticulum proteins and disposal by the proteosome. Evidence that HMG-CoA reductase is polyubiquitinated prior to proteolysis has been provided for the yeast and mammalian enzymes.

What is the signal for accelerated HMG-CoA reductase degradation? In mammalian cells, both mevalonate-derived non-sterols and sterols are required. That is, in cholesterol-depleted cells, the addition of sterols leads to accelerated HMG-CoA reductase degradation only when the sterols are accompanied by a mevalonate-derived non-sterol signal. Regulated degradation has also been demonstrated through in vitro experiments. HMG-CoA reductase degradation is more rapid in endoplasmic reticulum membranes isolated from mevalonate- or sterol-treated cells and in hepatic microsomes prepared from mevalonate-treated rats. Both the sterol and non-sterol signals are blocked by cycloheximide, indicating that regulated turnover of HMG-CoA reductase requires ongoing protein synthesis.

There is strong evidence that the non-sterol isoprenoid signal for HMG-CoA reductase degradation is derived from farnesyl-PR. This evidence has been recently reviewed and includes the following. Intracellular farnesol levels increase significantly after mevalonate addition to cells. Accelerated

HMG-CoA reductase degradation can be induced in cells incubated with mevalonate, farnesyl-PR or farnesol but not with a non-hydrolyzable analog of farnesyl-PP. Furthermore, inhibition of the enzyme farnesyl pyrophosphatase blocks the mevalonate-dependent, sterol-accelerated degradation of the enzyme.

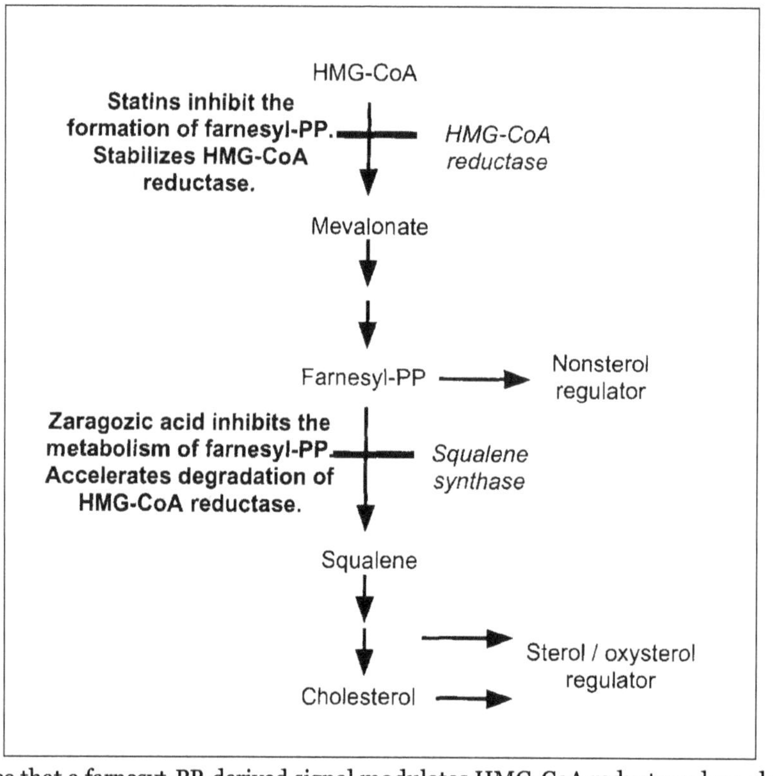

Figure: Evidence that a farnesyt-PP-derived signal modulates HMG-CoA reductase degradation. In yeast, conditions that decrease farnesyl-PP levels stabilize HMG-CoA reductase levels whereas conditions that increase farnesyl-PP levels accelerate HMH-CoA reductase degradation.

Many, but not all, aspects of regulated HMG-CoA reductase degradation are conserved among eukaryotes. In yeast, modulation of Hmg2p stability by a farnesyl-PPderived signal has been shown using pharmacologic and genetic approaches. Conditions chosen to increase farnesyl-PP levels (inhibition of squalene synthase by zaragozic acid or down-regulation of squalene synthase) accelerated Hmg2p ubiquitination and degradation, whereas conditions chosen to decrease farnesyl-PP levels (inhibition of HMG-CoA reductase by statins, farnesyl-PP synthase down-regulation, or squalene synthase overexpression) stabilized Hmg2p. One difference between the yeast and mammalian systems is the requirement for ongoing protein synthesis, which has not been shown in yeast.

Regulation of mammalian HMG-CoA reductase turnover requires both sterols and non-sterol isoprenoids, In contrast, the yeast isozyme can be suppressed by farnesyl-PP alone, although sterols enhance the non-sterol regulation. Pharmacologic and genetic approaches have again been extremely informative, showing that endogenously produced oxysterols serve as a positive signal for Hmg2P degradation in yeast. Oxysterols have long been known to accelerate HMG-CoA reductase degradation when added to the medium of cultured mammalian cells; however, an endogenously produced oxysterol regulator of mammalian HMG-CoA reductase degradation has not yet been identified.

How do sterol or non-sterol regulators influence HMG-CoA reductase degradation? They could have a direct effect on the enzyme itself, as an allosteric regulator or by binding to the membrane domain. Alternatively, they could have an effect on the biophysical properties of the endoplasmic reticulum membrane (i.e. fluidity, membrane organization) or affect the proteolytic machinery. These possible effects could be direct or through interaction with an effector protein.

The half-life of HMG-CoA reductase is also influenced by the enzyme's oligomerization state and expression level. Oligomerization of HMG-CoA reductase through its cytosolic domain appears to stabilize the protein, as does a higher expression level. Analysis of the crystal structure of the catalytic portion of human HMG-CoA reductase revealed that this domain forms tight tetramers. It was suggested that oligomerization of the catalytic domain may induce association of the membrane domains, which would decrease accessibility of the enzyme to proteases. Sterols may induce dissociation of the HMG-CoA reductase tetramer, resulting in accelerated proteolysis.

Sterol-Sensing Domain

How do HMG-CoA reductase and SCAP detect rising cellular cholesterol levels? A key feature of these proteins was revealed when SCAP was cloned and the deduced protein sequence compared with that of HMG-CoA reductase. A five transmembrane domain was found in SCAP with 25% identity and 55% similarity with a corresponding region in HMG-CoA reductase. This domain, termed the sterol-sensing domain, appeared to be critical for SCAP's cholesterol-regulated SREBP escort function since certain amino acid substitutions in the sterol-sensing domain led to constitutive activity. In HMG-CoA reductase, transmembrane span 2 (which is within the sterol-sensing domain) was shown to be necessary for regulated degradation. Chimeric mutants of HMG-CoA reductase were constructed that combined membrane-spanning domains from the hamster enzyme (which is subject to regulated degradation) with membrane-spanning domains from the sea urchin enzyme (which shares 62% amino acid sequence identity with the hamster enzyme in the membrane domain, but is not subject to regulated degradation). Analysis of regulated turnover of the chimeric molecules showed that hamster transmembrane span 2 was sufficient to confer regulated degradation upon the sea urchin enzyme.

Sterol-sensing domains with a high degree of sequence similarity are found in several other proteins with obvious connections to cholesterol homeostasis. One is the biosynthetic enzyme 7-dehydrocholesterol Δ7-reductase. The function of the sterolsensing domain in 7-dehydrocholesterol Δ7-reductase is not clear; however, amino acid substitutions in that region cause the loss of 90% of catalytic activity. Another sterol-sensing domain-containing protein is NPC1, a 1278-amino-acid glycoprotein found in late endosomes and lysosomes. NPCI is hypothesized to play a role in trafficking of cholesterol, gangliosides and other cargo from late endosomes to destinations throughout the cell. Mutations in NPC 1 lead to the predominant form of Niemann-Pick C disease, a human genetic disease characterized by progressive neurodegeneration. The biological function of NPC 1 is still not clear, but structure/function analysis indicates that the sterol-sensing domain is important. Mutations that cause amino acid substitutions within the sterolsensing domain lead to a rapidly progressing, infantile form of the disease, whereas amino acid substitutions throughout the rest of the protein cause the classical juvenile presentation. NPC1L1 is a Golgi protein with 42% identity and 51% similarity with NPC 1. NPC 1L 1 has a sterol-sensing domain, but the function of this NPC l-like protein is unknown.

Other proteins with a sterol-sensing domain have a more tenuous link to cholesterol homeostasis. Two proteins, Patched and dispatched, are involved in developmental patterning. Patched is the receptor for the morphogen, Sonic Hedgehog, which is the only known protein with a covalently attached cholesterol moiety. Dispatched is the plasma membrane protein required for secretion of cholesterol modified Hedgehog. Hedgehog binding to Patched leads to a signal transduction cascade that activates transcription of specific genes. Mutations in Patched cause basal cell nevus syndrome, which is characterized by developmental abnormalities and basal cell carcinomas. A role for Patched and dispatched in cholesterol metabolism has not been established; however, it is intriguing that exposure of embryos to inhibitors of cholesterol biosynthesis, such as Triparanol, AY-9944 or BM 15.766, cause profound developmental defects that resemble those in Sonic Hedgehog mutant embryos.

Metabolism of Cholesterol

Cellular cholesterol levels are regulated, not only by feedback inhibition of cholesterol synthesis, but also by feed forward regulation of cholesterol metabolism. Excess cholesterol is metabolized to oxysterols. In addition to blocking SCAP-facilitated proteolysis of SREBP and thereby down-regulating endogenous cholesterol synthesis and LDL receptor levels, oxysterols also activate bile acid synthesis and cholesterol esterification, which further reduces the cellular content of unesterified cholesterol.

Oxysterols

Oxysterols are potent suppressors of cholesterol synthesis. Their effectiveness has been attributed to their ability to diffuse into and through cells to activate regulatory processes, thus bypassing the need for receptor-mediated entry. It was long assumed that cholesterol was the natural regulator and that oxysterols were contaminants found in commercial supplies of cholesterol or formed upon storage of stock cholesterol solutions. Now it is known that there are many naturally occurring oxysterols that have diverse actions on cellular lipid metabolism. 25-Hydroxycholesterol is the most studied oxysterol; however, other oxysterols are as, or more, physiologically important.

Oxysterols can be formed by the action of at least three distinct hydroxylases. The mitochondrial sterol 27-hydroxylase participates in an alternative pathway of bile acid biosynthesis, hydroxylating cholesterol and several other intermediates in the bile acid synthetic pathway. 27-Hydroxycholesterol formed in peripheral tissues is a potent inhibitor of endogenous cholesterol synthesis. It is also thought to be secreted into the bloodstream and transported to the liver, where 27-hydroxycholesterol binds the liver X receptor (LXR) nuclear hormone receptor. LXR forms a heterodimer with the retinoid X receptor (RXR) and activates transcription of genes encoding bile acid biosynthetic enzymes. The physiological significance of sterol 27-hydroxylase is illustrated by the genetic disease cerebrotendinous xanthomatosis, which is caused by mutations in the sterol 27-hydroxylase gene. The absence of this critical hydroxylase activity precludes the mobilization of excess cholesterol from peripheral tissues and leads to cholesterol deposition and xanthoma development.

24-Hydroxylase is an endoplasmic reticulum enzyme predominantly expressed in brain. Bjorkhem and colleagues have provided strong evidence that 24-hydroxylase maintains cholesterol homeostasis in the brain, which cannot participate in highdensity lipoprotein-mediated reverse cholesterol transport. 24- Hydroxycholesterol is secreted from the brain into the circulation, taken up by the liver and metabolized into bile acids. 25-Hydroxylase is an endoplasmic reticulum and Golgi enzyme with low-level expression in most tissues. 25-Hydroxycholesterol is a potent regulator of SREBP proteolytic processing. Given that the enzyme resides in the same subcellular compartment as SREBP and SCAP, 25-hydroxycholesterol may be a physiological regulator of cholesterol synthesis.

Oxysterol binding to LXR activates transcription of several genes that play key roles in maintaining bodily cholesterol homeostasis. One is the gene encoding the ATPbinding cassette transporter ABCA 1, a 2201-amino-acid plasma membrane protein that stimulates cholesterol and phospholipid efflux. Cholesterol effluxed from peripheral cells by the action of ABCA1 is transferred by plasma high-density lipoproteins to the liver in a process called reverse cholesterol transport. A second example is LXR-activated transcription of the gene encoding cholesteryl ester transfer protein, which promotes transfer of cholesteryl esters from high-density lipoproteins to very low-density lipoproteins for clearance by the liver. Finally, oxysterol binding to LXR also stimulates expression of SREBP-lc, but not SREBP-Ia or SREBP-2. Therefore, increased cellular cholesterol should lead to oxysterol formation, which would increase expression of SREBP-Ic and increase fatty acid synthesis. Indeed, administration of an LXR selective agonist to mice led to increased lipogenesis and higher plasma triacylglycerol and phospholipid levels.

Another protein that binds oxysterols with high affinity was first reported by A.A. Kandutsch et al. and called oxysterol-binding protein (OSBP). At the time the protein was purified and cDNA-cloned, OSBP was expected to be a cytosolic protein that translocated into the nucleus and repressed transcription of cholesterogenic genes when oxysterols were present. Given our current knowledge of transcriptional control, we might expect OSBP to bind to the SREBP precursor or SCAP in the endoplasmic reticulum or site-1 protease or site-2 protease in the Golgi to interfere with SREBP proteolytic processing. However, a direct role for OSBP in transcriptional control has not been demonstrated. OSBP is a high-affinity 25-hydroxycholesterol-binding protein (K_d 10 nM) that translocates from cytosol and vesicles to the Golgi when ligand is bound.

How might OSBP transduce signals'? One reasonable hypothesis is that when cellular cholesterol levels are high, cholesterol hydroxylation occurs. The resultant oxysterol then binds to OSBR which translocates to the Golgi and signals suppression of cholesterol synthesis. However, OSBP responds paradoxically to cholesterol rather than oxysterols. When cells are cholesterol replete, OSBP moves to the cytosol and vesicles, not to the Golgi. OSBP moves to the Golgi when cells are cholesterol-depleted. Thus, it has been difficult to establish the identity of OSBP's endogenous ligand and which downstream events are mediated by OSBE The story is made more complex by the finding that the human OSBP family has at least five members, with sequence similarity in the C-terminal ligand-binding domain, whereas Saccharomyces cerevisiae has six related proteins.

Cholesterylester Synthesis

Excess cholesterol can also be metabolized to cholesteryl esters. ACAT is the endoplasmic reticulum enzyme that catalyzes the esterification of cellular sterols with fatty acids. In vivo, ACAT plays an important physiological role in intestinal absorption of dietary cholesterol, in intestinal and hepatic lipoprotein assembly, in transformation of macrophages into cholesteryl ester laden foam cells, and in control of the cellular free cholesterol pool that serves as substrate for bile acid and steroid hormone formation. ACAT is an allosteric enzyme, thought to be regulated by an endoplasmic reticulum cholesterol pool that is in equilibrium with the pool that regulates cholesterol biosynthesis. ACAT is activated more effectively by oxysterols than by cholesterol itself, likely due to differences in their solubility. As the fatty acyl donor, ACAT prefers endogenously synthesized, monounsaturated fatty acyl-CoA.

The cloning of the human ACAT gene and its orthologs, as well as the subsequent generation of ACAT-deficient mice, led to the realization that two ACAT isozymes must contribute to the enzyme activity. Human ACAT-I was cloned using an expression cloning strategy (C.C.Y. Chang, 1993). The gene encodes an integral membrane protein of 550 amino acids that is present in almost all cells and tissues examined. Orthologs were identified in other mammalian species, as well as Drosophila melanogaster and Caenorrhabditis elegans. The first indication of multiple ACATs came from the cloning of two ACAT-related enzymes (AREI and ARE2) from S. cerevisiae. The inactivation of both yeast genes was required to eliminate sterol esterification. In addition, ACAT-1-deficient mice showed the expected depletion of cholesteryl esters in adrenals, ovaries, testes and macrophages, but no changes in intestinal cholesterol absorption or hepatic cholesterol esterification. This result indicated that a second ACAT must be present in those mouse tissues.

The cloning of ACAT-2 revealed a protein of similar size to ACAT- 1, with a novel N-terminus but a C-terminus highly similar to ACAT-1. In adult humans, ACAT-2 is confined to the apical region of intestinal enterocytes, with low levels also expressed in hepatocytes. Disruption of the ACAT-2 gene in mice led to dramatic reduction in cholesterol absorption and prevention of hypercholesterolemia. The data suggest that, in humans, ACAT-1 plays a critical role in foam-cell formation and cholesterol homeostasis in extrahepatic tissues, whereas ACAT-2 has an important role in absorption of dietary cholesterol. ACAT-1 is the major isozyme in hepatocytes, although the total pool of cholesteryl esters produced by both enzymes regulates very low-density lipoprotein synthesis and assembly.

Cortisol Biosynthesis

In humans, cortisol, a glucocorticoid class of steroid hormone, is produced by the zona fasciculata of the cortex region of adrenal gland. It is also biosynthesized from cholesterol. The cortical region of adrenal gland especially zona glomerulosa produces aldosterone, zona reticularis produces sex hormones while medullary region of the adrenal gland lies beneath the cortex, primarily secretes the catecholamines namely: epinephrine (adrenaline) and norepinephrine (noradrenaline) under considerate stimulus. The cortisol biosynthesis is invigorated by adrenocorticotropic hormone (ACTH) secreted by the anterior lobe of the pituitary gland. The production of ACTH is vitalized by corticotrophin releasing hormone (CRH) discharge by the hypothalamus. Increase in ACTH concentration, increases the cholesterol level in the inner mitochondrial membrane through the regulation of the steroidogenic acute regulatory (STAR) protein thereby stimulating the major rate limiting step (cholesterol conversion to pregnenolone) in cortisol biosynthesis catalyzed by Cytochrome P450 side chain cleavage enzyme (CYP450 SCC).

As far as metabolism of cortisol is concerned, it gets metabolized via 11-β hydroxysteroid dehydrogenase system (11-β HSD) possessing two isoform of the enzymes: 11-β HSD1 and 11-β HSD2.

- The former 11-β HSD1 make use of NADPH and metamorphose it to biologically inert cortisone followed by final transformation to active cortisol.
- 11-β HSD2 exploit NAD+ to change cortisol to cortisone.

By and large, 11-β HSD1 dole out to augment the limited concentrations of cortisol and 11-β HSD2 decreases the restricted concentrations of cortisol.

Furthermore, cortisol is also metabolized to 5α tetrahydrocortisol (5α THF) and 5β tetrahydrocortisol (5β THF) via catalytic activity of 5α reductase and 5β reductase, the enzymes adjudged as the rate-limiting factors. 5β reductase is also acts as the rate-limiting factor in the conversion of cortisone to tetrahydrocortisone. It has also been suggested that modification of 11-β HSD1 play an important role in the pathogenesis of hypertension, insulin resistance and obesity, while 11-β HSD2 alteration has been implicated in essential hypertension and is acknowledged to escort to the syndrome of apparent mineralocorticoid excess (SAME).

Testosterone Biosynthesis

Similar to that of other steroidal hormones, testosterone is also gleaned from cholesterol. The step by step biosynthesis is described as under:

- It engross the oxidative cleavage of cholesterol side chain by CYP11A (mitochondrial cytochrome P450 oxidase) with the trouncing of 6C units to bestow pregnenolone.
- In ER, 2C atoms get obliterated by CYP17A enzyme to yield an assortment of C19 steroids. In toting up, the 3-hydroxyl group is oxidized by 3β HSD to generate androstenedione.
- The final step is the rate limiting step where 17β hydroxysteroid dehydrogenase reduces androstenedione (C17 keto group) to yield testosterone.

In men, the testis (leydig cells) produces high amount (>95%) of testosterone while in women it is also synthesized in extremely lesser quantity by by placenta, ovarian thecal cells as well as by the cortical region (zona reticularis) of adrenal gland in both sexes. Sertoli cells which are a part of the male generative glands also require testosterone for spermatogenesis. Similar to that of other hormones, in blood testosterone is afforded to target tissues where a large amount of it is elated to sex hormone binding globulin (SHBG, a specific plasma protein).

TAGs Biosynthesis

Triacylglycerol also known as triacylglyceride (TAG) or Triglyceride (TG) is synthesis from a molecule of glycerol and three molecules of fatty acids. As an important and vital component of blood, TAG assist in the bidirectional transmission of glucose and adipose fat from the liver. Based on the source of oil, TAG may either be saturated or highly unsaturated. Saturated are those either having single bonds (C-C) between the carbon atoms or where carbon atoms are bonded to hydrogen (C-H) or unsaturated are those having double bonds (C=C) between carbon moieties plummeting the integer of hydrogen atoms bonded to carbon units.

TAG being the chief constituent of human skin oils is synthesized by coalescing an alcohol (glycerol) with a hydroxyl (HO-) group and organic acids (3 fatty acid) with a carboxyl (-COOH) group. Both organic acid and alcohol together form esters through the formation of ester bonds. Usually the 3 moieties of fatty acids are different and the chain length also vary with most of them have 16, 18, or 20 carbon moieties. Even number of carbon atoms is present in fatty acids of animals and plants shimmering the origin of the biosynthetic pathway from 2C (Acetyl CoA). In prokaryotes (e.g. bacteria) odd and branched chain fatty acids synthesis takes place. A complex mixture of idiosyncratic triglycerides is present in natural fats and due to this reason they liquefy over a wide temperature range. Extraordinarily, cocoa butter consists of a small number of triglycerides obtained from stearic, oleic and palmitic acids. The simplest TAGs are those where all the 3 fatty acids are indistinguishable. The TAGs stearin is obtained from stearic acid and palmitin is derived from palmitic acid. The TAGs can be acquired as polymorphs (three forms) such as α, β, and β' vary in melting points.

In duodenum, lipids or fats cannot be absorbed in TAG form and hence hydrolysed by the pancreatic lipase that acts on ester bond resulting in the release of fatty acid and glycerol. One molecule of glycerol and one fatty acid (monoglycerides), sometimes diglycerides gets absorbed by the duodenum as TAGs have been broken down. In the intestine, after the secretion of bile and lipases, TAGs are ripped into free fatty acids and monoacylglycerol (MAG) by a procedure named as lipolysis. After lipolysis, TAGs consequently enthused to enterocytes and form chylomicrons together with protein and cholesterol. Through lymphatic system and blood, a number of tissues incarcerate the chylomicrons releasing TAGs to be utilized as energy source and stored in liver. As soon as energy requirement arises, glucagon signals the TAG breakdown by hormone sensitive lipase to discharge free fatty acids. The brain cannot exploit fatty acids as an energy source so it gets converted to a ketone and the glycerol gets transformed into glucose (gluconeogenesis) via Dihydroxyacetone phosphate (DHAP) and Glyceraldehyde 3-phosphate (G3P). Similarly, TAGs cannot pass through membranes liberally, hence breakdown by special enzyme (lipoprotein lipases) into glycerol and free fatty acids which further be taken up by cells via the fatty acid transporter (FAT).

TAGs are also vital constituent of chylomicrons and very-low-density lipoprotein (VLDL) and participate in metabolism as transporter and source of energy. A high level of TAGs in human blood is related to atherosclerosis, stroke and heart disease. The National Cholesterol Education Program has set up some guidelines for TAGs levels. The TAGs level can be tested after 8-12 h of fasting and it was observed that the level was higher after diet intake or eating.

TAGs Level	mg per dL	Mmol per L
Normal to Low Risk	< 150	< 1.70
Slightly Over Normal	150-199	1.70-2.25
Some Risk	200-499	2.26-5.65
High to Very High Risk	500 or higher	> 5.65

As recommended by the American Heart Association, an optimal TAGs level of 1. 1 mmol/L (100 mg/dL) or lesser perk up heart health. TAGs level also increases with carbohydrate rich diet (60%) and it correlates with a higher body mass index (BMI >28), insulin resistance and hyper triglyceridemia. In women also high glycemic index increases insulin production and boost TAGs level. Alcohol intake also bumps up TAGs level in blood. TAGs level can be reduced by the intake of flax seed oil and ώ-3 fatty acid from fish and other sources and through moderate exercise. Carnitine also lowers the TAGs level in blood and at some instance; fibrates have been used to fetch TAGs downward. Epanova, and ώ-3 carboxylic acid is developed by AstraZeneca for treating high TAGs level.

In oil paints, Linseed and related oils, rich in di and tri unsaturated fatty acid form, are used for coatings. In presence of oxygen, these oils also known as drying oils tend to generate heat and consolidate by polymerization offensive to the carbon backbone. During biodiesel manufacturing, TAGs undergo trans-esterification and the ensuing esters of fatty acid are used as fuel. The glycerin has been used in the pharmaceutical production and food manufacturing. Lysochromes (dye) are used for staining lipids, TAGs, fatty acids and lipoproteins. Other dye used include: Sudan Black B, Sudan IV, and Oil Red O.

Biosynthesis of Triacylglycerols

Three main pathways for triacylglycerol biosynthesis are known, the sn-glycerol-3-phosphate and dihydroxyacetone phosphate pathways, which predominate in liver and adipose tissue, and a monoacylglycerol pathway in the intestines. In maturing plant seeds and some animal tissues, a fourth pathway has been recognized in which a diacylglycerol transferase is involved. The most important route to triacylglycerols is the sn-glycerol-3-phosphate or Kennedy pathway, first described by Professor Eugene Kennedy and colleagues in the 1950s, by means of which more than 90% of liver triacylglycerols are produced.

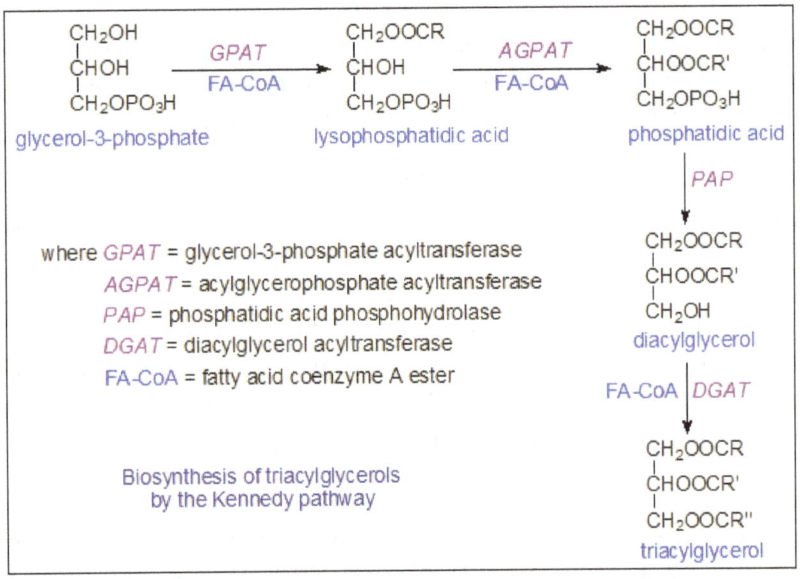

In this pathway, the main source of the glycerol backbone has long been believed to be sn-glycerol-3-phosphate produced by the catabolism of glucose (glycolysis) or to a lesser extent by the action of the enzyme glycerol kinase on free glycerol. However, there is increasing evidence that a significant proportion of the glycerol is produced de novo by a process known as glyceroneogenesis via pyruvate. Indeed, this may be the main source in adipose tissue.

Subsequent reactions occur primarily in or at the endoplasmic reticulum. First, the precursor sn-glycerol-3-phosphate is esterified by a fatty acid coenzyme A ester in a reaction catalysed by a glycerol-3-phosphate acyltransferase (GPAT) at position sn-1 to form lysophosphatidic acid, and this is in turn acylated by an acylglycerophosphate acyltransferase (AGPAT) in position sn-2 to form a key intermediate in the biosynthesis of all glycerolipids - phosphatidic acid. Numerous isoforms of these enzymes are known; they are expressed with specific tissue and membrane distributions and they are regulated in different ways.

Then, the phosphate group is removed by a family of enzymes - phosphatidic acid phosphohydrolases

(PAPs or 'phosphatidate phosphatases' or 'lipid phosphate phosphatases'). PAPs are also important as they produce sn-1,2-diacylglycerols as essential intermediates in the biosynthesis not only of triacylglycerols but also of phosphatidylcholine and phosphatidylethanolamine (and of monogalactosyldiacylglycerols in plants). This is a key branch-point in lipid biosynthesis as it may dictate the flow of lipids for storage or membrane biogenesis.

Much of this phosphatase activity leading to triacylglycerol biosynthesis in animals resides in three related cytoplasmic proteins, termed lipins, i.e. lipin-1, lipin-2 and lipin-3, which have tissue-specific roles in glycerolipid synthesis. Unusually, these were characterized and named before the nature of their enzymatic activities were determined. Each of the lipins appears to have distinctive expression and functions, but lipin-1 (PAP1) in three isoforms (designated 1α, 1β and 1γ) accounts for most of the PAP activity in adipose tissue and skeletal muscle in humans. Lipin 2 is the most abundant lipin in liver, but is also expressed substantially in the small intestine, macrophages and some regions of the brain, while lipin 3 activity overlaps with that of lipin 1 and lipin 2. Lipins are cytosolic enzymes but associate transiently with membranes to access their substrate, i.e. they are translocated to the endoplasmic reticulum in response to elevated levels of fatty acids within cells, although they do not have trans-membrane domains. Lipin-1 activity requires Mg^{2+} ions and is inhibited by N-ethylmaleimide, whereas the membrane-bound activity responsible for synthesising diacylglycerols as a phospholipid intermediate is independent of Mg^{2+} concentration and is not sensitive to the inhibitor.

Perhaps surprisingly, lipin-1 has a dual role in that it operates in collaboration with known nuclear receptors as a transcriptional coactivator to modulate lipid metabolism (lipin 1α) while lipin 1β is associated with induction of lipogenic genes such as fatty acid synthase, stearoyl CoA desaturase and DGAT. They can have profound effects on signalling in a variety of cell types. Abnormalities in lipin-1 expression are known to be involved in some human disease states that may lead to the metabolic syndrome and inflammatory disorders. Lipin 2 is a similar phosphatidate phosphohydrolase, which is present in liver and brain and is regulated dynamically by fasting and obesity (in mice), while lipin 3 is found in the gastrointestinal tract and liver.

In the final step in this pathway, the resultant 1,2-diacyl-sn-glycerol is acylated by diacylglycerol acyltransferases (DGAT), which can utilize a wide range of fatty acyl-CoA esters to form the triacyl-sn-glycerol. In fact there are two DGAT enzymes, which are structurally and functionally distinct. In animals, DGAT1 is located mainly in the endoplasmic reticulum and is expressed in skeletal muscle, skin and intestine, with lower levels of expression in liver and adipose tissue. It is believed to have dual topology contributing to triacylglycerol synthesis on both sides of the membrane of the endoplasmic reticulum, but esterifying only pre-formed fatty acids of exogenous origin. Perhaps surprisingly, DGAT1 is the only one present in the epithelial cells that synthesise milk fat in the mammary gland. Also DGAT1 can utilize a wider range of substrates, including monoacylglycerols, long-chain alcohols (for wax synthesis) and retinol, and it is reported to have an important role in protecting the endoplasmic reticulum from the lipotoxic effects of high-fat diets. Orthologues of this enzyme are present in most eukaryotes, other than yeasts, and it is especially important in plants.

DGAT2 is the main form of the enzyme in hepatocytes and adipocytes (lipid droplets), although it is expressed much more widely in tissues. It is associated with distinct regions of the endoplasmic reticulum, at the surface of lipid droplets and in mitochondria, and it esterifies fatty acids of both endogenous and exogenous origin. DGAT2 is believed to have a targeting domain that enables it to tether between the endoplasmic reticulum and lipid droplet thereby channelling triacylglycerols from the synthesis site in the endoplasmic reticulum to the nascent lipid droplet, where they accumulate and lead to the expansion of the latter (see below). Both enzymes are important modulators of energy metabolism, although DGAT2 appears to be especially important in controlling the homeostasis of triacylglycerols in vivo. As the glycerol-3-phosphate acyltransferase (GPAT) has the lowest specific activity of these enzymes, this step may be the rate-limiting one. However, DGATs are the dedicated triacylglycerol-forming enzymes, and they are seen as the best target for pharmaceutical intervention in obesity and attendant ailments; clinical studies of DGAT1 inhibitors are at an early stage.

In a second pathway for triacylglycerol biosynthesis, dihydroxyacetone-phosphate in peroxisomes or endoplasmic reticulum can be acylated by a specific acyltransferase to form 1-acyl dihydroxyacetone-phosphate, which is reduced by dihydroxyacetone-phosphate oxido-reductase to lysophosphatidic acid, which can then enter the pathway above to triacylglycerols. The precursor dihydroxyacetone-phosphate is important also as part of the biosynthetic route to plasmalogens, and neutral plasmalogens can be significant components of cytoplasmic droplets in many mammalian cells types but not adipose tissue.

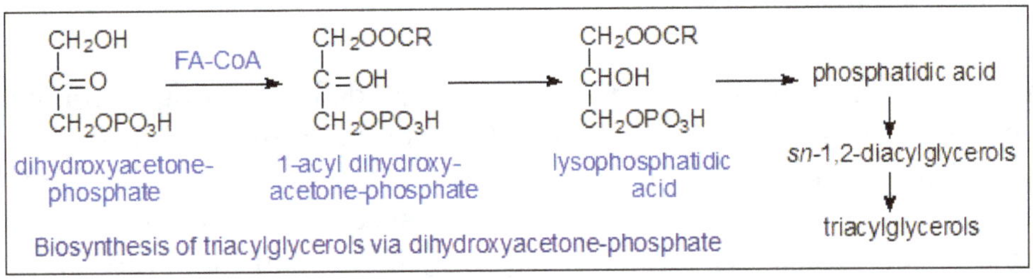

In the enterocytes of intestines after a meal, up to 75% of the triacylglycerols are formed via a monoacylglycerol pathway. 2-Monoacyl-sn-glycerols and free fatty acids released from dietary triacylglycerols by the action of pancreatic lipase within the intestines (see below) are taken up by the enterocytes. There, the monoacylglycerols are first acylated by an acyl coenzyme A:monoacylglycerol acyltransferase with formation of sn-1,2-diacylglycerols mainly as the first intermediate in the process, though some sn-2,3-diacylglycerols (~10%) are also produced (DGAT1 can also acylate monoacylglycerols). 1-Monoacylglycerols can also be synthesised by the acylation of glycerol and these can also be acylated. There are three isoforms of the monoacylglycerol acyltransferase in humans of which MGAT2 is most active in the intestines and liver and MGAT1 in adipose tissue. Finally, the acyl coenzyme A: diacylglycerol acyltransferase (DGAT1) reacts with the sn-1,2-diacylglycerols only to form triacylglycerols.

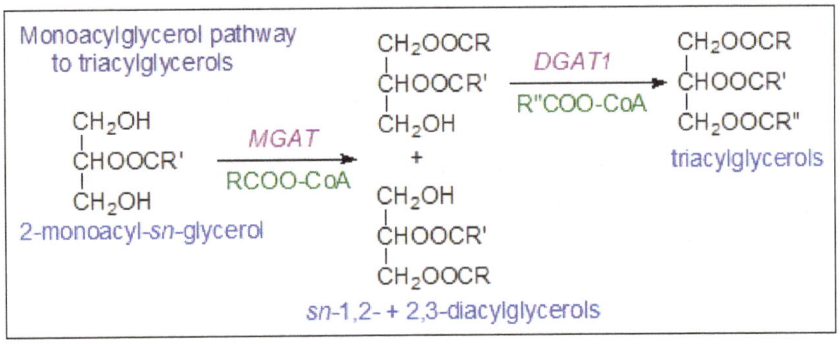

In a fourth biosynthetic pathway, which is less well known, triacylglycerols are synthesised by a transacylation reaction between two racemic diacylglycerols that is independent of acyl-CoA. The reaction was first detected in the endoplasmic reticulum of intestinal micro villus cells and is catalysed by a diacylglycerol transacylase. Both diacylglycerol enantiomers participate in the reaction with equal facility to transfer a fatty acyl group with formation of triacylglycerols and a 2-monoacyl-sn-glycerol. A similar reaction has been observed in seed oils.

Triacylglycerol biosynthesis via diacylglycerol transacylases

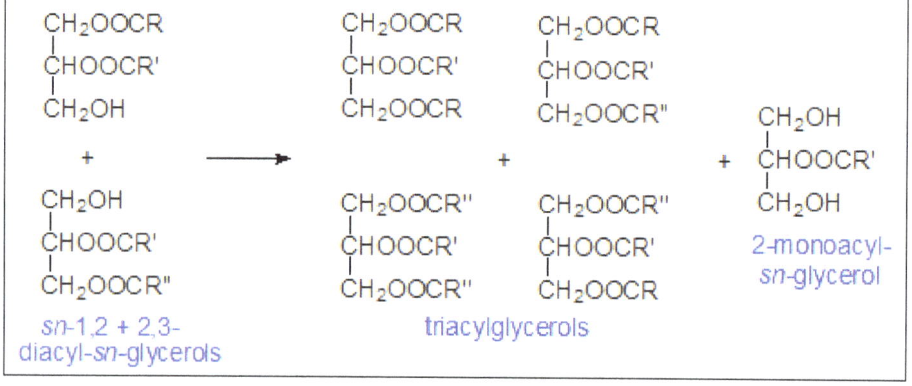

It has been suggested that this enzyme may function in remodelling triacylglycerols post synthesis, especially in oil seeds, and it is possible that it may be involved in similar processes in the liver and adipose tissue, where extensive hydrolysis/re-esterification is known to occur. There is evidence for selectivity in the biosynthesis of different molecular species in a variety of tissues and organisms, which may be a consequence of the varying biosynthetic pathways. Also in adipose tissue, fatty acids synthesised de novo are utilized in different ways from those from external sources in that they enter positions sn-1 and 2 predominantly, while a high proportion of the oleic acid synthesised in the tissue by desaturation of exogenous stearic acid is esterified to position sn-3.

In prokaryotes, the glycerol-3-phosphate pathway of triacylglycerol biosynthesis only occurs, but in yeast both glycerol-3-phosphate and dihydroxyacetone-phosphate can be the primary precursors and synthesis takes place in cytoplasmic lipid droplets and the endoplasmic reticulum. In plants, the glycerol-3-phosphate pathway is most important.

Among other potential routes to the various intermediates, lysophosphatidic acid and phosphatidic acid can be synthesised in mitochondria, but must then be transported to the endoplasmic reticulum before they enter the pathway for triacylglycerol production. 1,2-Diacyl-sn-glycerols are also produced by the action of phospholipase C on phospholipids.

In the glycerol-3-phosphate and other pathways, the starting material is of defined stereochemistry and each of the enzymes catalysing the various steps in the process is distinctive and can have preferences for particular fatty acids (as their coenzyme A esters) and for particular fatty acid combinations in the partially acylated intermediates. It should not be surprising, therefore, that natural triacylglycerols exist in enantiomeric forms with each position of the sn-glycerol moiety esterified by different fatty acids.

While triacylglycerols are essential for normal physiology, an excessive accumulation in human adipose tissue and other organs results in obesity and other health problems, including insulin resistance, steatohepatitis and cardiomyopathy. Accordingly, there is considerable pharmaceutical interest in drugs that affect triacylglycerol biosynthesis and metabolism.

Triacylglycerol Metabolism in the Intestines, Liver and Mammary Gland

Fat comprises up to 40% of the energy intake in the human diet in Western countries, and a high proportion of this is triacylglycerols. The process of fat digestion is begun in the stomach by acid-stable gastric or lingual lipases, the extent of which depending on species but may be important for efficient emulsification. However, this is insignificant in quantitative terms in comparison to the reaction with pancreatic lipase, which occurs in the duodenum. Entry of triacylglycerol degradation products into the duodenum stimulates synthesis of the hormone cholecystokinin and causes the gall bladder to release bile acids, which are strong detergents and act to emulsify the hydrophobic triacylglycerols so increasing the available surface area. In turn, cholecystokinin stimulates the release of the hydrolytic enzyme pancreatic lipase together with a co-lipase, which is essential for the activity of the enzyme. Pancreatic lipase, co-lipase, bile salts and calcium ions act together in a complex at the surface of the emulsified fat droplets to hydrolyse triacylglycerols. The process is regiospecific and results in the release of the fatty acids from the 1 and 3 positions with formation of 2-monoacyl-sn-glycerols. Isomerization of the latter to 1(3)-monoacyl-sn-glycerols occurs to some extent, and these can be degraded completely by the enzyme to glycerol and free fatty acids. Other lipases hydrolyse the phospholipids and other complex lipids in foods at the same time.

Hydrolysis of triacylglycerols in the intestines

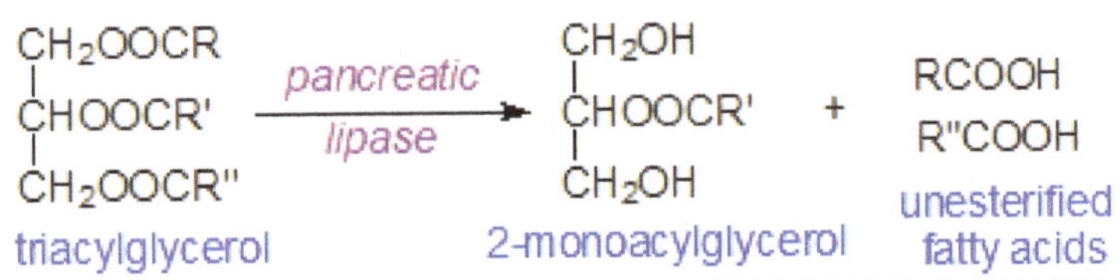

This process is somewhat different in neonates and young infants, in whom pancreatic lipase is less active but is effectively replaced by lipases in breast milk and by an acid gastric lipase (pH optimum 4-6).

There is evidence that the regiospecific structure of dietary triacylglycerols has an effect on the uptake of particular fatty acids and may influence further the lipid metabolism in humans. In particular, incorporation of palmitic acid into the position sn-2 of milk fat may be of benefit to the human infant (as a source of energy for growth and development), although it increases the atherogenic potential for adults. In addition, 2-monoacylglycerols and 2-oleoylglycerol especially have a signalling function in the intestines by activating a specific G-protein coupled receptor GPR119, sometimes termed the 'fat sensor'. When stimulated, this causes a reduction in food intake and body weight gain in rats and regulates glucose-stimulated insulin secretion. The free fatty acids released have a similar effect, though by a very different mechanism, via the receptor GPR40. Overall, it has become evident that triacylglycerol metabolism in the intestine has regulatory effects on the secretion of gut hormones and on systemic lipid metabolism and energy balance.

The free fatty acids and 2-monoacyl-sn-glycerols are rapidly taken up by the intestinal cells, from the distal duodenum to the jejunum, via specific carrier molecules but also by passive diffusion. A specific fatty acid binding protein prevents a potentially toxic build-up of unesterified fatty acids and targets them for triacylglycerol biosynthesis. The long-chain fatty acids are converted to the CoA esters and esterified into triacylglycerols by the monoacylglycerol pathway as described above. In contrast, short and medium-chain fatty acids (C12 and below) are absorbed in unesterified form and pass directly into the portal blood stream, where they are transported to the liver to be oxidized.

Subsequently, the triacylglycerols are incorporated into lipoprotein complexes termed chylomicrons in the enterocytes. In brief, these consist of a core of triacylglycerols together with some cholesterol esters that is stabilized and rendered compatible with an aqueous environment by a surface film consisting of phospholipids, free cholesterol and one molecule of a truncated form of apoprotein B (48 kDa). These particles are secreted into the lymph and thence into the plasma for transport to the peripheral tissues for storage or structural purposes. Adipose tissue in particular exports appreciable amounts of the enzyme lipoprotein lipase, which binds to the luminal membrane of endothelial cells facing into the blood, where it rapidly hydrolyses the passing triacylglycerols at the cell surface releasing free fatty acids, most of which are absorbed into the adjacent adipocytes and re-utilized for triacylglycerol synthesis within the cell.

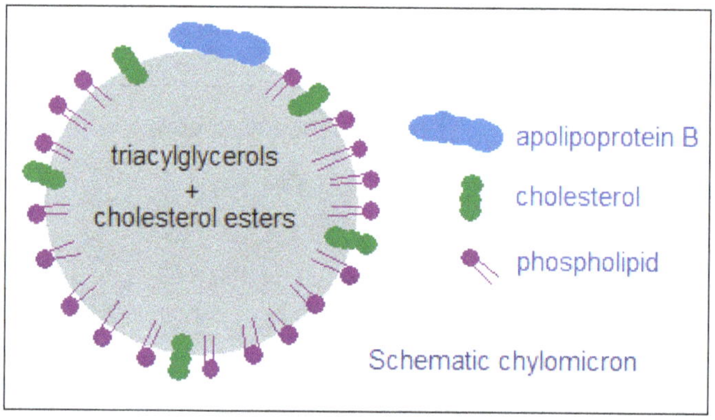

The chylomicrons remnants eventually reach the liver, where the remaining lipids are hydrolysed at the external membranes by a hepatic lipase and absorbed. The fatty acids within the liver can be utilized for a variety of purposes, from oxidation to the synthesis of structural lipids, but a proportion is re-converted into triacylglycerols, and some of this is stored as lipid droplets within the cytoplasm of the cells. In addition, phosphatidylcholine from the high-density lipoproteins is taken up by the liver, and a high proportion of this is eventually converted to triacylglycerols. In healthy liver, the levels of triacylglycerols are low (<5% of the total lipids), because the rates of acquisition of fatty acid from plasma and synthesis de novo within the liver are balanced by rates of oxidation and secretion into plasma. On the other hand, excessive accumulation of storage triacylglycerols is associated with fatty liver, insulin resistance and type 2 diabetes.

Most of the newly synthesised triacylglycerols are exported into the plasma in the form of very-low-density lipoproteins (VLDL), consisting again of a triacylglycerol and cholesterol ester core, surrounded by phospholipids and free cholesterol, together with one molecule of full-length apoprotein B (100 kDa), apoprotein C and sometimes apoprotein E. These particles in turn are transported to the peripheral tissues, where they are hydrolysed and the free acids absorbed. Eventually, the remnants are returned to the liver.

In the mammary gland, triacylglycerols are synthesised in the endoplasmic reticulum and large lipid droplets are produced with a monolayer of phospholipids derived from this membrane. These are transported to the plasma membrane and bud off into the milk with an envelope comprised of the phospholipid membrane to form milk fat globules as food for the newborn. The process is thus very different from that involved in the secretion of triacylglycerol-rich lipoproteins from other organs.

Triacylglycerol Synthesis and Catabolism

Adipose tissue and the adipocytes are characterized by accumulations of triacylglycerols, which act as the main energy store for animals, although they also cushion and insulate the body. Thus, triacylglycerols stored when there is a surplus of nutrients are mobilized for energy production during starvation. Adipose tissue also functions as a reserve of bioactive lipids, such as eicosanoids and lipid-soluble vitamins, and when required provides structural components, including fatty acids, cholesterol and retinol, for membrane synthesis and repair. Large fat depots occur around internal organs such as the liver, and also subcutaneously.

Similarly, within most other animal cells, even ganglia in the brain, a proportion of the fatty acids taken up from the circulation is converted to triacylglycerols as described above and incorporated into cytoplasmic lipid droplets (also termed 'fat globules', 'oil bodies', 'lipid particles', 'adiposomes', etc). By buffering against fatty acid accumulation that might exceed their capacity, non-adipose cells defend themselves in this way against lipotoxicity while providing a rapid source of energy and essential metabolites. Acting in concert with other cellular organelles, they function in many different metabolic processes. The triacylglycerol droplets together with cholesterol esters and other neutral lipids are surrounded by a protective monolayer that includes phospholipids, cholesterol and hydrophobic proteins. The phospholipid component of the monolayer consists mainly of phosphatidylcholine and phosphatidylethanolamine derived from cytosolic leaflets of the endoplasmic reticulum and plasma membrane. Lipid droplets do not occur only in the cytoplasm of animal cells but are present in the cytoplasm of some prokaryotes and in the plastids and other organelles of plants.

Among the proteins are many that function directly in lipid metabolism, and they include acyltransferases, lipases, perilipins, caveolins and the Adipose Differentiation Related Protein (ADRP or adipophilin). In adipocytes, the lipid droplets can range to up to 200 μm in diameter, while other cell types contain smaller lipid droplets of the order of 50 nm in diameter. Cytosolic lipid droplets with similar metabolic activities are found in the fly Drosophila melanogaster, and in higher plants and yeasts. Like adipose tissue cells, lipid droplets have a major function in that they sense and respond rapidly to changes in systemic energy balance. Within cells, lipid droplets facilitate the coordination and communication between different organelles and act as vital hubs of cellular metabolism.

1. Lipid droplet assembly: This process takes place in sub-domains of the endoplasmic reticulum, where at least one isoform of each of the enzymes of triacylglycerol biosynthesis, from acyl-CoA synthetases through to glycerol-3-phosphate acyltransferases, is located probably in a protein assembly or 'interactome'. Triacylglycerols accumulate and so attract perilipins and other proteins that allow lipid droplets to grow in patches of the membrane as lens-like swellings between the two membrane leaflets. A non-enzymatic protein seipin stabilizes the nascent droplets with minimal disruption to the membrane and enables them to mature by a mechanism that is still uncertain

but may involve regulation of protein and lipid trafficking into the droplet (and in yeast, nematodes, and plants). The growing lipid droplets bud toward the cytosol, a process that is believed to be directed and aided by surface proteins such as perilipin, while the triacylglycerol core attracts and is largely surrounded by phospholipids from the outer leaflet of the endoplasmic reticulum. Growth continues through an extended endoplasmic reticulum/droplet junction or bridge until finally with the involvement of membrane curvature-inducing coat proteins, fat storage-inducing transmembrane ('FIT') proteins that bind diacylglycerols, lysophospholipids and phosphatidic acid (non-bilayer forming lipids), the droplets bud off into the cytoplasm with their surface monolayer of phospholipids and proteins, including the enzymes of triacylglycerol biosynthesis.

Subsequently, mitochondria, peroxisomes and other organelles may contribute or exchange lipids at inter-organelle contact sites and effect changes in protein composition, although the lipid droplets remain close to the endoplasmic reticulum, and presumably this enables a dynamic response to any change in metabolic status sensed by the this organelle. Seipin appears to be essential for maintaining functional junctions between lipid droplets with other organelles in addition to its role in the initial steps of their formation; it is essential for lipid droplet formation in nuclei. In adipocytes, lipid droplets can also grow by fusion of smaller droplets, although again the mechanism is not fully understood.

Some of the surface proteins on lipid droplets can extend long helical hairpins of hydrophobic peptides deep into the lipid core. For example, perilipins constitute a family of at least five phosphorylated proteins that bind to droplets in animals and share a common region, the so-called 'PAT' domain, named for the three original members of the family that include perilipin and ADRP. Proteins related evolutionarily to these are found in more primitive organisms, including insects, slime moulds and fungi, but not in the nematode Caenorhabditis elegans. In mammals, perilipin A (or 'PLIN1' or more accurately the splice variant 'PLIN1a') is a well-established regulator of lipolysis in adipocytes, and it is believed to be involved in the formation of the large lipid droplets in white adipose tissue. The perilipins PLIN1 and PLIN2 have functions in triacylglycerol metabolism in tissues other than adipocytes, and PLIN2 in particular is the main perilipin in hepatocytes; PLIN5 operates in tissues that oxidize fatty acids such as the heart. Other surface proteins of lipid droplets are enzymes intimately involved in triacylglycerol metabolism, although there is a suggestion that cytoplasmic droplets may act as a storage organelle for hydrophobic proteins whose function is elsewhere in the cell.

On the basis of profiling of the surface proteins and phospholipids, lipid droplets in cells are now considered to be complex, metabolically active organelles that function in the supply of fatty acids for various purposes, including membrane trafficking and possibly in the recycling of both simple and complex lipids. Although they originate in the endoplasmic reticulum, lipid droplets can associate with most other cellular organelles through membrane contact sites in a highly dynamic manner. For example, within the liver, triacylglycerols are stored as lipid droplets in the cytoplasm adjacent to the endoplasmic reticulum where a triacylglycerol hydrolase can affect lipolysis to di- and monoacylglycerols that are more soluble in the membrane, which they are able to cross. They are then available for re-synthesis into triacylglycerols by luminally oriented acyltransferases before assembly into nascent lipoprotein complexes. Similar organelles can be found in most eukaryotic cells and in bacteria, and they provide a reservoir not only for triacylglycerols and their fatty acid constituents but also for eicosanoids, and cholesterol and retinol esters, for example. Lipid droplets even occur in the nuclei of cells where their constituents interact with numerous proteins and are involved in the regulation of nuclear events.

2. Lipolysis: When fatty acids are required by other tissues for energy or other purposes, they are released from the triacylglycerols by the sequential actions of three cytosolic enzymes at neutral pH, i.e. adipose triacylglycerol lipase (ATGL), hormone-sensitive lipase (HSL) and monoacylglycerol lipase, which cycle between the cytoplasmic surfaces of the endoplasmic reticulum and the surface layer of lipid droplets. Simplistically, ATGL hydrolyses triacylglycerols to diacylglycerols, which are hydrolysed by HSL to monoacylglycerols before these are hydrolysed by the monoacylglycerol lipase to complete the process. A protein perilipin (PLIN1) has been described as "the gatekeeper of the adipocyte lipid storehouse". Thus, the lipolytic process is regulated by perilipin, which acts as a barrier to lipolysis in non-stimulated cells, but on β-adrenergic stimulation as during fasting it is phosphorylated by the cAMP-protein kinase. This changes its shape and reduces its hydrophobicity, and in the process activates lipolysis. An isoform, perilipin A, is the main regulatory factor in white adipose tissue. However, many other proteins interact with the three enzymes to modulate their activity, location and stability.

The adipose triacylglycerol lipase, which initiates the process, was discovered surprisingly recently. It is structurally related to the plant acyl-hydrolases in that it has a patatin-like domain in the NH2-terminal region (patatin is a non-specific acyl-hydrolase in potato) and is located on the surface of the lipid droplet both in the basal and activated states. This lipase is specific for triacylglycerols containing long-chain fatty acids, preferentially cleaving ester bonds in the sn-1 or sn-2 position (but not sn-3), and yields diacylglycerols and free fatty acids as the main products, with low activity only towards diacylglycerols, and none to monoacylglycerols and cholesterol esters. However, it also has transacylase and phospholipase activities, and it hydrolyses retinol esters in hepatic stellate cells. Adipose triacylglycerol lipase can be activated at the same time as hormone-sensitive lipase and is now believed to be rate limiting for the first step in triacylglycerol hydrolysis.

Regulation of the enzymatic activity is a complex process, and for example, a lipid droplet protein designated Gene identification-58 (CGI-58 or ABHD5), is known to be an important activating factor and is required for hydrolysis of fatty acids from position sn-1. In the resting state this protein binds to perilipin (PLIN1), but on hormonal stimulation, the latter is phosphorylated leading to dissociation and interaction of CGI-58 with phosphorylated ATGL to commence the first step in triacylglycerol hydrolysis. Mutations in adipose triacylglycerol lipase or CGI-58 are believed to be responsible for a syndrome in humans known as 'neutral lipid storage disease'. A second protein (GoS2) inhibits the enzyme.

Hormone-sensitive lipase is regulated by the action of the hormones insulin and noradrenaline by a mechanism that ultimately involves phosphorylation of the enzyme by cAMP-protein kinase (as with perilipin), thereby increasing its activity and causing it to translocate from the cytosol to the lipid droplet to in initiate the second step in hydrolysis. In addition to its activity towards triacylglycerols, hormone-sensitive lipase will rapidly hydrolyse diacylglycerols, monoacylglycerols, retinol esters and cholesterol esters. In fact, diacylglycerols are hydrolysed ten times as rapidly as triacylglycerols. Within the triacylglycerol molecule, hormone-sensitive lipase preferentially hydrolyses ester bonds in the sn-1 and sn-3 positions, leaving free acids and 2-monoacylglycerols as the main end products, although it is capable of hydrolysing a broad spectrum of other lipids. The activity of the enzyme is regulated further by a variety of proteins that include PLINs and fatty acid binding proteins (FABP).

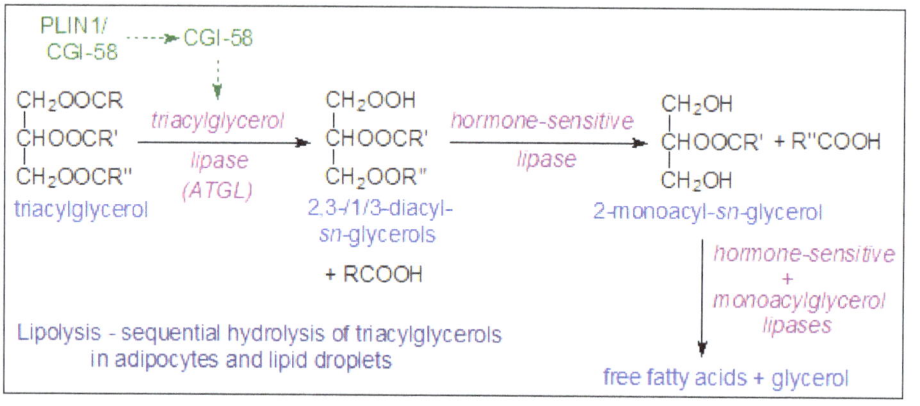

The monoacylglycerol lipase is believed to be the rate-limiting enzyme in monoacylglycerol hydrolysis, i.e. the final step in triacylglycerol catabolism releasing free glycerol and fatty acids, and is found in the cytoplasm, the plasma membrane, and in lipid droplets. It is specific for monoacylglycerols and has no activity against di- or triacylglycerols. As it is the enzyme mainly responsible for deactivation of the endocannabinoid 2-arachidonoylglycerol and is highly active in malignant cancers, it is attracting pharmaceutical interest. Further lipolytic enzymes, including carboxylesterases, are believed to operate against triacylglycerols in cytoplasmic lipid droplets in the liver.

Free fatty acids released by the combined action of these three lipases are exported into the plasma for transport to other tissues in the form of albumin complexes, while the glycerol released is transported to the liver for metabolism by either glycolysis or gluconeogenesis. Eventually, the whole organelle can disappear, including the proteins, when they undergo a process of autophagy ('lipophagy'), i.e. the delivery of the organelles to lytic compartments for degradation. This process is especially important during starvation and is also relevant to tumorigenesis and cancer metastasis, and while it is mechanistically distinct from lipolysis, there is cross-talk between the two.

Not only does the adipocyte provide a store of energy but it manages the flow of energy through the formation of the hormone leptin, which stimulates secretion of various factors that communicate with other tissues, including cytokines, adiponectin and resistin. The synthesis of leptin is tightly controlled by adipocytes and its main function is believed to be the provision of information on the state of fat stores to other tissues. Lipid droplets may play a role in this process, since perilipin is required for the sensing function. As caveolae, which contain the proteins caveolins (and presumably sphingolipids) and are particularly abundant in adipocytes, modulate the flux of fatty acids across the plasma membrane and are involved in signal transduction and membrane trafficking pathways, it is evident that they have a major role in this aspect of lipid metabolism. In addition, insulin is the main hormone that affects metabolism and its receptor at the plasma membrane is located in caveolae. Release of proinflammatory cytokines can stimulate lipolysis and cause insulin resistance, in turn leading to dysfunction of adipose tissue and systemic disruption of metabolism. Thus, adipose tissue metabolism has profound effects on whole-body metabolism, and defects in these processes can have severe implications for such severe pathological conditions as diabetes, obesity, cardiovascular disease, fatty liver disease and cancer in humans.

3. Functions other than energy management: Lipid droplets accumulate within many cell types other than adipocytes, including leukocytes, epithelial cells, hepatocytes and even astrocytes, especially during infections, cancer and other inflammatory conditions. They are important for the cellular storage and release of hydrophobic vitamins, signalling precursors and other lipids that are not related to energy homeostasis, while reducing the dangers of lipotoxicity. A variety of enzymes are associated with lipid droplets, including protein kinases, which are involved in many different aspects of lipid metabolism, such as cell signalling, membrane trafficking and control of the production of inflammatory mediators like the eicosanoids. Lipolysis enables secretion of lipid species termed lipokines (more generally 'adipokines') from adipocytes that may signal in a hormone-like fashion to other tissues, thereby modulating gene expression and physiological function, including food intake, insulin sensitivity, insulin secretion and related processes. These include palmitoleic acid (9-16:1) and fatty acid esters of hydroxy fatty acids (FAHFA), though the circulating proteins adiponectin and leptin have been studied more intensively. Adiponectin is a powerful insulin sensitizer and suppressor of apoptosis and inflammation with anti-diabetic and anti-atherosclerotic functions, often operating through its effects on sphingolipids, while leptin exerts most of its effects on the brain to trigger behavioral, metabolic, and endocrine responses to control the body's fuel reserves.

Indeed, there are now suggestions those lipid droplets in all cell types are essential for the response mechanisms to cellular stress, including inflammation and immunity, and act as hubs to integrate metabolic and inflammatory processes. Via their lipolytic machinery, they regulate the availability of fatty acids for the activation of signalling pathways and for the production of oxylipins from polyunsaturated fatty acids. For example, triacylglycerols in cytoplasmic lipid droplets of human mast cells, which are potent mediators of immune reactions and influence many inflammatory diseases, have a high content of arachidonic acid and this can be released by adipose triacylglycerol lipases as a substrate for production of specific eicosanoids when the cells are stimulated appropriately. Similarly, triacylglycerols in lipid droplets of the skin are a highly specific source of the linoleic acid that is required for the formation of the O-acylceramides, which are essential for epidermal barrier function. An organelle termed the midbody in dividing cells in humans and rodents contains a unique triacylglycerol that is a single molecular species consisting of three fatty acids 16:1-12:0-18:1, but its function is not known.

Vitamin E (tocopherols) and vitamin A in the form of retinyl esters are stored in cytoplasmic lipid droplets, and the latter are present in appreciable concentrations in the stellate cells of liver, for example. In endocrine cells of the gonads and adrenals, cholesterol esters stored in lipid droplets are an important source of cholesterol for the mitochondrial biosynthesis of various steroid hormones. In the nucleus of the cell, in addition to providing a reservoir of fatty acids for membrane remodelling, lipid droplets can sequester transcription factors and chromatin components and generate the lipid ligands for certain nuclear receptors.

In addition to their role in lipid biochemistry, lipid droplets participate in protein degradation and glycosylation. Their metabolism can be manipulated by pathogenic viruses and bacteria such as Mycobacterium tuberculosis with unfortunate consequences for the host, but they also serve as reservoirs for proteins that fight intracellular pathogens. In consequence, such lipid droplets and their enzyme systems may be markers for disease states and are also considered to be targets for pharmaceutical intervention.

Brown Adipose Tissue

Most adipose tissue depots ('white fat') serve primarily as storage and endocrine organs that provide a reservoir of nutrients for release when the food supply is low. However, a second specialized form of adipose tissue, brown fat, is multilocular, highly vascularized and rich in mitochondria and the iron-containing pigments that transport oxygen and give the tissue its colour and name. Brown adipocytes arise from progenitor cells that are closer to those of skeletal muscle than white adipocytes. In humans, these depots tend to be located in specific anatomical regions such as subcutaneous areas around the neck, where its function may be to supply warm venous blood directly to the spinal cord and brain, and elsewhere to the heart, kidney, pancreas and liver. Brown fat is able to oxidize fat so rapidly that heat is generated ("non-shivering thermogenesis"), and it is especially important in young animals and in those recovering from hibernation.

In brief, during cold exposure, release of noradrenaline and stimulation of β-adrenergic receptors in the nervous system initiate a catabolic program that commences with rapid breakdown of cellular triacylglycerol stores and release of unesterified fatty acids and transient activation of a co-activator of peroxisome proliferator-activated receptor gamma (PPARγ). These set in motion a signalling process those results in the efficient β-oxidation of fatty acids to produce heat. The key molecule is believed to be the uncoupling protein-1 (UCP1), which acts as a valve to uncouple electron transport in the respiratory chain from ATP production with a highly exothermic release of chemical energy, i.e. as heat rather than as ATP. Although many aspects of the mechanism are uncertain, it is clear that proton conductance by UCP1 is highly regulated and inducible. It is activated by free long-chain fatty acids and inhibited by purine nucleotides, i.e. fatty acids are not only the substrate for thermogenesis, but act also as self-regulating second messengers. The mitochondrial phospholipid cardiolipin, which is intimately involved in oxidative phosphorylation, is indispensable for stimulating and sustaining the function of thermogenic fat. There is evidence also that acylcarnitines produced in the liver from fatty acids released from white adipose tissue in response to cold exposure are transported in plasma to brown adipose tissue and can serve as a substrate for thermogenesis. Upon activation of brown adipose tissue, the dense vasculature increases the delivery of fatty acids and glucose to the brown adipocytes and warms the blood passing through the tissue.

The activities of acyl-CoA synthetases and acyl-CoA thioesterases determine the availability of substrates for β-oxidation and consequently the thermogenic capacity. Synthesis of the lipokine octadecanoid 12,13-dihydroxy-9Z-octadecenoic acid (12,13-diHOME), is induced by cold also, and this stimulates the activity of brown adipose tissue by promoting the uptake of fatty acids. Similarly, there are suggestions that n-3 polyunsaturated fatty acids may promote adaptive thermogenesis, for example through the activity of the 12-lipoxygenase metabolite 12-hydroxyeicosapentaenoic acid (12-HEPE). Peroxisomal synthesis of plasmenyl-phospholipids is believed to regulate adipose tissue thermogenesis by mediating mitochondrial fission.

In hibernating mammals, brown adipose tissue is especially important metabolically, and even in laboratory animals such as mice, it can consume about 50% of dietary lipids and glucose when the animals are exposed to cold. Similarly, in even humans, the activity of brown adipose tissue is induced acutely by cold and is stimulated via the sympathetic nervous system, and the relevance of this tissue to human metabolism is now becoming apparent. For example, there are suggestions that brown adipose tissue can behave as an endocrine system to secrete endocrine factors ('batokines') that may be favourable towards cardiovascular risk. For obvious reasons, there are efforts to determine whether sustained activation of brown fat by pharmaceutical means could be beneficial towards a number of human disease states, including obesity, diabetes and cardiovascular disease.

Such research has been stimulated by the observation that clusters of distinct adipocytes with thermogenic capacity in addition to their storage function can be present in white adipose tissue and emerge in response to various physiological signals, especially reactive oxygen species. They are termed 'beige or brite' adipocytes and arise from multipotent preadipocytes. Brown adipose tissue is a discrete organ in animals, but beige adipose tissue is interspersed with white and the two forms have different developmental origins. In adult humans, most of the deposits once thought to be classical brown adipose tissue do not contain the genetic markers for this tissue and are now recognized to be beige/brite fat, although brown adipose tissue per se is present in significant amounts in human newborns and infants. However, beige adipocytes utilize the same machinery to release heat by oxidation of fatty acids under β-adrenergic stimulation.

Other Functions of Triacylglycerol Depots

Subcutaneous depots act as a cushion around joints and serve as insulation against cold in many terrestrial animals, as is obvious in the pig, which is surrounded by a layer of fat, and it is especially true for marine mammals such as seals. Those adipocytes embedded in the skin differ from the general subcutaneous depots and support the growth of hair follicles and regenerating skin, and they may also have a defensive role both as a physical barrier and by responding metabolically to bacterial infection.

In marine mammals and fish, the fat depots are less dense than water and so aid buoyancy with the result that less energy is expended in swimming. More surprisingly perhaps, triacylglycerols together with the structurally related glyceryl ether diesters and wax esters are the main components of the sonar lens used in echo-location by dolphins and toothed whales. The triacylglycerols are distinctive in that they contain two molecules of 3-methylbutyric (isovaleric) acid with one long-chain fatty acid. It appears that the relative concentrations of the various lipids in an organ in the head of the animals (termed the 'melon') are arranged anatomically in a three-dimensional topographical pattern to enable them to focus sound waves.

In cold climates, many insects do not feed over winter and must manage their energy stores to meet the energetic demands of development and reproduction in the spring. Some insect species that are tolerant of freezing produce triacylglycerols containing acetic acid, and these remain liquid at low temperatures; by interacting with water, they may play a role in cryoprotection.

Triacylglycerol Metabolism in Plants and Yeasts

1. Seed oils: Fruit and seed oils are major agricultural products with appreciable economic and nutritional value to humans. The mesocarp of fruits is a highly nutritious energy source that attracts animals that help to disperse the seeds, and in plants such as the oil palm and olive trees a high proportion of the fruit flesh contains triacylglycerols. Similarly in seeds, triacylglycerols are the main storage lipid and can comprise as much as 60% of their weight. Fruit lipids are not intended

for use by the plant per se and are stored in lipid droplets in large irregular structures that break down readily, but seed lipids are required for the development of the plant embryo so their metabolism is of particular importance. However, triacylglycerol biosynthesis and metabolism are required also for pollen viability and to maintain lipid homeostasis in chloroplasts.

In plants, fatty acids synthesised in the plastid compartment are stored in the embryo or endosperm tissues of seeds as triacylglycerols in lipid droplets with a coherent surface layer of proteins and lipids. In addition to the common range of fatty acids synthesised in plastids, some plant species produce novel fatty acids, including medium- and very-long-chain components and those with oxygenated and other functional moieties. Some very specific means of diverting these two seeds and triacylglycerol production must exist to prevent disruption of the plant membranes. Seed development occurs in three stages - rapid cell division with no accumulation of storage material, rapid deposition of triacylglycerols and other energy-rich metabolites, and finally desiccation.

During the period of oil accumulation in seeds, there must be a mechanism to hydrolyse the newly formed ACP esters of fatty acids and export the unesterified fatty acids to the endoplasmic reticulum, where they are converted to the CoA esters and triacylglycerols are synthesised by the Kennedy and other pathways described above. In yeast and plants, diacylglycerol esterification is the only committed step in triacylglycerol production and this occurs by mechanisms that can be both dependent or independent of acyl-CoA esters. The acyl-CoA-dependent route is catalysed by diacylglycerol:acyl-CoA acyltransferases (DGATs) with acyl-CoA and diacylglycerols as substrates. In plants, two membrane-bound enzymes (DGAT1 and DGAT2) and a cytosolic enzyme (DGAT3) are known, while the acyl-CoA-independent reaction utilizes a phospholipid:diacylglycerol acyltransferase with phospholipids as acyl donor and diacylglycerol as acyl acceptor to produce triacylglycerols and lysophospholipids. DGAT2 is especially important in those plant species with unusual fatty acid compositions.

In addition, a substantial proportion in some species is synthesised by a flux through the membrane phospholipid phosphatidylcholine, produced by both the eukaryotic and prokaryotic pathways with differing positional distributions (see our web-page on galactosyldiacylglycerols), in which diacylglycerols are generated by the action of a phosphatidate phosphatase as an intermediate. Direct transfer of one fatty acid from phosphatidylcholine to diacylglycerol by the action of the phospholipid:1,2-diacyl-sn-glycerol-acyltransferase (PDAT) enzyme also occurs, possibly to remove damaged or unusual membrane components. As phosphatidylcholine undergoes extensive remodelling and its fatty acid components are subject to modification, for example by desaturation to form linoleic and linolenic acids, the compositions and especially the positional distributions of triacylglycerols produced in this way can be very different from those synthesised by the 'classical' pathways. Phosphatidylcholine may also function as a carrier for the trafficking of acyl groups between organelles and membrane subdomains, and it has been suggested that an assembly of interacting enzymes may facilitate the transfer of polyunsaturated fatty acids from this phospholipid to triacylglycerols in seeds.

As triacylglycerol synthesis continues, oil droplets accumulate between the leaflets of the endoplasmic reticulum and are surrounded by a monolayer of phospholipids and proteins, which in Arabidopsis include oleosins, a caleosin, a steroleosin, a putative aquaporin and a glycosylphosphatidylinositol-anchored protein. These eventually "bud off" from the endoplasmic reticulum with their monolayer of phospholipids and proteins and are released into the cytosol by a yet-to-be defined mechanism.

At the onset of germination, water is absorbed and lipases are activated. The process of lipolysis begins at the surface of oil bodies, where the oleosins, which are the most abundant structural proteins, are believed to serve to assist the docking of lipases. They also control the size and stability of lipid droplets in seeds. A number of lipases have been cloned from various plant species and are typical α/β-hydrolases, with a conserved catalytic triad of Ser, His, and Asp or Glu as in patatin (an especially abundant lipolytic protein in potatoes), which are able to hydrolyse triacylglycerols but not phospho- or galactolipids. The most important of these is believed to be the 'sugar-dependent lipase 1 (SDP1)', which is a patatin-like lipase similar to the mammalian adipose triacylglycerol lipase discussed above, and is located on the surface of the oil body. This is active mainly against triacylglycerols to generate diacylglycerols, but presumably works in conjunction with di- and monoacylglycerol lipases to generate free fatty acids and glycerol.

The lipid droplets in seeds exist in close proximity with glyoxysomes (broadly equivalent to peroxisomes). These are the membrane-bound organelles that contain most of the enzymes required to oxidize fatty acids derived from the triacylglycerols via acetyl-CoA to four-carbon compounds, such as succinate, which are then converted to soluble sugars to provide germinating seeds with energy to fuel the growth of the seedlings and to produce shoots and leaves. In addition, they supply structural elements before the seedlings develop the capacity to photosynthesise. How the products of lipolysis are transported to the glyoxysomes for further metabolism has still to be determined, but a specific 'ABC' transporter is required to import fatty acids into the glyoxysomes in Arabidopsis. The free acids are converted to their coenzyme A esters by two long-chain acyl-CoA synthetases located on the inner face of the peroxisome membrane before entry into the β-oxidation pathway. All of these processes are controlled by an intricate regulatory network, involving transcription factors that crosstalk with signalling events from the seed maturation phase through to embryo development. After about two days of the germination process, the glycoxysomes begin to break down, but β-oxidation can continue in peroxisomes in leaf tissue.

2. Lipid droplets - plastoglobules: Triacylglycerol-rich lipid droplets (LD) have been observed in most cell types in vegetative tissues of plants as well as in seeds, and although their origin and function are poorly understood, they contain all the enzymes required for triacylglycerol metabolism together with phospholipases, lipoxygenases and other oxidative enzymes. Rather than oleosins, these lipid droplets in plants and algae contain a family of ubiquitously expressed 'LD-associated proteins' on the surface, together with a monolayer of phospholipids (mainly phosphatidylcholine), galactolipids such as sulfoquinovosyldiacylglycerol and in some species betaine lipids. As in yeast and humans, seipins (three in Arabidopsis) are necessary for normal LD biogenesis. Again, LD form within the bilayer of the endoplasmic reticulum and pinch off into the cytoplasm. Abiotic stresses can induce remodelling of lipid membranes through lipase action with the formation of toxic lipid intermediates, and these can be sequestered by triacylglycerols in lipid droplets to inhibit membrane damage and potentially prevent cell death. While they are believed to be involved mainly in stress responses, lipid droplets may have other specialized roles, for example in anther and pollen development, where triacylglycerols serve as a source of fatty acids for membrane biosynthesis. Fatty acids derived from triacylglycerols in lipid droplets are believed to be subjected to peroxisomal β-oxidation to produce the ATP required for stomatal opening and no doubt many other purposes.

In addition, lipid droplets that have been termed 'plastoglobules' are produced by localized accumulation of triacylglycerols and other neutral lipids between the membrane leaflets of the thylakoid cisternae and then pinch off into the stroma, where they are involved in a wide range of biological functions from biogenesis to senescence via the recruitment of specific proteins. During senescence for example, lipid droplets accumulate rapidly in leaves of A. thaliana. In reproductive tissues, may have a more direct function by recruiting and transporting proteins, both for organ formation and successful pollination. Antifungal compounds such as 2-hydroxy-octadecatrienoic acid and other oxylipins are produced from α-linolenic acid in these organelles, and it has been suggested that the latter function as intracellular factories to produce stable metabolites via unstable intermediates by concentrating the enzymes and hydrophobic substrates in an efficient manner. Plastoglobules are also implicated in the biosynthesis and metabolism of vitamins E and K.

3. Microalgae and yeasts: Triacylglycerol metabolism in lipid droplets in microalgae is under intensive study because of their potential for biodiesel production. It seems that similar processes occur as in higher plants, but with a simpler genome encoding few redundant proteins. Lipid droplets in yeast are considered to be a highly dynamic and functionally diverse hub that ensures stress resistance and cell survival by promoting membrane and organelle homeostasis. As most of the important biosynthetic and catabolic enzymes involved in triacylglycerol metabolism are conserved between yeasts and mammals, the former are proving to be useful models for the study of triacylglycerol production. The size and triacylglycerol content of lipid droplets in yeasts change appreciably in different stages of growth and development, and Saccharomyces cerevisiae contains a single phosphatidic acid phosphatase (Pah1), which has an essential role in this process. During vegetative growth, Pah1 in the cytosol is phosphorylated by multiple protein kinases, and this enables the synthesis of phospholipids rather than triacylglycerols. As cells progress into stasis, the Pah1 is dephosphorylated and translocates to the endoplasmic reticulum, which ultimately leads to triacylglycerol synthesis for storage in lipid droplets. Some fatty acids derived from phospholipids are utilized for triacylglycerol biosynthesis at the inner nuclear membrane, and this is important for nuclear integrity.

Triacylglycerol Metabolism in Prokaryotes

Study of the biosynthesis of triacylglycerols in bacteria has been stimulated by an awareness of the role of this lipid class in the pathogenesis of Mycobacterium tuberculosis and the relationship with antibiotic biosynthesis by Streptomyces coelicolor. For example, triacylglycerols are believed to be an energy reserve for the long-term survival of M. tuberculosis during the persistence phase of infection as well as a means by which unesterified fatty acids are detoxified. Increasing numbers of bacterial species, for example from the genera Mycobacterium, Nocardia, Rhodococcus, Micromonospora, Dietzia and Gordonia, are now known to produce triacylglycerols (sometimes wax esters), and these can be stored in lipid droplets in the organisms. The first three steps in triacylglycerol biosynthesis are catalysed by GPAT, LPAT and PAP enzymes comparable to those in other organisms. However, it has become apparent that the DGAT can be a dual-function CoA-dependent acyltransferase known as wax ester synthase/diacylglycerol acyltransferase, which accepts a broad diversity of acyl-CoA substrates for esterification of diacylglycerols or long-chain fatty alcohols for the synthesis of triacylglycerols or wax esters, respectively, depending on which intermediates are present in the organisms. Bacteria that lack such an enzyme are unable to produce these non-polar lipids.

Biosynthesis of Phospholipids

Phospholipids are produced by the esterification of an alcohol (serine, ethanolamine and choline) to the phosphate of 1, 2 diacylglycerol 3 phosphate (phosphatidic acid). The alcohols contain nitrogen except glycerol and inositol and the fatty acid on C1 is saturated and on C2 is unsaturated of the glycerol backbone. The main classifications of phospholipids are: Phosphatidylcholine (PC), Phosphatidylethanolamine (PE), Phosphatidylserine (PS), Phosphatidylinositol (PI), Phosphatidylglycerol (PG), Cardiolipin (diphosphatidylglycerol, DPG). There are two known mechanism for the synthesis of phospholipids.

- CDP activated polar head group for the attachment to the phosphate group of phosphatidic acid.
- CDP activated 1, 2 diacylglycerol and an inactivated polar head group.

Phosphatidylcholine (PC)

Also known as lecithins, PC are neutral zwitterions at physiological pH. At C1 position, PC contain either stearic or palmitic acid and at C2 position linolenic or linoleic acid is present. Dipalmitoyllecithin is the chief (80%) phospholipid of pulmonary alveoli that hold palmitate at both C1 and C2 of glycerol. The synthesis occurs in the following steps:

- Activation of choline by phosphorylation followed by coupling to CDP preceding its attachment to the phosphatidic acid.
- Totting up of choline to CDP activated 1, 2 diacylglycerol.
- Conversion of either PE or PS to PC. The switching of PS to PC first entails decarboxylation of PS to yield PE followed by a series of methylation reactions employing S-adenosylmethionine (SAM) as the methyl group donor.

Phosphatidylethanolamine (PE)

Like PC, at physiological pH, PE molecules are also neutral zwitterions. At C1 position, PE have either stearic or palmitic and at C2 position, a long chain unsaturated fatty acid (22:6, 20:4 and 18:2) is present. The PE synthesis crop up by two pathways:

- Activation of ethanolamine by phosphorylation followed by coupling to CDP which is further transmitted from CDP ethanolamine to phosphatidic acid yielding PE.
- Decarboxylation of PS to yield PE.

Phosphatidylserines (PS)

At physiological pH, PS carries a net charge of −1 and is poised of fatty acids alike to PE. The PS biosynthetic pathway engrosses a substitution of serine for ethanolamine in PE. This replacement arises when PE is in the membrane lipid bilayer. In addition, PS via decarboxylation reaction also dish up as a source of PE.

Phosphatidylinositol (PI)

At C1 position, PI exclusively contains stearic acid and at C2 position arachidonic acid is present. At physiological pH, PIs utterly compiled of inositol (non-phosphorylated) demonstrated a net charge of −1. PIs stay alive in membranes with an assortment of phosphate level esterified to the hydroxyl group of inositol.

Phosphorylated inositol containing molecules are expressed as polyphosphoinositides (vital intracellular signal transducers originated from the plasma membrane). The PI synthesis rivets CDP activated 1, 2 diacylglycerol condensation with myo inositol followed by the phosphorylations of the hydroxyl group of inositol escorting to the polyphosphoinositides production. Phosphatidylinositol 4, 5 bisphosphate (PIP_2) is an essentially imperative phospholipid of the membrane involved in the transmission signals of the cell growth and differentiation.

Phosphatidylglycerols (PG)

At physiological pH, PG displays a net charge of −1 and present in the mitochondrial membrane in high amount as a constituent of pulmonary surfactant. PG is also a well-known predecessor for cardiolipin synthesis and itself synthesized from CDP diacylglycerol and Glycerol 3 Phosphate. In addition to the above facts, an imperative role of PG is to dole out as the precursor for diphosphatidylglycerols (DPGs) synthesis.

Diphosphatidylglycerols (DPG)

At physiological pH, DPG show a net charge of −2 and are very acidic in nature. DPG is present as components of pulmonary surfactant principally in the inner mitochondrial membrane. Cardiolipins are one of the key class of DPGs synthesized by the condensation of CDP diacylglycerol with PG.

At C1 and C2 position of glycerol, fatty acid allocation within phospholipids is recurrently in flux due to the degradation and the incessant remodeling of phospholipid that transpires within the membranes. Phospholipases degrades phospholipid and there are different phospholipases that demonstrate substrate specificities for phospholipids at different positions. In most of the cases, the acyl group transferred to glycerol by Acyl transferases, is different from that of present in phospholipid when it inhabits within a membrane. In phospholipids, the acyl group remodeling is the consequence of phospholipase A_1 (PLA_1) and phospholipase A_2 (PLA_2) action. The resultant products are called Lysophospholipids and serve as substrates for Acyl transferases employing different Acyl CoA groups. In an swapping reaction catalyzed by lysolecithin: lecithin acyltransferase (LLAT), lysophospholipids also accepts Acyl groups from other phospholipids.

PLA_2, an important class of enzyme (30 enzymes reported), acts on the C2 position of membrane phospholipids and releases arachidonic acid which serve as a substrate for eicosanoids biosynthesis. PLA_2 exist in ten isozymic forms in the secretory pathway and abbreviated as $sPLA_2$. $sPLA_2$ are Ca^{2+} -dependent, low molecular weight proteins involved in various processes such as inflammation, host defense and modification of eicosanoid generation. Another family phospholipases, the $cPLA_2$ (cytosolic PLA_2 family) embraces 3 isozymes with $cPLA_2$-α being an indispensable factor of the initiation of arachidonic acid metabolism. Similar to that of $sPLA_2$ enzymes, the $cPLA_2$

are tightly regulated by Calcium ions and phosphorylation processes. Two PLA_2 isozymes (iPLA2) are also recognized as Ca^{2+} -independent for activity and are mainly involved with phospholipid remodeling. PAF, another class of PLA_2 enzymes with four members, are being involved in the inactivation and hydrolysis of platelet activating factor. Like $iPLA_2$ family, PAF hydrolyzing PLA_2 isozymes are Ca^{2+} - independent and the unique activity was called PAF acetylhydrolase (PAF AH). They also oxidized phospholipids and found to be allied with lipoprotein particles as lipoprotein associated PLA_2 (Lp PLA_2) family.

Plasmalogens

They are glycerol ether phospholipids with two basic types namely alkenyl ether (–O–CH=CH–) and alkyl ether ($–O–CH_2–$). DHAP serves as the precursor for its synthesis and 3 most important classes of plasmalogens have been acknowledged: Serine, Ethanolamine and Choline plasmalogens. In cardiac tissue, choline plasmalogen is abundant and 1-O-1'-enyl-2-acetyl-sn-glycero-3-phosphocholine (platelet activating factor, PAF) has been recognized as an enormously powerful biological mediator, accomplished of stirring up cellular responses at low concentrations ($< 10^{-11}M$). PAF functions as a moderator of anaphylactic shock, acute inflammatory reactions and hypersensitivity. In retort to antigen-IgE complexes formation, PAF is synthesized on the surfaces of basophils, eosinophils, macrophages, monocytes and neutrophils. In cells, the PAF synthesis and discharge leads to platelet aggregation and the liberation of serotonin. PAF also generates response in heart, liver, lung, smooth muscle and uterine tissues. In myelin tissue, ethanolamine plasmalogen is ubiquitous.

References

- Cholesterol-biosynthesis: ijs.si, Retrieved 03, March 2020
- The-lipid-web: lipidmaps.org, Retrieved 19, July 2020

4

Lipid Oxidation

The oxidative degradation of lipids is termed as lipid oxidation. It is a highly complex set of reactions occurring between fatty acids and oxygen resulting in the oxidative degradation of lipids. The process takes place by a free radical chain reaction mechanism. This chapter closely examines the key concepts of lipid oxidation to provide an extensive understanding of the subject.

'Lipid oxidation' is a term explaining different types of reactions, having both positive and negative implications on the human body. In the body lipid oxidation is important for several physiological reactions, for instance when utilizing fatty acids for the production of energy through β-oxidation. Oxidation is also involved in the production of signalling substances called eicosanoids. These are formed from the omega-3 fatty acid eicosapentaenoic acid (EPA) and the omega-6 fatty acid arachidonic acid (AA) by the action of specific enzyme systems. Lipid oxidation could also refer to uncontrolled oxidative degradation of lipids initiated by free radicals stealing electrons, which is the first step in the formation of several cytotoxic and mutagenic substances in the body. Uncontrolled oxidative damage also affects food products, influencing the overall quality.

The main lipids of the body are basically triglycerides (neutral fats), the phospholipids and the sterols which are the principal and most efficient source of stored energy because of high fuel value (9.3 calorie/g). These fat reserves are constantly broken down and re-synthesized. An optimal nutrition must contain a certain amount of fat along with essential fatty acids. Previously carbohydrates were attributed for major source of energy but increasing evidences had established the fact that fat is the chief source of energy for many tissues and perhaps the main source of energy for the body as a whole.

Lipids: Digestion, Absorption and Secretion

Daily intake of lipids is approximately 80g of which more than 90% is normally triacylglycerol (TAG) or triglyceride while remaining consists of cholesterol, its esters, phospholipids and unsterilized (free) fatty acid.

Digestion and Emulsification

The lipids digestion begins in the stomach by lingual lipase (acid-stable) present on the back of the tongue and gastric lipase present in the gastric mucosa. The lipases mainly digest TAG with short or medium chain length (≤ C12) as present in milk. The emulsification of lipids crop up in the duodenum where the surface area of the hydrophobic lipid molecules increases for effective digestion by enzymes via two modes i.e. through the mechanical mixing and detergent property of bile salts resulting from peristalsis. In liver, bile salts are synthesized and stored in the gall bladder are actually imitative of cholesterol. Emulsification actually stabilizes the smaller individual particles and prevents them from coalescing.

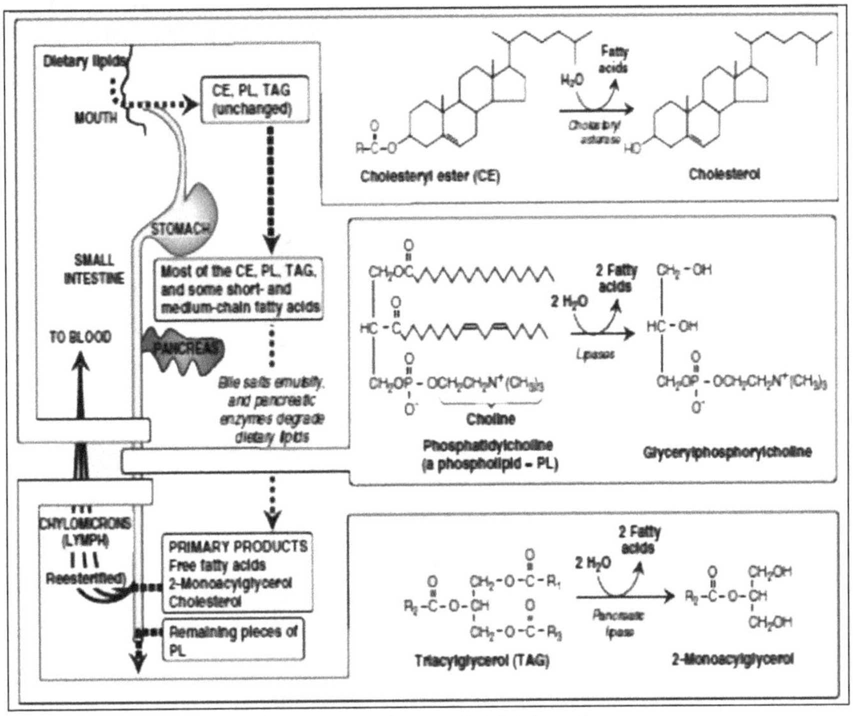

Figure: Overview of Lipid Digestion.

Dilapidation of Lipids

The dietary lipids are degraded by enzymes of pancreas and the secretions of these pancreatic enzymes are hormonally regulated.

TAG

TAG are large enough to be degraded by mucosal cells of intestinal villi, thus they are first converted to monoacylglycerol and free fatty acid by esterase and pancreatic lipase.

Cholesteryl Ester

Overall dietary cholesterol is present mostly in free non-esterified form while only 10-15% occurs in esterified form. Cholesteryl ester is degraded by pancreatic cholesteryl ester hydrolase (or known as cholesterol esterase) which produces free fatty acid and cholesterol. In presence of bile salts, the enzyme activity gets increased.

Phospholipids

Phospholipids are acted upon by activated pancreatic phospholipase A_2 and ultimately forming fatty acid and lysophospholipid. The lysophospholipid is acted upon by lysophospholipase leaving a glyceryl phosphoryl base that may be excreted in the feces or it can be further degraded or absorbed by the body.

Hormonal Control

Cholecystokinin (CCK) which is a small peptide hormone is secreted by mucosal cells of lower duodenum and jejunum. It acts on gallbladder, exocrine cells of pancreas and small intestine. Another hormone, secretin produced by intestinal cells helps in balancing pH for optimum activity.

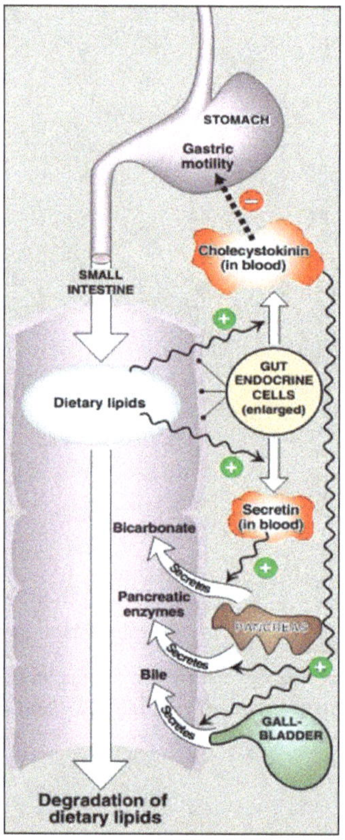

Figure: Hormonal Control of Lipid Digestion in the Small Intestine.

Absorption

Absorption of lipids is done mainly by enterocytes (mucosal cells of intestine). Before absorption, these lipids are ultimately breakdown in jejunum into 2-monoacylglycerol, cholesterol and free fatty acids which are the key products. After that micelles are formed in the intestinal lumen and then these lipid droplets are absorbed in the brush border membrane of the enterocytes. In enterocytes, re-synthesis of TAG and cholesteryl esters occurs.

Secretion

The newly re-synthesized cholesteryl esters and TAGs are incredibly hydrophobic in nature therefore it is obligatory to package it as lipid droplet surrounded by thin layer of phospholipid, Apolipoprotein-B and unesterified cholesterol. This helps in stabilization and increase solubility.

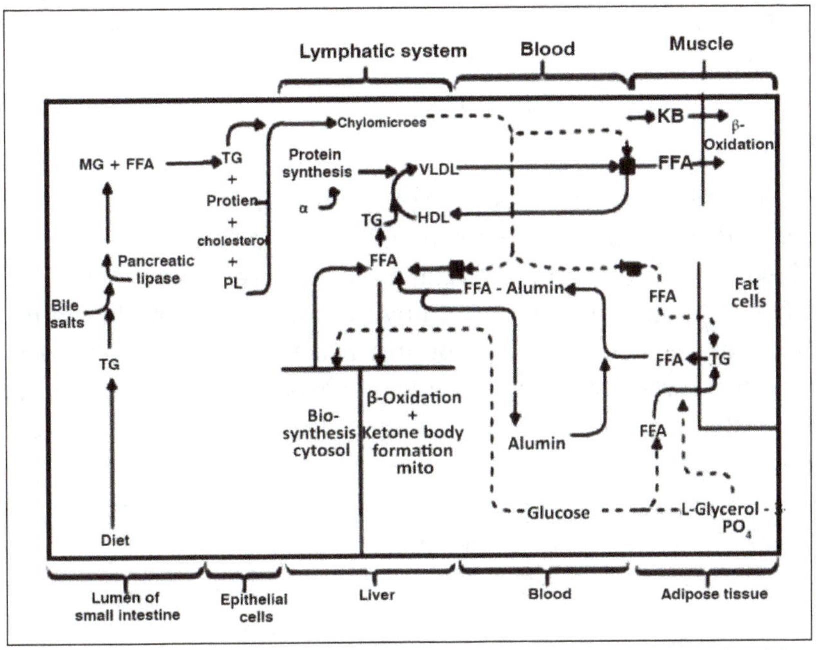

Figure: Utilization of Lipids in Different Compartments of Animals. TG: triacylglycerol, MG: Monoacylglycerol, FFA: Free fatty acid, PI: Phospholipids, KB: Ketone bodies, VLDL: Very low density lipoproteins, HDL: High density lipoproteins.

These particles known as chylomicrons are released from enterocytes into the lacteals (lymphatic system present in small intestine villi).

Use of Released Chylomicrons by Tissues

In adipose tissue and skeletal muscle mainly, but also in kidney, heart, liver and lung; these chylomicrons are broken down to TAG which are further degraded into free fatty acid and glycerol by lipoprotein lipase. The paucity of lipoprotein lipase or its coenzyme apolipoprotein-C II also known as 'Familial lipoprotein lipase' deficiency or 'Type I hyper-lipoproteinemia' is an exceptional autosomal recessive disorder.

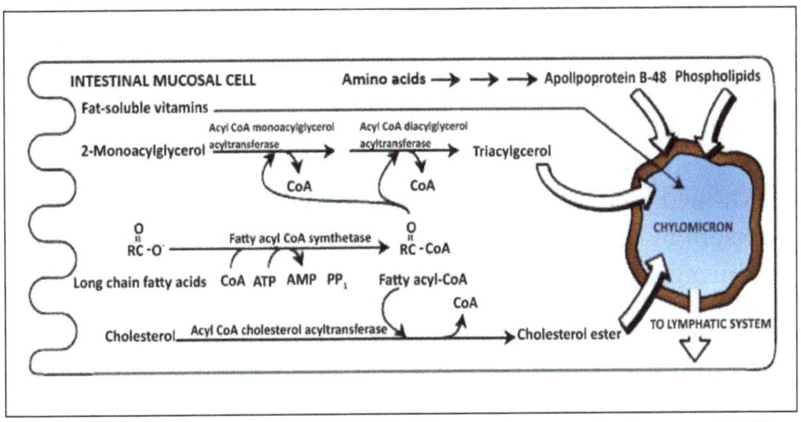

Figure: Chylomicrons: Assembly and Secretion by Intestinal Mucosal Cells.

Destiny of Free Fatty Acid

The free fatty acids may enter adipocytes or muscle cells or may be transported in blood in conjugation with serum albumin until taken up by peripheral tissues. Adipocytes can again produce TAG from free fatty acids which are stored until needed. In adipocytes, hormone-sensitive lipases hydrolyze TAGs and release the fatty acids (free fatty acids) into the blood, wherever they bind with serum albumin.

Albumin (Mr 66,000) which is more or less half of the total serum protein binds non-covalently to as much as 10 fatty acids per monomer of protein. The inexplicable fatty acids have been passed to tissues such as heart, renal cortex and skeletal muscle. To serve as a fuel in these target tissues, fatty acids secede from albumin and are moved by the transporter of plasma membrane into the cells. Numerous proteins other than serum albumin also act as transporters to carry the absorbed TAGs from intestine to different organs and tissues such as liver, heart muscle and adipose tissues via blood. These proteins wrap around the lipid molecule to facilitate their movement across the membranes in cells.

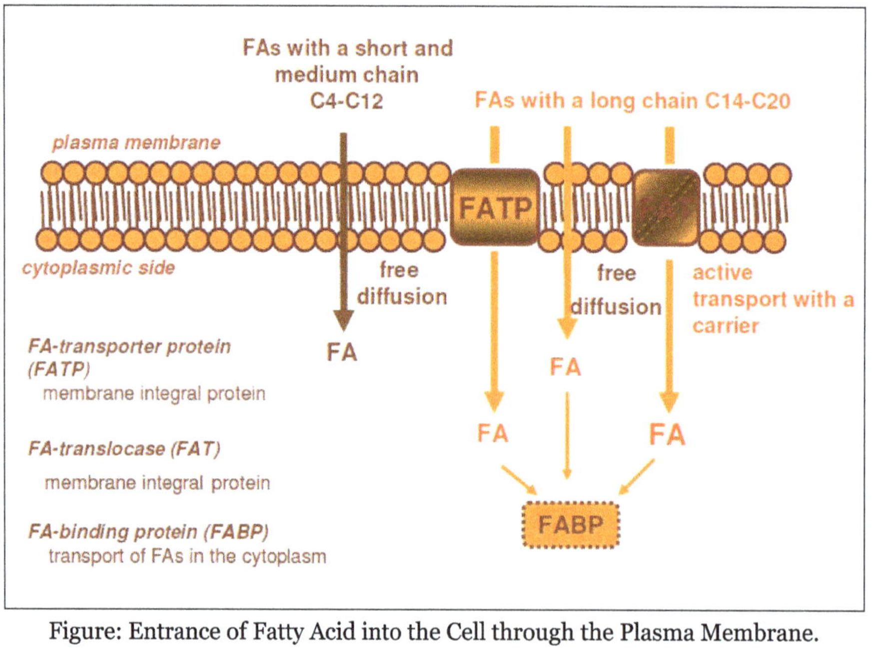

Figure: Entrance of Fatty Acid into the Cell through the Plasma Membrane.

The fatty acid binding proteins (FABPs) present in the cytosol of diverse animal tissues have low molecular weight (14-15 kDa). In the cytosolic compartment, it functions as carrier of fatty acids. FABPs may also be implicated in the intracellular storage, the delivery of fatty acids and/or cellular uptake of fatty acids to the sites of their consumption. In fatty acid metabolism, the significance of FABPs is sustained by the reduction in exploitation and uptake of long chain fatty acids in knockout mice deficient in heart FABE. These animals develop cardiac hypertrophy and exhibit exercise intolerance during aging. Nevertheless, the molecular mechanism of FABP function remains to be explicated.

Destiny of Glycerol

Used exclusively by liver to produce glycerol-3-phosphate entering glycolysis or gluconeogenesis.

About 95% of the biologically accessible energy of TAGs resides in their long chain fatty acids and merely 5% is supplied by the glycerol moiety. The released glycerol obtained by the action of lipase is phosphorylated by glycerol kinase and the ensuing glycerol 3-phosphate (G3P) is oxidized to dihydroxyacetone phosphate (DHAP). The triose phosphate isomerase, which is a glycolytic enzyme converts G3P to glyceraldehyde 3-phosphate which is auxiliary oxidized through the glycolysis pathway.

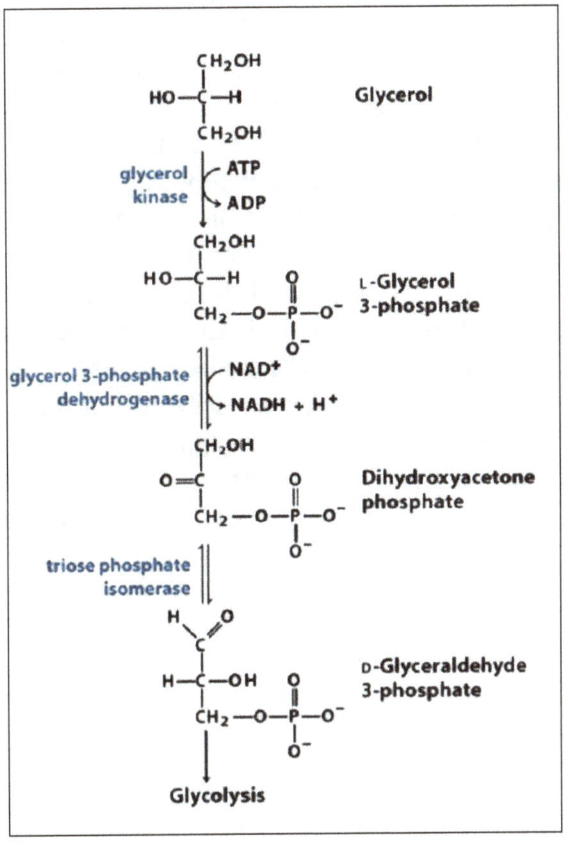

Destiny of Remaining Portion of Chylomicron

After maximum TAGs had been removed, the remaining portion (containing cholesteryl esters, phospholipids, fat soluble vitamins, apolipoprotein and few TAG) present on liver cells binds to the receptors and are endocytosized where they are degraded. If this endocytosis is decreased then they accumulate in the plasma and the condition is clinically known as 'Type III hyper-lipoproteinemia' or 'Familial betalipoproteinemia'.

Oxidation of Fatty Acids

The lipids of metabolic significance in the mammalian organisms include triacylglycerols (= triglycerides, neutral fats), phospholipids and steroids, together with products of their metabolism such as long-chain fatty acids, glycerol and ketone bodies. An overview of their metabolic interrelationships and their relationship to carbohydrate metabolism is depicted.

At least 10 to 20% of the body weight of a normal animal is due to the presence of lipids, a major part of which is in the form of triglycerides which are uncharged esters of glycerol. Body lipids are distributed in varying amounts in all organs and stored in highly specialized connective tissues called depot. In these depots, a large part of the cytoplasm of the cell is replaced by droplets of lipids. Body lipids serve as an important source of chemical potential energy.

Fats (or triacylglycerols) are highly concentrated stores of metabolic energy. They are the best heat producers of the three chief classes of foodstuffs. Carbohydrates and proteins each yield 4.1 kilocalories (4.1 kcal) of heat for every gram oxidized in the body ; whereas fats yield 9.3 kcal more than twice as much. The basis of this large difference in caloric yield is that fats contain relatively more carbon and hydrogen in relation to oxygen as compared to proteins or carbohydrates. In other words, fats are compounds that are less completely oxidized to begin with and therefore can be oxidized further and yield more energy. Furthermore, triacylglycerols are very nonpolar and so they are stored in a nearly anhydrous form, whereas carbohydrates and proteins are much more polar and more highly hydrated. In fact, a gram of dry glycogen (a carbohydrate) binds about 2 grams of water. Consequently, a gram of nearly anhydrous fat stores more than 6 times as much energy as a gram of hydrated glycogen, which is the reason that triacylglycerols, rather than glycogen, were selected in evolution as the major energy reservoir. A normal man, weighing 70 kg, possesses fuel reserves of 10,000 kcal in triacylglycerols, 25,000 kcal in proteins (mostly in muscles), 600 kcal in glycogen and 40 kcal in glucose. As mentioned, triacylglycerols constitute about 11 kg of his total body weight. If this amount of energy were stored in glycogen, his total body weight would be 55 kg greater.

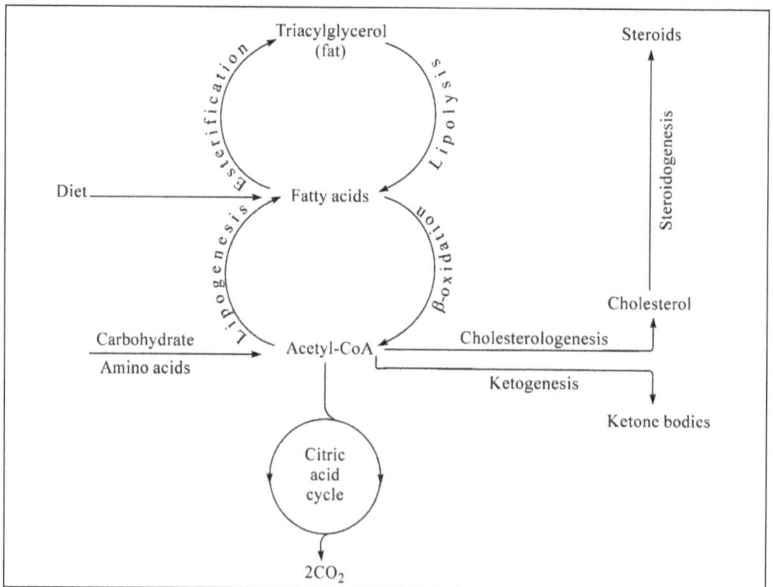

Figure: An overview of the principal pathways of lipid metabolism.

The normal animal contains a greater quantity of easily mobilized lipids than of carbohydrates or proteins. About 100 times more energy is stored as mobilizable lipids than as mobilizable carbohydrate in the normal human being. In times of caloric insufficiencies, an animal can meet the endogenous requirements necessary for the maintenance of life by drawing on its lipid depots. In addition, neutral lipids serve as insulators of delicate internal organs of the body. This function is best exemplified in marine animals, whose water environment is both colder than body temperature and a far better thermal conductor than air. Lipids also serve as shock absorbers in protecting joints, nerves and other organs against mechanical trauma.

An overview of the intermediary metabolism with special emphasis on fatty acids and triglycerides is given in figure below:

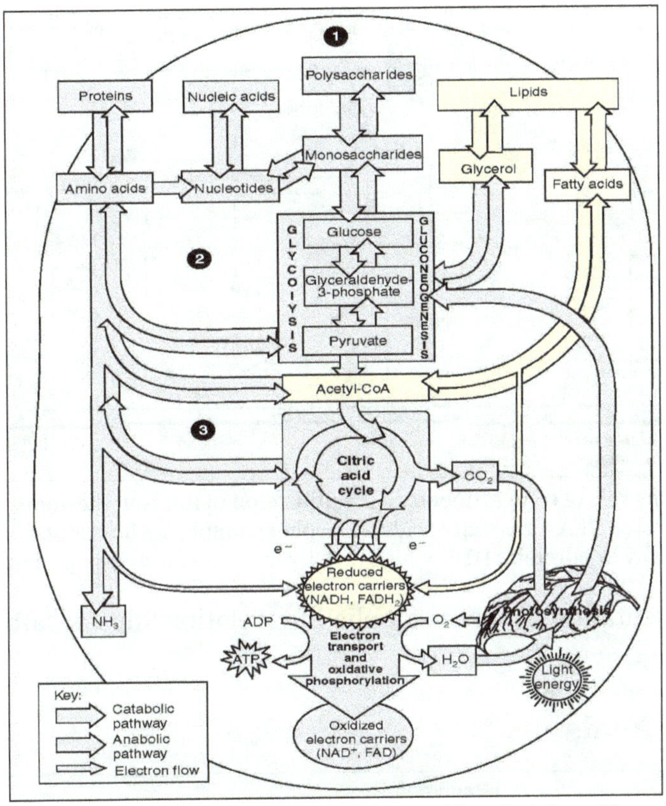

Figure: An overview of intermediary metabolism with fatty acid and triglyceride pathways highlighted.

When food intake exceeds caloric utilization, the excess energy is invariably stored as fat for the body cannot store any other form of food in such large amounts. The capacity of the animal to store carbohydrates (such as glycogen) is strictly limited, and there is no provision for the storage of excess proteins. Moreover, in an adult organism in which active growth has ceased, nitrogen output is more or less geared to nitrogen intake, and the organism shows no tendency to store surplus proteins from the diet. Plants differ from animals in that the energy reserves needed for reproduction are stored in the form of carbohydrates (as in corn or wheat) or as a combination of reserve proteins and oils (as in oil seeds, flax seed, safflower seed or sunflower seed).

In mammals, the major site of accumulation of triacylglycerols is the cytoplasm of adipose cells (= fat cells). Droplets of triacylglycerol coalesce to form a large globule, which may occupy most of the cell volume. Adipose cells are specialized for the synthesis and storage of triacylglycerols and for their mobilization into fuel molecules that are transported to other tissues by the blood. More than 99% of the lipid of human adipose tissue is triacylglycerol, regardless of anatomical location. In general, depot lipid is richer in saturated fatty acids than liver lipid. The more nearly saturated a sample of lipid, the higher the energy yield available from oxidation.

Presents, schematically, the flow of lipids in the body. Three important compartments are the liver, blood and adipose tissue. Both liver and adipose tissue are the principal sites of metabolic activity while the blood serves as a transport system. Other compartments, such as cardiac and skeletal muscle, are important utilizers of fatty acids and ketone bodies.

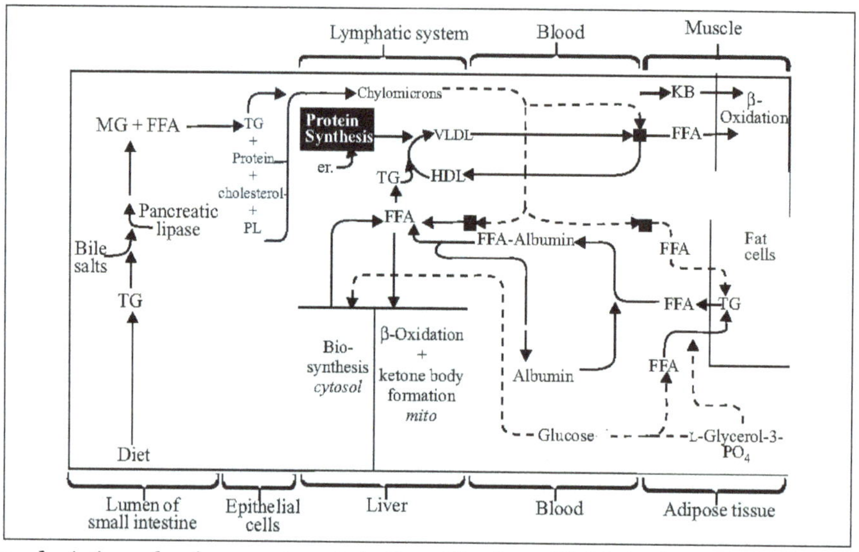

Figure: Scheme depicting role of compartments in the utilization of lipids in the animals TG = triacylglycerol, MG = monoacylglycerol, FFA = free fatty acids, PL = phospholipids, KB = ketone bodies, VLDL = very low-density lipoprotein, HDL = high density lipoprotein, ■= lipoprotein lipase.

A diagrammatic representation of the metabolic interrelationships of fatty acids is presented in figure below.

Oxidation of Fatty Acids

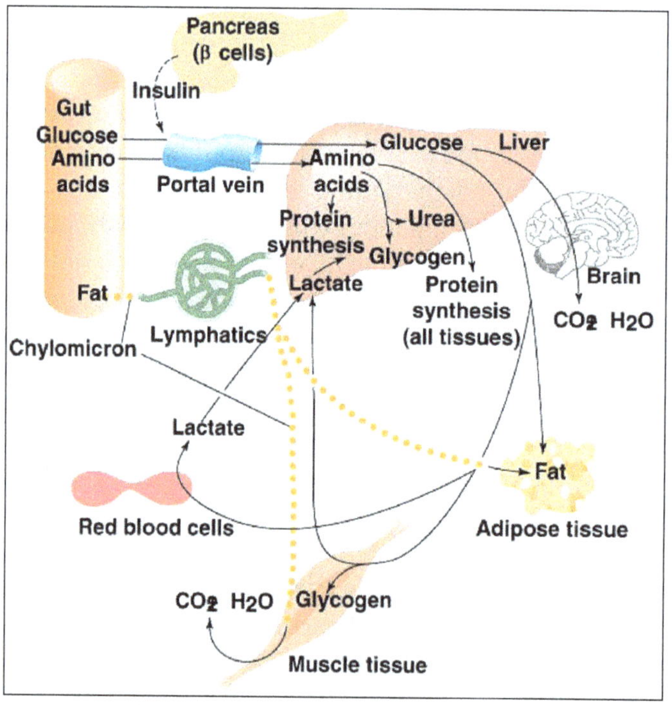

Figure: Metabolic interrelationships of fatty acids in the human.

The importance of oxidation of fatty acids is not limited to the obese or to devotees of greasy foods ; it is a critical part of the metabolic economy in the lean as well as the lardy. The oxidation of long-chain fatty acids to acetyl-CoA is a central energy-yielding pathway in animals, many protists and some bacteria. Complete combustion or oxidaton of a typical fatty acid, palmitic acid, yields 2,380 kcal per mole.

$$CH_3{-}(CH_2)_{14}{-}COOH + 23O_2 \rightarrow\rightarrow\rightarrow\rightarrow 16CO_2 + 16H_2O + 2380 \text{ kcal/mole}$$

In some organisms, acetyl-CoA produced by fatty acid oxidation has alternative fates. In vertebrate animals, acetyl-CoA may be converted in the liver into ketone bodies, which are water-soluble fuels exported to the brain and other tissues when glucose becomes unavailable. In higher plants, acetyl-CoA from fatty acid oxidation serves primarily as a biosynthetic precursor and only secondarily as fuel. Although the biological role of fatty acid oxidation differs from one organism to another, the mechanism is essentially the same.

In 1904, Franz Knoop elucidated the mechanism of fatty acid oxidation. He fed dogs straightchain fatty acid in which the ω-carbon atom was joined to a phenyl group. Knoop found that the urine of these dogs contained a derivative of phenylacetic acid when they were fed phyenylbutyrate. In contrast, a derivative of benzoic acid was formed when they were fed phenylpropionate. In fact, benzoic acid was formed whenever a fatty acid containing an odd number of carbon atoms was fed, whereas phenylacetic acid was produced whenever a fatty acid containing an even number of carbon atoms was fed. Knoop deduced from these findings that fatty acids are degraded by oxidation at the β-carbon. In other words, the fatty acids are degraded in two-carbon units and that the obvious two-carbon unit is acetic acid. This finding later came to be known as Knoop's hypothesis. These experiments are a landmark in Biochemistry because they were the first to use a synthetic level to elucidate synthetic mechanisms. Deuterium and radioisotopes came into Biochemistry several decades later.

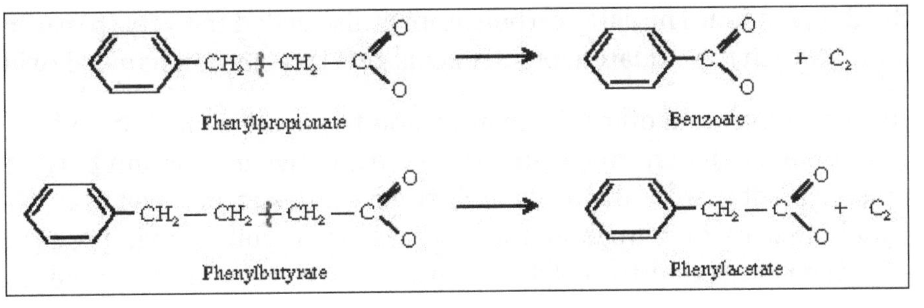

The complete combustion of fatty acids to CO_2 and H_2O occurs in the mitochondria, where the transfer of electrons from the fatty acids to oxygen can be used to generate ATP. The combustion occurs in 2 stages:

- The fatty acid is sequentially oxidized so as to convert all of its carbons to acetyl-coenzyme A.
- The acetyl-coenzyme A is oxidized by the reactions of the citric acid cycle. Both stages generate ATP by oxidative phosphorylation.

Activation of a Fatty Acid

Because of their hydrophobicity and extreme insolubility in water, triacylglycerols are segregated into lipid droplets, which do not raise osmolarity of the cytosol and, unlike polysaccharides, do not contain extra weight as water of solvation. The relative chemical inertness of triacylglycerols allows their extracellular storage in large quantities without the risk of undesired chemical reactions with other cellular constituents.

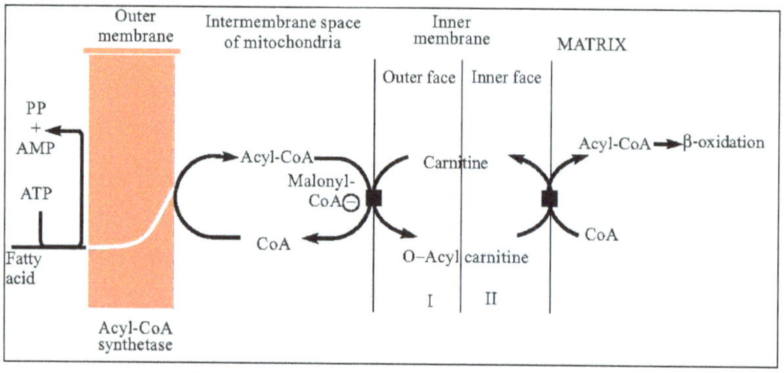

Figure: Transport mechanism of fatty acids from cytosol to the β-oxidation site in the mitochondrion.

Carnitine : acyl-CoA transferase I (outer face) and carnitine : acyl-CoA transferase II (inner face), the two distinct enzymes that catalyze the same reaction; malonyl - CoA ⊖ indicates inhibition of transferase I.

But the same properties that make triacylglycerols good storage compounds present problems in their role as fuels. Because of their insolubility in water, the ingested triacylglycerols must be emulsified before they can be digested by water-soluble enzymes in the intestine, and triacylglycerols absorbed in the intestine must be carried in the blood by proteins that counteract their insolubility. The relative stability of the C—C bonds in a fatty acid is overcome by activation of the carboxyl group at C-1 by attachment to coenzyme A, which allows stepwise oxidation of the fatty acyl group at the C-3 position. This later carbon atom is also called the beta (β) carbon in common nomenclature, from which the oxidation of fatty acids gets its common name: β oxidation.

An unusual property of liver and other tissue mitochondria is their inability to oxidize fatty acids or fatty acyl-CoA's unless (–)- carnitine (3-hydroxy-4-trimethyl ammonium butyrate) is added in catalytic amounts. Evidently, free fatty acids or fatty acyl-CoA's cannot penetrate the inner membranes of liver and other tissue mitochondria, whereas acyl carnitine readily passes through the membrane and is then converted to acetyl-CoA in the matrix. Fig. outlines the translocation of acetyl-CoA from outside the mitochondrion to the internal site of the β oxidation system. The key enzyme is carnitine acetyl-CoA transferase.

Reactions of Fatty Acid Oxidation

The free fatty acids that enter the cytosol from the blood cannot pass directly through the mitochondrial membranes, but must first undergo a series of 3 enzymatic reactions. These are described as under:

First Reaction: It is catalyzed by a series of family of isozymes present in the outer mitochondrial membrane, acyl-CoA synthetases (also called fatty acid thiokinases), which promote the general reaction.

$$\text{Fatty acid} + \text{CoA} + \text{ATP} \underset{\Delta G' = \sim 0\,\text{Kcal/mole}}{\rightleftharpoons} \text{Fatty acyl} - \text{CoA} + \text{AMP} + \text{PPi}$$

Three different acyl-CoA synthetases occur in the cell and act on fatty acids of short, intermediate and long carbon chains, respectively. One type of synthetase activates acetate and propionate to corresponding thioesters, another activates medium chain fatty acids from C_4 to C_{11}, and the third activates fatty acids from C_{10} to C_{20}. Acetyl-CoA synthetase catalyzes the formation of thioester linkage between the fatty acid carboxyl group and the thiol (—SH) group of coenzyme A to yield a fatty acyl-CoA; simultaneously, ATP undergoes cleavage to AMP and PPi. The reaction actually takes place in 2 steps:

$$\text{Fatty acid} + \text{ATP} + \text{Enzyme} \rightleftharpoons \text{Enzyme} - \text{acyllladeny late} + \text{PPi}$$

$$\text{Enzyme-acyladney late} + \text{CoA-SH} \rightleftharpoons \text{Enzyme} + \text{Fatty1} - \text{acyl} - \text{S} - \text{CoA} + \text{AMP}$$

Fatty acyl-CoAs, like acetyl-CoA, are high-energy compounds ; their hydrolysis to free fatty acid and CoA has a large negative standard free-energy change ($\Delta G^{o\prime} \approx -31$ kJ/mol). The formation of fatty acyl-CoAs is made more favourable by the hydrolysis of 2 high energy bonds in ATP; the pyrophosphate formed in the activation reaction is immediately hydrolyzed by a second enzyme, inorganic pyrophosphatase, which pulls the preceding activation reaction in the direction of the formation of fatty acyl-CoA. The overall reaction is:

$$\text{Fatty acid} + \text{CoA} + \text{ATP} \rightarrow \text{Fatty acyl} - \text{CoA} + \text{AMP} + 2\text{P}_i$$

$$\Delta G^{o\prime} = -32.5\,\text{kJ / mol}$$

Second Reaction: Fatty acyl-CoA esters, formed in the outer mitochondrial membrane, do not cross the inner mitochondrial membrane intact. Instead, the fatty acyl group is transiently attached to the hydroxyl group of carnitine and the fatty acyl-carnitine is carried across the inner mitochondrial membrane by a specific transporter. In this enzymatic reaction, carnitine acyltransferase I, present on the outer face of the inner membrane, catalyzes transesterification of the fatty acyl group from coenzyme A to carnitine. The fatty acyl-carnitine ester crosses the inner mitochondrial membrane into the matrix by facilitated diffusion through the acyl-carnitine/carnitine transporter.

Third Reaction: In this final step of the entry process, the fatty acyl group is enzymatically transferred from carnitine to intramitochondrial coenzyme A by carnitine acyltransferase II. This isozyme is located on the inner face of the inner mitochondrial membrane, where it regenerates fatty acyl-CoA and releases it, along with free carnitine, into the matrix. Carnitine reenters the space between the inner and outer mitochondrial membranes via the acyl-carnitine/carnitine transporter. Once inside the mitochondrion, the fatty acyl-CoA is ready for the oxidation of its fatty acid component by a set of enzymes in the mitochondrial matrix.

Oxidation of Even-Chain Saturated Fatty Acids (Knoop's β Oxidation Pathway)

Mitochondrial oxidation of fatty acids takes place in 3 stages:

Stage 1: A long-chain fatty acid is oxidized to yield acetyl residues in the form of acetyl-CoA.

Stage 2: The acetyl residues are oxidized to CO_2 via the citric acid cycle.

Stage 3: Electrons derived from the oxidations of Stages 1 and 2 are passed to O_2 via the mitochondrial respiratory chain, providing the energy for ATP synthesis by oxidative phosphorylation.

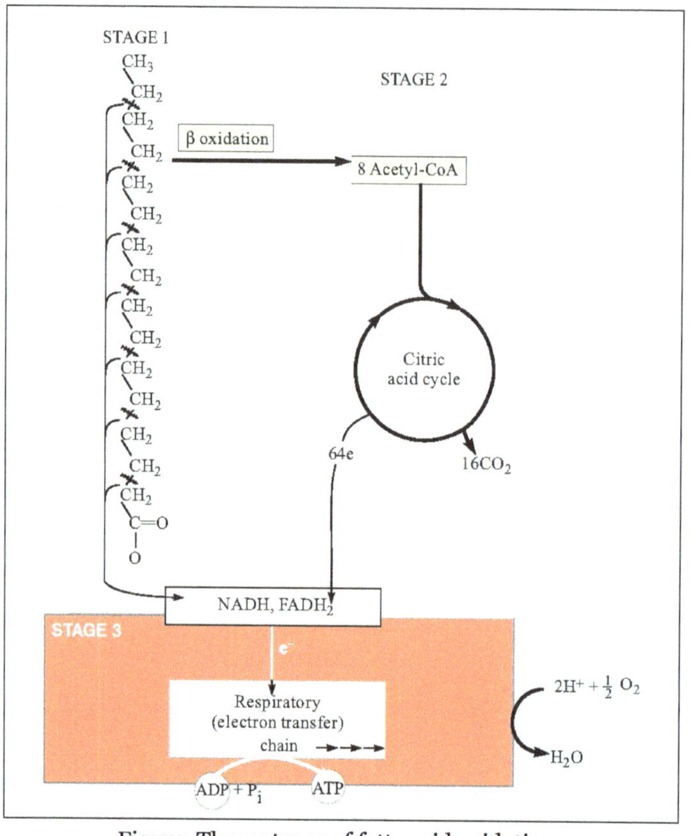

Figure: Three stages of fatty acid oxidation.

First Stage: β oxidation pathway: In this stage, the fatty acids undergo oxidative removal of successive two-carbon units in the form of acetyl-CoA, starting from the carboxyl end of the fatty acyl chain. For example, the C16 fatty acid palmitic acid (palmitate at pH 7) undergoes 7 passes through this oxidative sequence, in each pass losing two carbons as acetyl-CoA. At the end of seven cycles, the last two carbons of palmitate (originally C-15 and C-16) are left as acetyl-CoA. The overall result is the conversion of 16-carbon chain of palmitate to 8 two-carbon acetyl-CoA molecules. Formation of each molecule of acetyl-CoA requires removal of 4 hydrogen atoms (two pairs of electrons and 4 H^+) from the faty acyl moiety by the action of dehydrogenases.

Second Stage: Citric acid cycle: In this stage of fatty acid oxidation, the acetyl residues of acetylCoA are oxidized to CO_2 via the citric acid cycle, which also takes place in the mitochondrial matrix. Acetyl-CoA derived from fatty acid oxidation, thus, enters a final common pathway of oxidation along with acetyl-CoA derived from glucose via glycolysis and pyruvate oxidation.

Third Stage: Mitochondrial respiratory chain: The first two stages of fatty acid oxidation produce the electron carriers, NADH and FADH2, which in the third stage donate electrons to the mitochondrial respiratory chain, through which electrons are carried to oxygen. Coupled to this flow of electrons is the phosphorylation of ADP to ATP. Thus, energy released by fatty acid oxidation is conserved as ATP.

The first stage of fatty acid oxidation for the simple case of a saturated chain with an even number of carbons, and for the slightly more complicated cases of unsaturated and odd-number chains, will now be described in detail.

Four Steps of β Oxidation

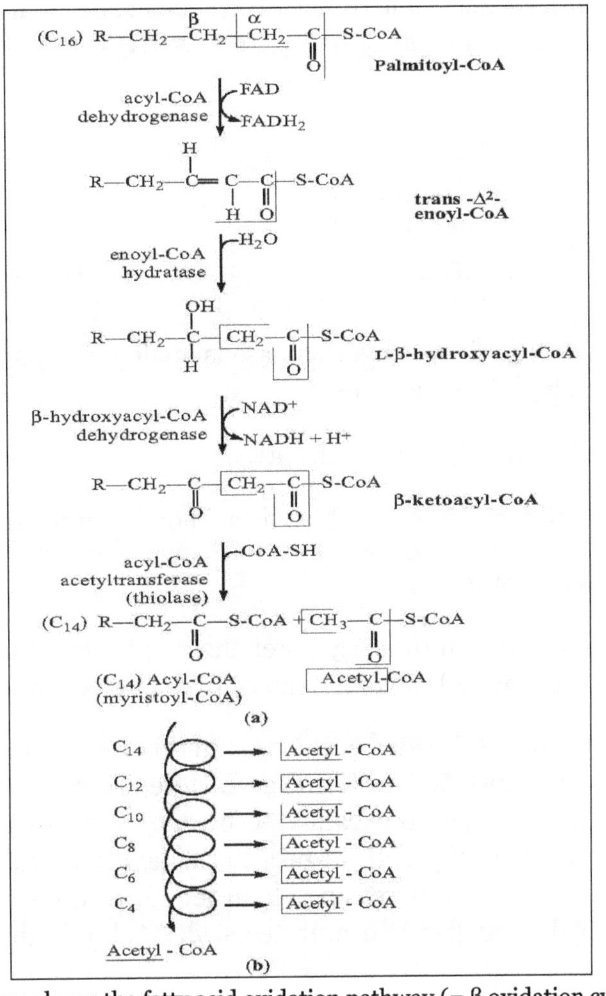

Figure: shows the fatty acid oxidation pathway (= β oxidation cycle).

(a) In each pass through this sequence, one acetyl residue (shaded) is removed in the form of acetyl-CoA from the carboxyl end of palmitate (C_{16}) which enters as palmitoyl-CoA.

(b) Six more passes through the pathway yield 7 more molecules of acetyl-CoA, the seventh arising from the last 2 carbon atoms of the C-16 chain. In all, 8 molecules of acetyl-CoA are formed.

β Oxidation of saturated fatty acids is accomplished by a 4-step mechanism, illustrated in Figure above. The four steps of the fatty acid spiral, as it is also called, are described below:

First Step: α, β dehydrogenation of acyl-CoA: In this step, acyl-CoA is oxidized by an acyl-CoA dehydrogenase to produce an enoyl-CoA with a trans double bond between α and β carbon atoms (C-2 and C-3). It is thus, better written as trans-Δ^2-enoyl-CoA (Recall that naturally occurring unsaturated fatty acids normally have their double bonds in the cis configuration).

$$R{-}C^{\gamma}H_2{-}C^{\beta}H_2{-}C^{\alpha}H_2{-}\overset{O}{\overset{||}{C}}{-}S{-}CoA + FAD \longrightarrow R{-}CH_2{-}CH = CH{-}\overset{O}{\overset{||}{C}}{-}S{-}CoA + FADH_2$$

$$\Delta G' = -4.8 \text{ kcal/mole}$$

Three acyl-CoA dehydrogenases (E.C. No. 1.3.99.3) are found in the matrix of mitochondria. They all have FAD as a prosthetic group. The first has a specificity ranging from C_4 to C_6 acylCoAs, the second from C_6 to C_{14} and the third from C_6 to C_{18}. The $FADH_2$ is not directly oxidized by oxygen but traces the following path:

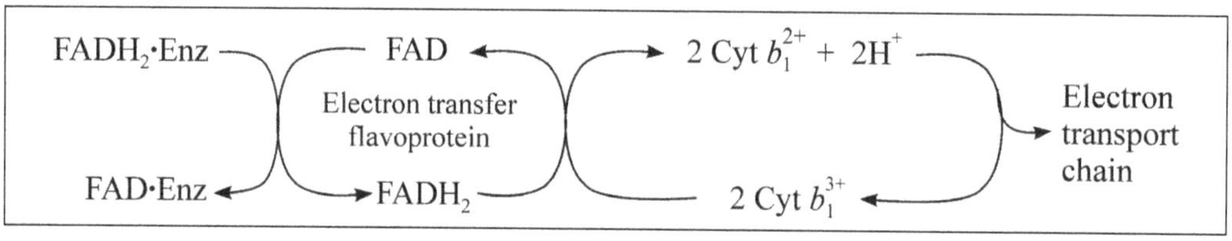

The oxidation catalyzed by acetyl-CoA dehydrogenase is analogous to succinate dehydrogenation in the citric acid cycle, as in both the reactions:

- The enzyme is bound to the inner membrane.
- A double bond is introduced into a carboxylic acid between the α and β carbons.
- FAD is the electron acceptor.
- Electrons from the reaction ultimately enter the respiratory chain and are carried to O_2 with the concomitant synthesis of 2 ATP molecules per electron pair.

Second Step: Hydration of α, β-unsaturated acyl-CoAs: In this step, a mole of water is added to the double bond of the trans-Δ^2 -enoyl-CoA to form the L-stereoisomer of β-hydroxyacyl-CoA (also called 3-hydroxyacyl-CoA). The reaction is catalyzed by enoyl-CoA hydratase or crotonase (E.C. No. 4.2.1.17), and has broad specificity with respect to the length of the acyl group. However, its activity decreases progressively with increasing chain length of the substrate (It may be noted that the enzyme will also hydrate α, β-cis unsaturated acylCoA, but in this case D (–)-β-hydroxyacyl-CoA is formed).

$$R-CH_2-\overset{\beta}{C}H=\overset{\alpha}{C}H-\overset{\overset{O}{\|}}{C}-S-CoA+H_2O \underset{\Delta G' = -0.75\ \text{kcal/mole}}{\rightleftharpoons} R-CH_2-\underset{OH}{\underset{|}{C}H}-\underset{H}{\underset{|}{C}H}-\overset{\overset{O}{\|}}{C}-S-CoA$$

trans L (+) -β-hydroxyacyl-CoA

This reaction catalyzed by enoyl-CoA hydratase is formally analogous to the fumarase reaction in the citric acid cycle, in which water adds across an α-β double bond. The hydration of enoyl-CoA is, in fact, the prelude to the second oxidation reaction, i.e., Step 3.

Third Step: Oxidation of β-hydroxyacyl-CoA: In this step of fatty acid oxidation cycle, the L-β-hydroxyacyl-CoA is dehydrogenated (or oxidized) to form β-ketoacyl-CoA by the action of an enzyme, β-hydroxyacyl-CoA dehydrogenase (E.C. No. 1.1.1.35), which in absolutely specific for the L stereoisomer of the hydroxyacyl substrate. NAD^+ is the electron acceptor in this reaction and the NADH, thus formed, donates its electrons to NADH dehydrogenase (complex I), an electron carrier of the respiratory chain. Three ATP molecules are generated from ADP per pair of electrons passing from NADH to O^2 via the respiratory chain.

$$R-CH_2-\underset{\substack{|\\ OH}}{CH}=\underset{\substack{|\\ H}}{CH}-\overset{\substack{O\\ \|}}{C}-S-CoA + NAD^+ \rightleftharpoons R-CH_2-CO-CH_2-\overset{\substack{O\\ \|}}{C}-S-CoA + NADH + H^+$$

$\Delta G' = +3.75$ kcal/mole

β-ketoacyl-CoA

This reaction, catalyzed by β-hydroxyacyl-CoA, is closely analogous to the malate dehydrogenase reaction of the citric acid cycle. Thus, we see that the first three reactions in each round of fatty acid oxidation closely resemble the last steps in the citric acid cycle:

$$\begin{array}{ccccccc} \text{Aceyl}-\text{CoA} & \longrightarrow & \text{Enoyl}-\text{CoA} & \longrightarrow & \text{Hydroxyacyl}-\text{CoA} & \longrightarrow & \text{Ketoacyl}-\text{CoA} \\ \text{Succinate} & \longrightarrow & \text{Fumarate} & \longrightarrow & \text{Malate} & \longrightarrow & \text{Oxaloacetate} \end{array}$$

The net result of the first three reactions is the oxidation of methylene group at β (or C-3) position to a keto group of the substrate, acyl-CoA.

Fourth Step: Thiolysis or Thioclastic scission: Thiolysis is a splitting by thiol (–SH) group, aided by enzymatic catalysis. This is the final step and brings about the cleavage of β-ketoacyl-CoA by the thiol group of a second mole of CoA, which yields acetyl-CoA and an acyl-CoA, shortened by two carbon atoms. This thiolytic cleavage is catalyzed by the enzyme, acyl-CoA acetyltransferase (E.C. No. 2.3.1.16), which also has broad specificity. This enzyme is more commonly called β-ketothiolase or simply thiolase.

$$\underset{\substack{\beta\text{-ketoacyl-CoA}\\ \text{(n carbons)}}}{R-CH_2-CO-CH_2-\overset{\substack{O\\ \|}}{C}-S-CoA} + \underset{\text{Coenzyme A}}{CoA-SH} \rightleftharpoons \underset{\substack{\text{Acyl-CoA}\\ \text{(n-2 carbons)}}}{R-CH_2-CO-S-CoA} + \underset{\text{Acetyl-CoA}}{CH_3-\overset{\substack{O\\ \|}}{C}-CoA}$$

$\Delta G' = {}^{-}6.65$ kcal/mole

Although the overall reaction is reversible, the equilibrium position is greatly in the direction of cleavage.

As to the mechanism of thiolase action, the enzyme protein has a reactive thiol (–SH) group on a cysteinyl residue that is involved in the following series of reactions:

$$\underset{\beta\text{-ketoacyl-CoA}}{R-CH_2-CO-CH_2-\overset{\substack{O\\ \|}}{C}-S-CoA} + \underset{\text{Thiolase}}{Enz-SH} \rightleftharpoons \underset{\text{Acyl-S-Enz}}{R-CH_2-CO-S-Enz} + \underset{\text{Acetyl-CoA}}{CH_3-\overset{\substack{O\\ \|}}{C}-S-CoA}$$

$$R-CH_2-CO-S-Enz + CoA-SH \rightleftharpoons \underset{\text{Acyl-CoA}}{R-CH_2-CO-S-CoA} + Enz-SH$$

In summary, the shortening of a fatty acyl-CoA derivative by two carbon atoms can be represented by the equation:

$$R-CH_2-CH_2-CH_2-CO-S-CoA+FAD+NAD+CoA-SH \rightarrow$$
$$R-CH2-CO-S-CoA+CH^3-CO-S-CoA+FADH_2+NADH+H^+$$

The shortened acyl-CoA then undergoes another cycle of oxidation, starting with the reaction catalyzed by acyl-CoA dehydrogenase. Beta-ketothiolase, hydroxyacyl dehydrogenase and enoyl-CoA hydratase all have broad specificity with respect to the length of the acyl group. Thus, by repeated turns of the cycle, a fatty acid is degraded to acetyl-CoA molecules with one being produced every turn until the last cycle, wherein two are produced. The β-oxidation of fatty acids is presented in a cyclical manner in figure.

The β oxidative system is found in all organisms. However, in bacteria grown in the absence of fatty acids, the β oxidative system is practically absent but is readily induced by the presence of fatty acids in the growth medium. The bacterial β oxidation system is completely soluble and hence is not membrane-bound. Curiously, in germinating seeds possessing high lipid content, the β oxidation system is exclusively located in microbodies called glyoxysomes, but in seeds with low lipid content, the enzymes are associated with mitochondria.

Stoichiometry of β Oxidation

The energy yield derived from the oxidation of a fatty acid can be calculated. In each reaction cycle, an acyl-CoA is shortened by two carbons and one mole each of FADH2, NADH and acetylCoA are formed.

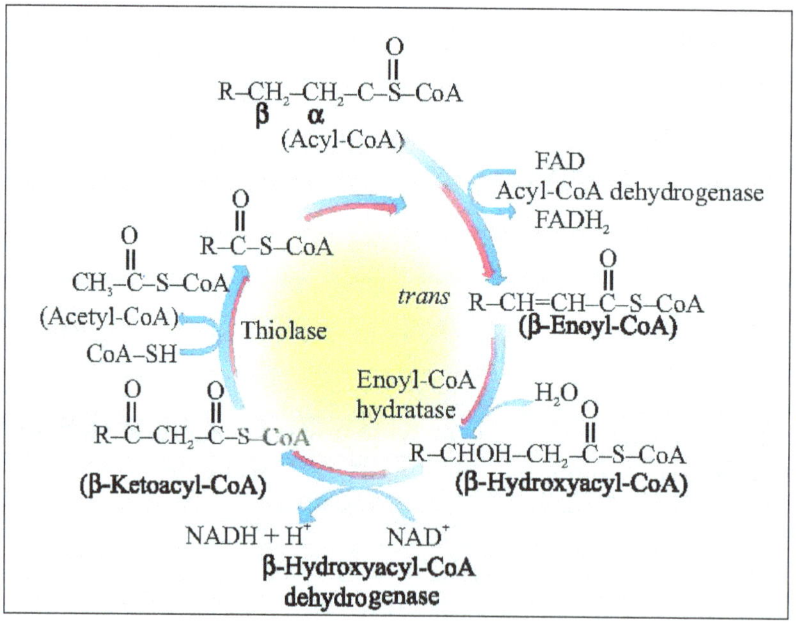

Figure: The β-oxidation cycle for fatty acids.

$$C_n - \text{acyl} - \text{CoA} + \text{FAD} + \text{NAD}^+ + \text{CoA} + H_2O$$
$$\rightarrow C_n - 2 - \text{acyl} - \text{CoA} + \text{FADH}_2 + \text{NADH} + \text{Acetyl} - \text{CoA} + H^+$$

The degradation of palmitoyl-CoA (C_{16}-acyl-CoA), for example, requires 7 reaction cycles. In the seventh cycle, the C_4 - ketoacyl-CoA is thiolyzed to 2 moles of acetyl-CoA. Hence, the stoichiometry of oxidation of palmitoyl-CoA is:

$$\text{Palmitoyl} - \text{CoA} + 7\text{FAD} + 7\text{NAD}^+ + 7\text{CoA} + 7H_2O$$
$$\rightarrow 8\,\text{Acetyl} - \text{CoA} + 7\text{FADH}_2 + 7\,\text{NADH} + H^+$$

Three ATP are generated when each of these NADH is oxidized by the respiratory chain, whereas two ATP are formed for each $FADH_2$ because their electrons enter the chain at the level of coenzyme Q. Recall that the oxidation of one mole of acetyl-CoA by the citric acid cycle yields 12 ATP molecules. Hence, the number of ATP moles formed in the oxidation of palmitoyl-CoA is 14 from the 7 $FADH_2$, 21 from the 7 NADH, and 96 from the 8 moles of acetyl-CoA. This totals to 131. Two high-energy phosphate bonds are consumed in the activation of palmitate, in which ATP is split into AMP and 2 Pi. Thus, the net yield from the complete oxidation of a mole of palmitate is 131 – 2 = 129 ATP molecules.

The efficiency of energy conservation in fatty acid oxidation can be estimated from the number of ATP formed and from the free energy of oxidation of palmitic acid to CO_2 and H_2O, as determined by calorimetry. The standard free energy of hydrolysis of 129 ATP is 129 × – 7.3 kcal = – 941.7 kcal or roughly – 942 kcal. The standard free energy of oxidation of palmitic acid is – 2,340 kcal.

$$CH_3(CH_2)_7 - COOH + 23O_2 \rightarrow 16CO_2 + 16H_2O$$
$$\Delta G^{o\prime} = -2,340\,kcal / mole$$

Hence, the efficiency of energy conservation of the in vivo oxidation of fatty acids, under standard conditions, is 942/2340 × 100 = 42.56%, a surprisingly high figure. This value is similar to those of glycolysis, the citric acid cycle and oxidative phosphorylation. In other words, the 129 ATP produced account for a conservation of 942 kcal of the 2,340 kcal released by the oxidation of one mole of palmitic acid, i.e., roughly 42% efficiency of energy conservation. The remaining energy is lost probably as heat. It, hence, becomes clear why, as a food, fat is an effective source of available energy. In this calculation, we neglect the combustion of glycerol, the other component of a triacylglycerol.

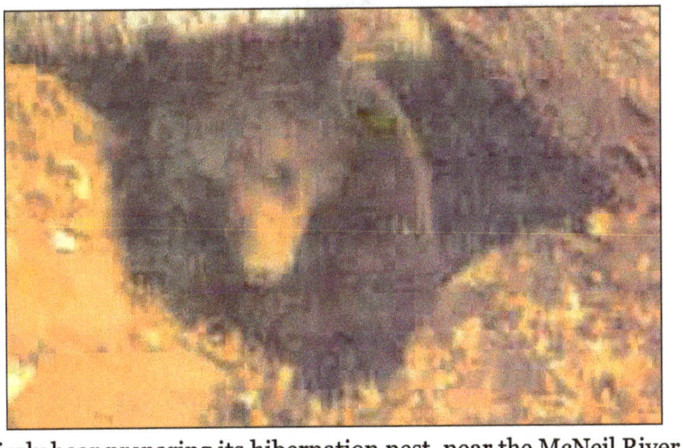

Figure: A grizzly bear preparing its hibernation nest, near the McNeil River in Canada.

In hibernating animals such as grizzly bear and the tiny dormouse, fatty acid oxidation provides metabolic energy, heat and water — all essential for survival of an animal that neither eats nor drinks for long periods. The camel, although not a hibernator, can synthesize and store triacylglycerols in large amounts in its hump, a metabolic source of both energy and water under desert conditions.

Oxidation of Unsaturated Fatty Acids

The fatty acid oxidation scheme described above operates only when the incoming fatty acid is a saturated one (having only single bonds) and possesses an even number of carbon atoms. However, most of the fatty acids in the triacylglycerols and phospholipids of animals and plants are unsaturated, having one or more double bonds in its carbon chain. These bonds are in cis configuration and cannot be acted upon by the enzyme, enoyl-CoA hydratase which catalyzes the addition of H_2O to the trans double bond of the Δ^2 - enoyl-CoA generated during β oxidation. However, by the action of two auxiliary enzymes, the fatty acid oxidation sequence described above can also break down the common unsaturated fatty acids. The action of these two enzymes, one isomerase and the other a reductase, will be illustrated by the following two examples:

Oxidation of Monounsaturated Fatty Acids

This requires only one additional enzyme, enoyl-CoA isomerase. Oleate, an abundant C–18 monounsaturated fatty acid with a cis double bond between C–9 and C–10 (denoted cis-Δ^9) is aken as an example. Oleate is converted into oleoyl-CoA which is transported through the mitochondrial membrane as oleoyl carnitine and then converted back into oleoyl-CoA in the matrix. Oleoyl-CoA then undergoes 3 passes through the β oxidation cycle to yield 3 moles of acetyl-CoA and the coenzyme A ester of a Δ^3, 12-carbon unsaturated fatty acid, cis-Δ^3 - dodecenoylCoA. This product cannot be acted upon by the next enzyme of the β oxidation pathway, i.e., enoyl-CoA hydratase, which acts only on trans double bonds. However, by the action of the auxiliary enzyme, enoyl-CoA isomerase, the cis-Δ^3 -enoyl-CoA is isomerized to yield the trans-Δ^2 -enoyl-CoA. The latter compound is now converted by enoyl-CoA hydratase into the corresponding L-β-hydroxyacyl-CoA (trans-Δ^2 -dodecenoyl-CoA). This intermediate is now acted upon by the remaining enzymes of β oxidation to yield acetyl-CoA and a C–10 saturated fatty acid as its coenzyme A ester (decanoyl-CoA). The latter undergoes 4 more passes through the pathway to yield altogether 9 acetyl-CoAs from one mole of the C–18 oleate.

Oxidation of Polyunsaturated Fatty Acids

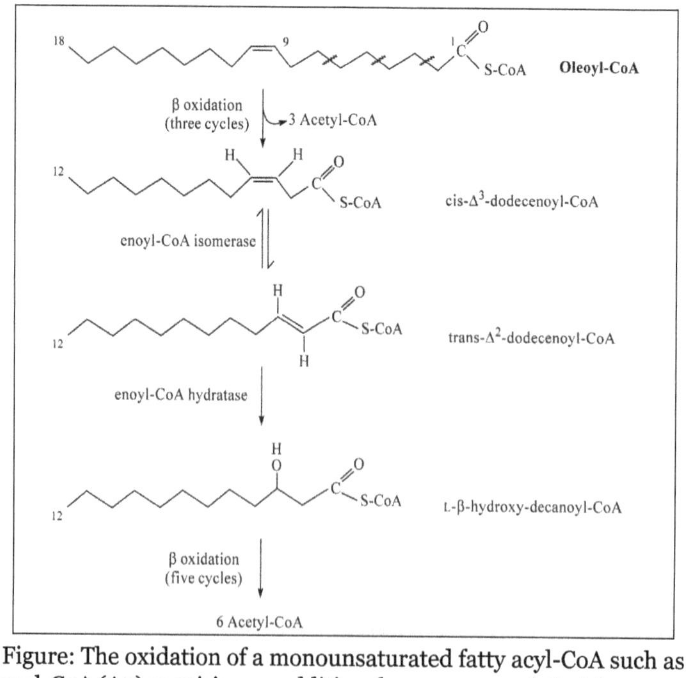

Figure: The oxidation of a monounsaturated fatty acyl-CoA such as oleoyl-CoA (Δ9) requiring an additional enzyme, enoyl-CoA isomerase.

This process requires two auxiliary enzymes, enoyl-CoA isomerase and 2, 4-dienoyl-CoAreductase. The mechanism is illustrated by taking linoleate, a C-18 polyunsaturated fatty acid with 2 cis double bonds at C_9 and C_{12} (denoted cis-Δ^9, cis-Δ^{12}), as an example. Linoleoyl-CoA undergoes 3 passes through the typical β oxidation sequence to yield 3 moles of acetyl-CoA and the coenzyme A ester of a C-12 unsaturated fatty acid with a cis-Δ^3, cis-Δ^6 configuration. This intermediate cannot be used by the enzymes of the β oxidation pathway; its double bonds are in the wrong position and have the wrong configuration (cis, not trans). However, the combined action of enoyl-CoA isomerase and 2, 4-dienoyl-CoA reductase allows reentry of this intermediate into the typical β oxidation pathway and its degradation to 6 acetyl-CoAs. The overall result is the conversion of linoleate to 9 moles of acetyl-CoA.

Note that the enzyme repositions the double bond converting the cis isomer to a trans isomer, a normal intermediate in β oxidation. Thus, both position and configuration of the double bond are shifted by the action of the enzyme.

Here is an excellent example of the beautiful economy of organization of metabolism. The introduction of 2 additional types of enzymes (an enoyl-CoA isomerase and a 3-hydroxyacyl-CoA racemase) makes it possible to handle any combination of double bonds found in an unsaturated chain through the same route used for saturated fatty acids.

The roles of the 3 additional enzymes which are necessary for the oxidation of a dienoic (or polyenoic) acid may be shown in outline below, where A is enoyl-CoA isomerase; B, enoyl-CoA hydratase; and C, 3-hydroxyacyl-CoA epimerase. Monoenoic and dienoic acids are oxidized at comparable rates.

$C_{18:2}$ (9 *cis*, 12 *cis*) $\xrightarrow{-3C_2}$ $C_{12:2}$ (3 *cis*, 6 *cis*) $\xrightarrow{A}$ $C_{12:2}$ (2 *trans*, 6 *cis*)

$\xrightarrow{-C_2}$ $C_{10:1}$ (4 *cis*) $\xrightarrow{-C_2}$ $C_{8:1}$ (2 *cis*) $\xrightarrow{B}$ 3-D-hydroxy $C_{8:0}$

$\xrightarrow{C}$ 3-L-hydroxy $C_{8:0}$ $\xrightarrow{-C_2}$ $C_{6:0}$ $\longrightarrow$ $3C_2$

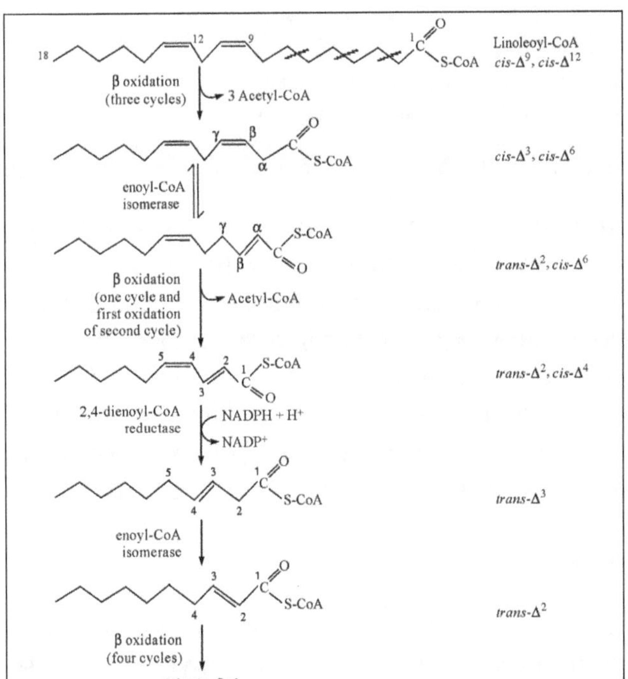

Figure: The oxidation of polyunsaturated fatty acids requiring two additional enzymes, enoylCoA isomerase and 2, 4-dienoyl-CoA reductase.

Note that the combined action of these two enzymes converts a trans-Δ^2, cis Δ^4 -dienoyl-CoA intermediate into the trans-Δ^2 -enoyl-CoA substrate, necessary for β oxidation.

Oxidation Of Odd-Chain Fatty Acids

Most naturally-occurring lipids contain fatty acids with an even number of carbon atoms, yet fatty acids with an odd number of carbon atoms are found in significant amounts in the lipids of many plants and some marine animals. Small quantities of C-3 propionate are added as a mould inhibitor to some breads and cereals, and thus propionate enters the human diet. Besides, cattle and other ruminants form large amounts of propionate during fermentation of carbohydrates in the rumen. The propionate so formed is absorbed into the blood and oxidized by the liver and other tissues.

A generalized scheme of the oxidation of an odd-chain fatty acid is presented in Figure below.

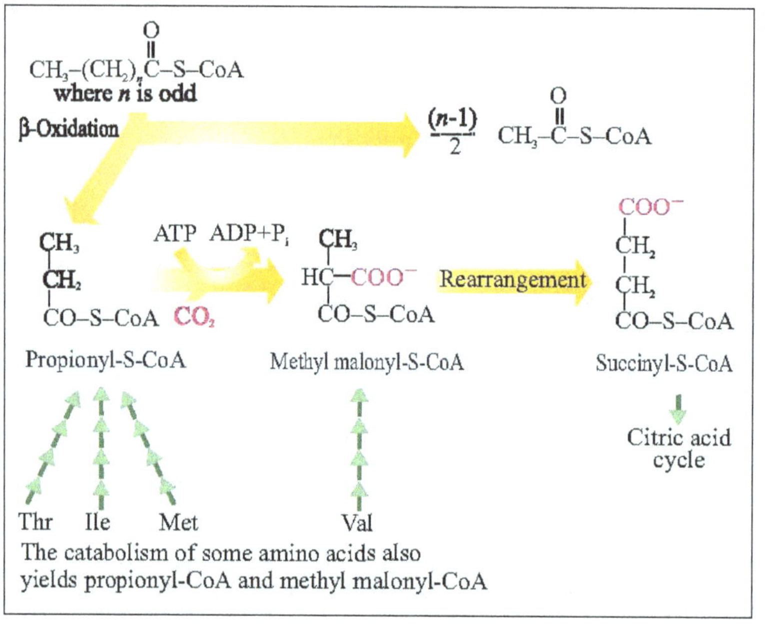

Figure: The oxidation of a fatty acid containing an odd number of carbon atoms.

The odd-carbon long-chain fatty acids are oxidized by the same pathway as the even-carbon fattty acids, starting at the carboxyl end of the chain. However, the substrate for the last pass through the β oxidation cycle is a fatty acyl-CoA, in which the fatty acid has 5 carbon atoms. When this is oxidized and finally cleaved, the products are acetyl-CoA and propionylCoA, rather than 2 moles of acetyl-CoA produced in the normal β oxidation cycle. The acetyl-CoA is, of course, oxidized via the citric acid cycle but the oxidation of propionylCoA presents an interesting problem, since at first glance the propionic acid (or propionylCoA) appears to be a substrate unsuitable for β oxidation. However, the substrate is held by two strikingly dissimilar pathways: methylmalonate pathway and β-hydroxy-propionate pathway.

Methylmalonate Pathway

This pathway is found only in animals and occurs in the mitochondria of liver, cardiac and skeletal muscles, kidney and other tissues. Propionate (or propionyl-CoA) is also produced by the oxidation of isoleucine, valine, methionine and threonine. Propionate is catalyzed by acetylCoA synthetase to produce propionyl-CoA. The propionyl-CoA is carboxylated to form the D stereoisomer of methylmalonyl-CoA by an enzyme propionyl-CoA carboxylase, which contains the cofactor biotin. In this reaction, as in pyruvate carboxylase reaction, the CO_2 (or its hydrated ion, HCO_3 –) is activated by attachment to biotin before its transfer to the propionate moiety. The formation of the carboxybiotin intermediate requires energy, which is provided by the cleavage of ATP to AMP and PPi. The dmethylmalonyl-CoA, thus formed, is enzymatically epimerized to L-methylmalonyl-CoA, by the action of methylmalonyl-CoA epimerase (The epimerase labilizes the α-hydrogen atom, followed by uptake of a proton from the medium, thus catalyzing interconversion of D- and L-methylmalonylCoA). The L-methylmalonyl-CoA undergoes an intramolecular rearrangement to form succinylCoA by the enzyme methylmalonyl-CoA mutase, which requires as its coenzyme deoxyadenosyl-cobalamin or coenzyme B_{12}. When [2– ^{14}C] methyl-malonyl-CoA was converted by the mutase enzyme, the label (marked by an asterisk, below) was found in the 3 position of succinyl-CoA, thus indicating an intramolecular transfer of the entire thioester group, –CO–S– CoA, rather than migration of the carboxyl carbon.

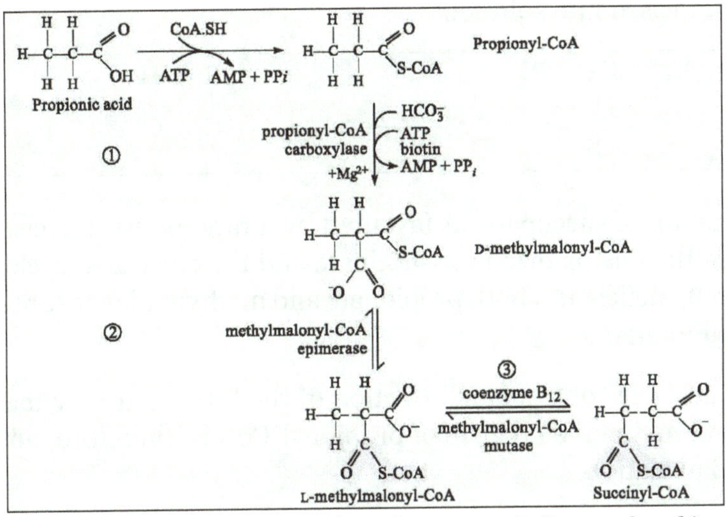

Figure: The methylmalonate pathway of propionate metabolism, as found in animals.

Note the third remarkable reaction in which substituents on adjacent carbon atoms exchange positions; the coenzyme B_{12} playing a key role in it.

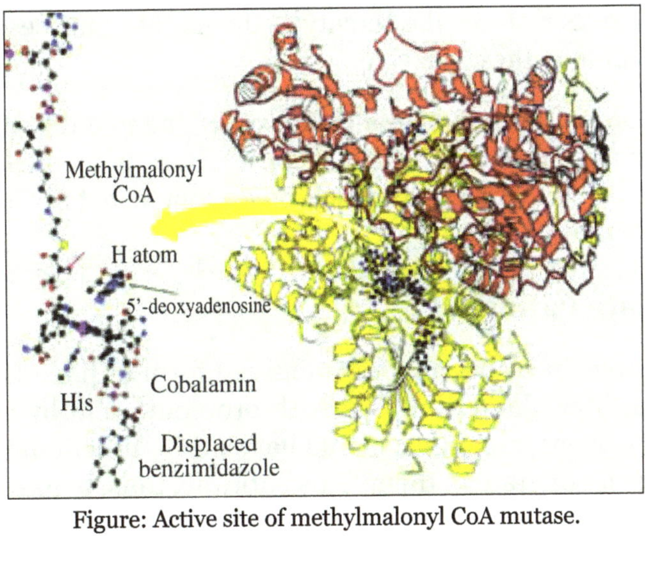

Figure: Active site of methylmalonyl CoA mutase.

The arrangement of substrate and coenzyme in the active site facilitates the cleavage of the cobalt-carbon bond and the subsequent abstraction of a hydrogen atom from the substrate.

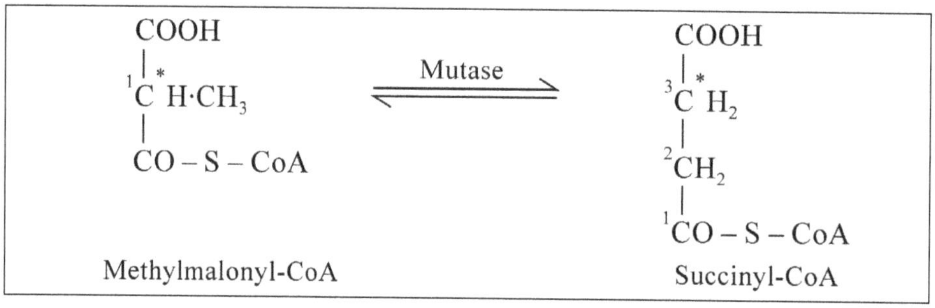

The role of the coenzyme B_{12} is to remove hydrogen from one carbon atom by transferring it directly to an adjacent carbon atom, simultaneously affecting the exchange of a second (R) substituent. The H and R are not released into solution.

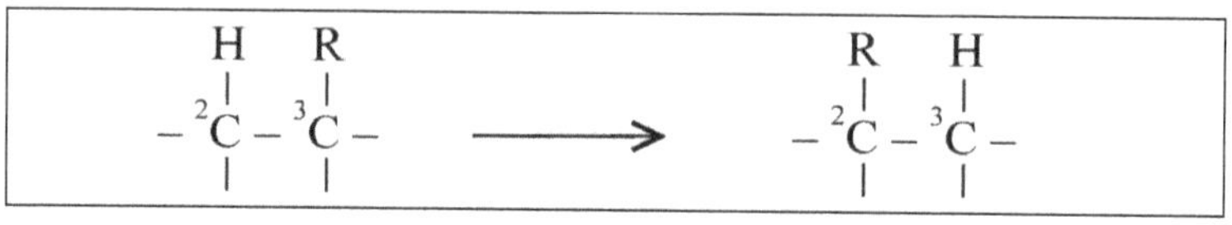

At equilibrium, formation of succinyl-CoA favoured by a ratio of 20 :1 over methylmalonyl-CoA. The succinyl-CoA can then be oxidized via succinate and the citric acid cycle to CO_2 and H_2O. In patients with vitamin B_{12} deficiency, both propionate and methylmalonate are excreted in the urine in abnormally large amounts.

The odd-chain fatty acids are only a small fraction of the total, and only the terminal 3 carbons appear as propionyl-CoA. The metabolism of propionyl-CoA is, therefore, not of quantitative significance in fatty acid oxidation.

Two inheritable types of methylmalonic acidemia (and aciduria) are associated in young children with failure to grow and mental retardness. In one type, the mutase protein is absent or defective since addition of coenzyme B_{12} to liver extracts does not restore the activity of the mutase. In the other type, feeding large doses of vitamin B_{12} relieves the acidemia and aciduria, and addition of coenzyme B_{12} to liver extracts restores the activity of the mutase ; in these cases, there is limited ability to convert the vitamin to the coenzyme.

Another inheritable disorder of propionate metabolism is due to a defect in propionyl-CoA carboxylase, resulting in propionic acidemia (and aciduria). Such individuals, as well as those with methylmalonic acidemia, are capable of oxidizing some propionate to CO_2, even in the absence of propionyl-CoA carboxylase.

β-Hydroxypropionate Pathway

This pathway is ubiquitous in plants and is a modified form of β oxidation scheme. It nicely resolves the problem of how plants can cope with propionic acid by a system not involving vitamin B_{12} as cobamide coenzyme. Since plants have no B_{12} functional enzymes, the methylmalonate pathway does not operate in them. This pathway, thus, bypasses the B_{12} barrier in an effective way.

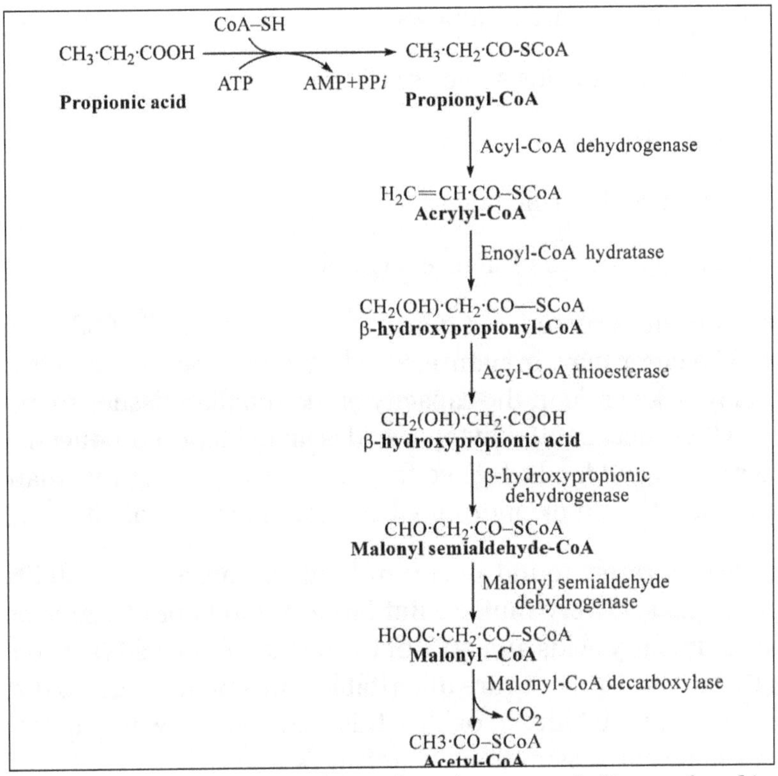

Figure: The β-hydroxypropionate pathway of propionate metabolism, as found in plants.

α Oxidation of Fatty Acids

Although β oxidation is major pathway for the oxidation of fatty acids, two other types of oxidation also occur, α and ω oxidation. α oxidation is the removal of one carbon atom (i.e., α carbon) at a time from the carboxyl end of the molecule. α oxidation was first observed in seeds and leaf tissues of plants. α oxidation of long-chain fatty acids to 2-hydroxy acids and then to fatty acids with one carbon atom less than the original substrate have been demonstrated in the microsomes of brain and other tissues also. Long-chain α hydroxy fatty acids are constituents of brain lipids, e.g., the C_{24} cerebronic acid (= 2 hydroxylignoceric acid), CH_3 $(CH_2)_{21}$. CH(OH). COOH. These hydroxy fatty acids can be converted to the 2-keto acids, followed by oxidative decarboxylation, resulting in the formation of long-chain fatty acids with an odd number of carbon atoms:

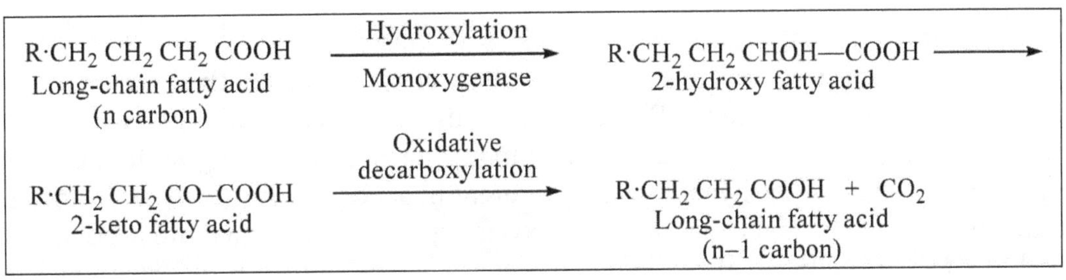

The initial hydroxylation reaction is catalyzed by a mitochondrial enzyme, monoxygenase that requires O_2, Mg^{2+}, NADPH and a heat-stable cofactor. Conversion of the α hydroxy fatty acid to CO_2 and the next lower unsubstituted acid appears to occur in the endoplasmic reticulum and to require O_2, Fe^{2+} and ascorbate.

The salient features of α oxidation are as follows:

- Only free long-chain fatty acids serve as substrates.
- Molecular oxygen is indirectly involved.
- It does not require CoA intermediates.
- It does not lead to generation of high-energy phosphates.

This mechanism explains the occurrence of α hydroxy fatty acids and of odd-numbered fatty acids in the biomolecules. The latter may, in nature, also be synthesized de novo from propionate. The α oxidation system plays a key role in the capacity of mammalian tissues to oxidize phytanic acid (=3,7,11,15-tetramethylhexadecanate). Phytanic acid is an oxidation product of phytol and is present in animal fat, cow's milk and foods derived from milk. The phytol presumably. originates from plant sources, as it is a substituent of chlorophyll and the side chain of vitamin K_2.

Normally, phytanic acid is rarely found in serum lipids because of the ability of normal tissue to degrade (or oxidize) the acid very rapidly. But large amounts of phytanic acid accumulate (as much as 20% of the serum fatty acids and 50% of the hepatic fatty acids) in the tissues and serum of individuals with Refsum's disease, a rare inheritable autosomal recessive disorder affecting the nervous system because of an inability to oxidize this acid. Diets low in animal fat and milk products appear to relieve some of the symptoms of Refsum's disease.

The presence of 3-methyl group in phytanic acid blocks β oxidation. In the mitochondria of normal individuals, α hydroxylation of phytanic acid by phytanate α hydroxylase is followed by oxidation by phytanate α oxidase to yield CO_2 and pristanic acid (= 2,6,10, 14-tetramethylpentadecanoic acid), which readily undergoes β oxidation after conversion to its CoA derivative. In Refsum's disease, there is a lack of the enzyme, phytanate α hydroxylase.

ω Oxidation of Fatty Acids

The biological oxidation of fatty acids at the omega (ω) carbon atom was first reported by Verkade and his group, who isolated from the urine dicarboxylic acids of the same chain length as those that were fed in the form of triglycerides. He proposed that certain acids were first oxidized at the ω carbon atom and then further metabolized by β oxidation proceeding from both ends of the dicarboxylic acid.

The ω oxidation scheme responsible for the oxidation of alkanes in both the animal and plant bacterial systems has been depicted in Fig. The mechanism involves an initial hydroxylation of the terminal methyl group to a primary alcohol. In animals, the cytochrome P_{450} system is the hydroxylase responsible for this alkane hydroxylation; whereas in bacteria, rubridoxin is the intermediate electron carrier which feeds electrons to ω hydroxylase system. The immediate product, RCH_2OH is oxidized to an aldehyde by an alcohol dehydrogenase, which in turn is oxidized to a carboxylic acid by an aldehyde dehydrogenase in both systems. Summarily, the $-CH_3$ group is converted to a $-CH_2OH$ group which subsequently is oxidized to $-COOH$, thus forming a dicarboxylic acid. Once formed, the dicarboxylic acid may be shortened from either end of the molecule, by the β oxidation sequence, to form acetyl-CoA.

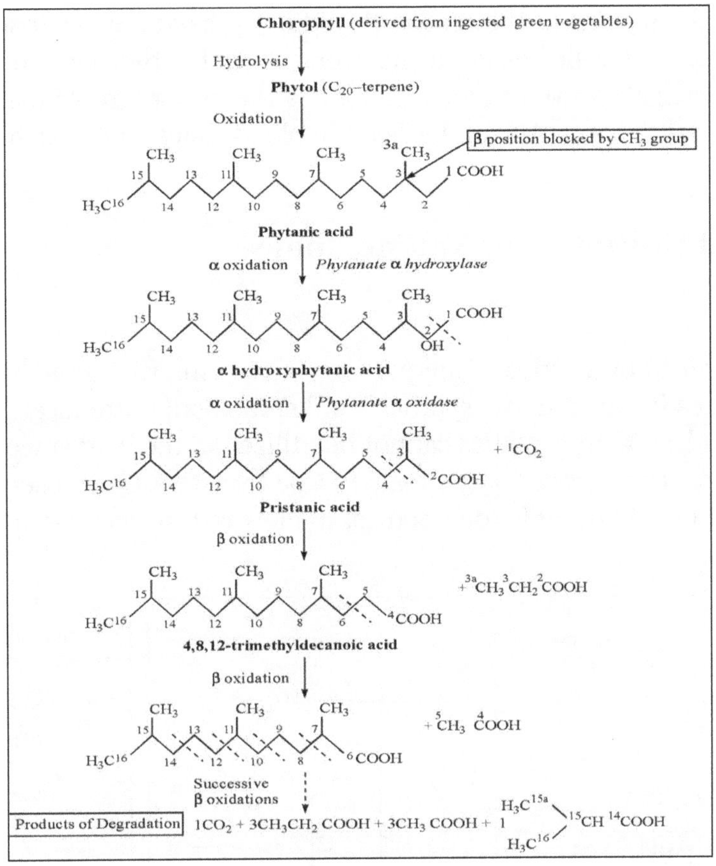

Figure: The metabolism of phytanic acid by the normal animal cell.

Dashed lines show the successive points of cleavage of the molecule. These series of reactions now have assumed an extremely important scavenging role in the bacterial biodegradation of both detergents derived from fatty acids and even more important the large amounts of oil spilled over the ocean surface. The rate of bacterial oxidation of floating oil under aerobic conditions is estimated as high as 0.5g/day per square metre of oil surface. The bacterial oxidation of oils is brought about primarily by ω oxidation mechanism.

Ketogenesis

The term ketogenesis means formation of ketone bodies. The acetyl-CoA, formed in fatty acid oxidation, enters the citric acid cycle if fat and carbohydrate degradation are approximately balanced. The molecular basis of the adage that fats burn in the flame of carbohydrates is now evident. The entry of acetyl-CoA into citric acid cycle depends on the availability of oxaloacetate for the formation of citrate. However, if fat breakdown predominates, the acetyl-CoA undergoes a different fate. This is because the concentration of oxaloacetate is lowered if carbohydrate is not available or else poorly utilized. Also, in fasting or in diabetes, the oxaloacetate is utilized in the formation of glucose and is thus unavailable for condensation with acetyl-CoA. Under these conditions, acetyl-CoA is diverted to the formation of acetoacetate (3-oxobutyrate, in systematic nomenclature) and D-3-hydroxybutyrate. Acetoacetate continually undergoes spontaneous decarboxylation to yield acetone, which is exhaled. The reaction is slow, but if the concentration of acetoacetate becomes high, enough acetone may be formed to make its characteristic odour detectable in the breath of the individuals. This is the part of the reason that the 3 substances (acetoacetate, D-3-hydroxy-butyrate and acetone) were collectively but inaccurately called the "ketone bodies" (or "acetone bodies") by early investigators even though acetone is the minor part of the total. The term now seems quaint, but it is still in use. And an increase in blood concentrations, of these compounds is called ketonemia.

Biosynthesis and Utilization of Ketone Bodies

Biosynthesis

Ketone bodies are formed by a series of unique reactions, primarily in the liver and kidney mitochondria. The enzymes involved in the synthesis of ketone bodies are localized primarily in liver and kidney mitochondria. Ketone bodies cannot be utilized in the liver since the key utilizing enzyme, β-ketoacyl : CoA transferase (= 3-oxoacid : Coa transferase) is absent in the tissue but is present in all tissues metabolizing ketone bodies, namely red muscle, cardiac muscle, brain and kidney.

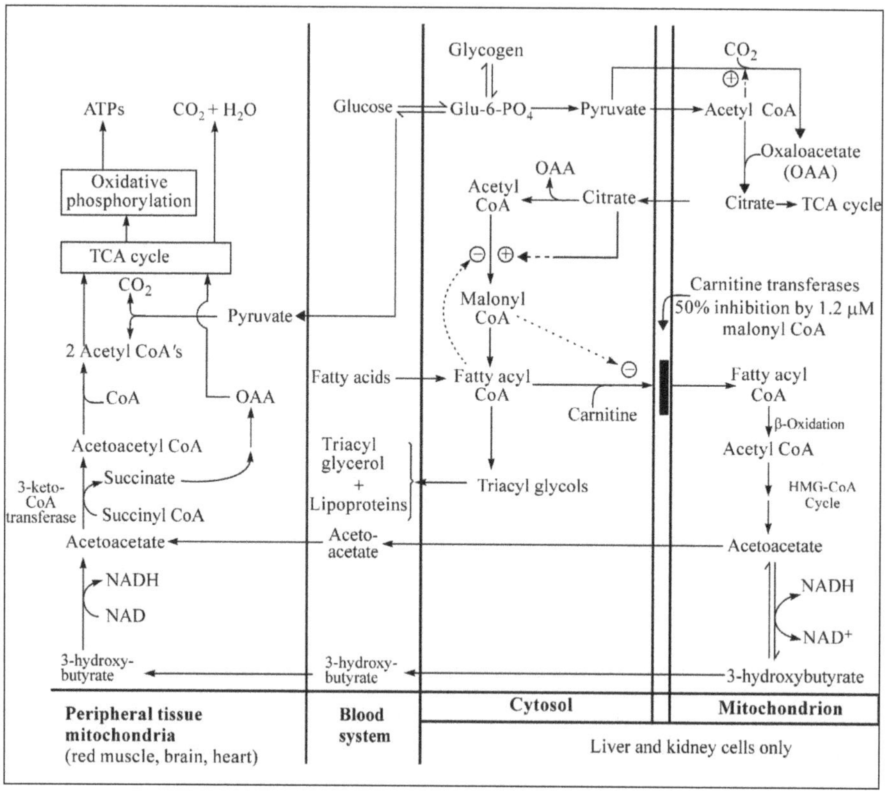

Figure: Biosynthesis of ketone bodies and their utilization.

Note that the C-6 compound, β-methylglutaryl-CoA (HMG-CoA) is also an intermediate of sterol biosynthesis, but the enzyme that forms HMG-CoA in that pathway is cytosolic. HMG-CoA lyase is present in the mitochondrial matrix but not in the cytosol.

Acetoacetate is produced from acetyl-CoA in the liver and kidneys by a simple three-step process. The first step in the formation of acetoacetate, one of the 3 ketone bodies, is the reversible head-to-tail condensation of 2 moles of acetyl-CoA, to produce a mole of acetoacetylCoA enzymatically by thiolase

(Step 1). This steps is simply the reversal of the last step of β oxidation, i.e., thiolysis. The acetoacetyl-CoA then condenses with acetyl-CoA and water to form β-hydroxy-β-methylglutaryl-CoA, HMG-CoA (Step 2). The unfavourable equilibrium in the formation of acetoacetyl-CoA is compensated for by the favourable equilibrium of this reaction, which is due to the hydrolysis of a thioester linkage.

If we think of acetoacetyl-coenzyme A as being analogous to oxaloacetate, we find that the condensation reaction (Step 2) in exactly analogous to the formation of citrate in the first step of the citric acid cycle. However, it is catalyzed by a quite different enzyme, 3-hydroxy-3- methylglutaryl-CoA synthetase. HMG-CoA is ultimately cleaved at a different point to yield acetoacetate and acetyl-CoA in an irreversible reaction (Step 3). The sum of these reactions is:

$$2\ \text{Acetyl} - \text{CoA} + H_2O \longrightarrow \text{Acetoacetate} + 2\ \text{CoA} + H^+$$

The acetoacetate so produced is reversibly reduced by a mitochondrial enzyme, D-βhydroxybutyrate dehydrogenase to produce D-β-hydroxybutyrate. This enzyme is specific for the D-stereoisomer; it does not act on L-β-hydroxyacyl-CoAs and is not to be confused with L-β-hydroxyacyl-CoA dehydrogenase, which participates in the β-oxidation pathway. The ratio of hydroxybutyrate to acetoacetate depends on the NADH/NAD+ ratio inside mitochondria. Accetoacetate is also easily decarboxylated to acetone, either spontaneously or by the action of acetoacetate decarboxylase.

Utilization

The acetoacetate and D-β-hydroxybutyrate diffuse from the liver mitochondria into the blood and are transported to peripheral tissues. George Cahill and others have shown that these two acetyl-CoA products are important molecules in energy production. Acetoacetate and D-β-hydroxybutyrate are normal fuels of respiration and are quantitatively important as source of energy. In fact, heart muscles and renal cortex use acetoacetate in preference to glucose. On the contrary, glucose is the major fuel for the brain in well-nourished persons on a balanced diet. However, brain adapts to utilization of acetoacetate during fasting, starvation and diabetes. In prolonged starvation, 75% of the fuel needs of the brain are met by acetoacetate.

Fatty acids are released by adipose tissue and converted into acetyl units by the liver, which then exports them as acetoacetate. Acetoacetate can, thus, be regarded as a water-soluble transportable form of acetyl units.

To sum up, ketone bodies are alternative substrates to glucose, for energy sources in muscle and brain. The precursors of ketone bodies, namely free fatty acids are toxic in high concentrations, have very limited solubility, and readily saturate the carrying capacity of the plasma membrane. On the other hand, the ketone bodies are low in toxicity, and tolerated at high concentrations, are very soluble, diffuse rapidly through membranes, and are rapidly metabolized to CO_2 and H_2O. The importance of this pathway is indicated by the fact that the normal human liver is potentially capable of making the equivalent of half its weight as acetoacetate each day!

In the extrahepatic tissues, D-β-hydroxybutyrate is oxidized to acetoacetate by an enzyme, Dβ-hydroxybutyrate dehydrogenase. Acetoacetate is activated to form its coenzyme A ester by transfer of CoA from succinyl-CoA, an intermediate of the citric acid cycle, in a reaction catalayzed by b-ketoacyl-CoA transferase. The acetoacetyl-CoA is then cleaved by thiolase to yield 2 moles of acetyl-CoA, which enter the citric acid cycle.

Ketogenic and Antiketogenic Substances

The ketogenic substances are, of course, all the fatty acids. In addition, at least 3 amino acids belong to this category: leucine, phenylalanine and tyrosine. Antiketogenic substances, in the sense of preventing the accumulation of ketone bodies, are the carbohydrates, the glycerol fraction of fat, and the following amino acids: glycine, alanine, valine, serine, threonine, cysteine, methionine, aspartic acid, glutamic acid, arginine, histidine, proline and ornithine. These are antiketogenic because their non-nitrogenous residues are convertible to glucose.

Regulation of Ketogenesis

Ketosis arises as a result of deficiency in available carbohydrate. This leads to the enhanced rate of ketogenesis. There are 3 crucial steps in the pathway of metabolism of free fatty acids (FFAs) that determine the magnitude of ketogenesis. These are:

1. It causes an imbalance between esterification and lipolysis in adipose tissue, with consequent of free fatty acids into the circulation. FFAs are the principal substrates for ketone body formation in the liver and therefore all factors, metabolic or endocrine, affecting the release of free fatty acids from adipose tissues, influence ketogenesis.

2. Upon entry of free fatty acids into the liver, the balance between esterification and oxidation of FFAs is influenced by the hormonal state of the liver and possibly by the availability of glycerol-3-phosphate or by the redox state of the tissue.

3. As the quantity of fatty acids presented for oxidation increases, more form ketone bodies and less form CO_2, regulated in such a manner that the total energy production remains constant.

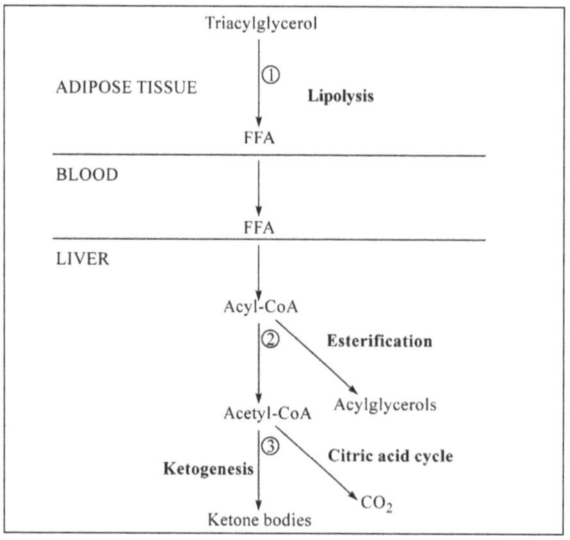

Figure: Regulation of ketogenesis.

The 3 crucial steps in the pathway of free fatty acids (FFAs) that determine the rate of ketogenesis are numbered as ①, ②, and ③.

Under circumstances of limited utilization of carbohydrates and/or substantial mobilization of fatty acids to the liver, there is a markedly diminished rate of operation of two of the three pathways for metabolizing acetyl-CoA, viz., the citric acid cycle and fatty acid synthesis. The result is a channeling of acetyl-CoA into β-hydroxy-β-methylglutaryl-CoA (HMG-CoA). This results in increased formation of acetoacetate, D-β-hydroxybutyrate and acetone, whereby elevating the concentration of the total ketone bodies in the blood and urine above normal. Higher than normal quantities of the ketone bodies present in the blood or urine constitute ketonemia (= hyperketonemia) and ketonuria, respectively. D-β-hydroxybutyrate is quantitatively the predominant ketone body present in the blood and urine in ketosis. Whenever a marked degree of ketonemia and ketonuria exists, the odour of acetone is likely to be detected in the exhaled air. The triad of ketonemia, ketonuria and acetone odour of the breath is collectively termed ketosis.

Table: Accumulation of ketone bodies in untreated diabetics with severe ketosis.

State of individual	Blood concentration* (mg/100 mL)	Urinary excretion (mg/24 hr)
Normal	<3	≤ 125
Extreme ketosis (untreated diabetic)	90	5,000

Since H^+ is produced along with oxybutyrates, ketosis is frequently accompanied by acidosis. Acidosis is the lowering of blood pH due to the rise in blood levels of acetoacetate and D-β-hydroxybutyrate. Ketosis and acidosis taken together are termed as ketoacidosis. The ketoacidosis is characterized by the continual excretion in quantity of the ketone bodies which entails some loss of buffer cation, thus progressively depleting the alkali reserve. Ketoacidosis is of frequent occurrence in the untreated juvenile diabetic because of failure to secrete sufficient insulin and may be fatal. In late-onset diabetes after age 40, severe ketoacidosis is less frequent.

Fatty Acid Oxidation in Peroxisomes

Although the major site of fatty acid oxidation in animal cells is the mitochondrial matrix, other compartment in certain cells also contain enzymes which are capable of oxidizing fatty acids to acetyl-CoA, by a pathway similar to, but not identical with, that in mitochondria. Peroxisomes are membrane-bound cellular compartments in animals and plants. The fatty acids are oxidized in peroxisomes to produce hydrogen peroxide which is then destroyed enzymatically. As in the oxidation of fatty acids in mitochondria, the intermediates are coenzyme A derivatives. The whole process consists of 4 steps.

(a) Dehydrogenation of long-chain acyl-coenzyme A.

(b) Addition of water to the resulting double bond.

(c) Oxidation of the β-hydroxyacyl-CoA to a ketone.

(d) Thiolytic cleavage by coenzyme A.

The fatty acid oxidation in glyoxysomes occurs by the peroxisomal pathway. The peroxisomal pathway differs from the mitochondrial pathway in 3 respects:

1. In peroxisomal pathway, the flavoprotein dehydrogenase, that introduces the double bond, passes electrons directly to O_2, producing H_2O_2. This strong and potentially damaging oxidant is immediately cleaved to H_2O and O_2 by catalase. Whereas in mitochondrial pathway, the electrons removed in first oxidative step pass through the respiratory chain to O_2, forming H_2O. This process

is accompanied by ATP synthesis. In peroxisomes, the energy released in the first oxidative step of fatty acid breakdown is dissipated as heat.

2. The NADH formed in β oxidation cannot be reoxidized, and the peroxisomes must export reducing equivalent to the cytosol (These equivalents are eventually passed on to mitochondria).

3. In mitochondria, the acetyl-CoA is further oxidized via the citric acid cycle. Acetyl-CoA produced by peroxisomes (and also glyoxysomes) is exported. The acetate from glyoxysomes serves as a biosynthetic precursor of polysaccharides, amino acids, nucleotides and some metabolic intermediates.

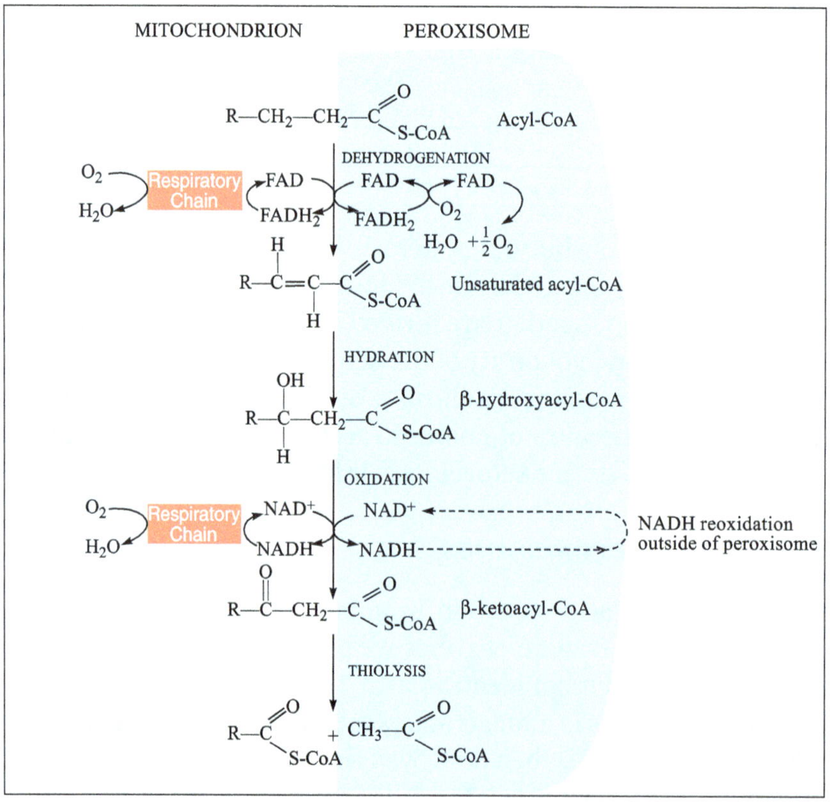

Figure: Comparison of β oxidation of fatty acids as it occurs in animal mitochondria and in animal and plant peroxisomes.

Metabolic Water

Another important biological feature of fatty acid oxidation (and of aerobic respiration, in general) is the production of metabolic water. For example, a mole of palmitic acid upon oxidation produces 16 moles of H_2O. This metabolic production of water is of significant importance to many organisms. In the case of a camel, for example, the lipids stored in its humps serve both as a source of energy and as a source of the water needed to help sustain the animal for extended periods of time in the desert. The kangaroo rat is an even more striking example of adaptation to an arid environment. This animal has been observed to live indefinitely without having to drink water. It lives on a diet of seeds, which are rich in lipids but contain little water. The metabolic water that the kangaroo rat produces is adequate for all its water needs. This metabolic response to arid conditions is usually accompanied by a reduced output of urine.

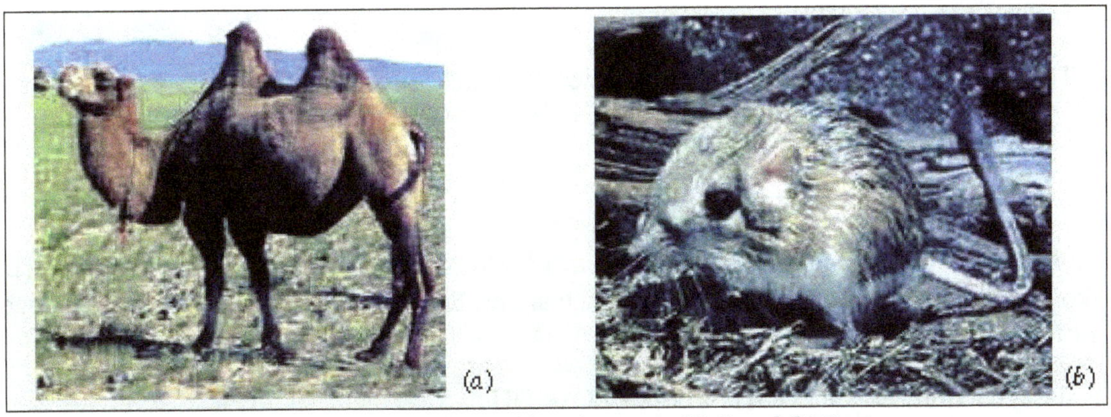

Figure: Adaptation of animals to arid environment (a) Camel (b) Kangaroo rat.

Oxidation of Lipids in Foods

Lipid oxidation is a major cause of deterioration in the quality of food and food products. Oxidation can occur in both triglycerides and phospholipids of food because lipids are divided into two main classes; polar lipids (phospholipids) and neutral lipids (triglycerides). Lipid oxidation has been long been recognized as a major problem in the storage of fatty acids in foods. Oxidation occurs by several molecular mechanisms such as generation of reactive oxygen precursors and free radicals. Oxidation affects many interactions among food constituents, leading to both desirable and undesirable products. Food lipids are the foods components that are most susceptible to oxidation, therefore oxidation reactions are one of the major sources of deterioration that occurs during manufacturing, storage, distribution and final preparation of foods. Lipid oxidation products are omnipresent in foods, although much variation exists in their kind and levels present. Although levels of these compounds are generally low, the problem of lipid oxidation severely compromises the quality of some food products and limits the shelf-life of others. Oxidative changes can cause rancidity such as off flavours, loss of colour, altered nutrient value, and may produce toxic compounds, which can be detrimental to the health of consumers. Antioxidants and chelating agents are the most helpful inhibitors of lipid oxidation.

Autoxidation

Autoxidation reaction leads to lipids breakdown and formation of wide range of oxidation products. When lipid substrate (LH) exposed to heat, light or metal ions their hydrogen atom of double bond is extracted and free or alkyl radical ($L^{\cdot}$) is formed. These free radicals reacts with oxygen and peroxy radical ($LOO^{\cdot}$) are formed with subtracting hydrogen atom from another unsaturated fatty acids. It leads to the formation of primary oxidation products called hydro-peroxides (LOOH) by the mechanisms of initiation, propagation and termination. The length of propagation cycle is directly proportional to the degree of lipid unsaturation. These primary oxidation products are not stable and further break down into carbonyl compounds such as aldehyde, ketones and alcohols.

- Initiation: $LH \rightarrow L^{\cdot} + H^{\cdot}$
- Propagation: $L^{\cdot} + O_2 \rightarrow LOO^{\cdot}$

- $LOO^{\cdot} + LH \rightarrow LOOH + L^{\cdot}$
- Termination: $LOO^{\cdot} + LOO^{\cdot} \rightarrow$ non-radical product
- $LOO^{\cdot} + L^{\cdot} \rightarrow$ non-radical product
- $L^{\cdot} + L^{\cdot} \rightarrow$ non-radical product

Further reaction of hydro-peroxide (LOOH) with ferrous ion (Fe^{2+}) yield peroxyl radical ($LOO^{\cdot}$) and alkoxyl radicals ($LO^{\cdot}$). This reaction will be fast and Fe2+ will be regenerated but the regeneration of Fe^{2+} will be slow:

- Fast reaction: $Fe^{2+} + LOOH \rightarrow Fe^{3+} + LO^{\cdot} + OH^{-}$
- Slow reaction: $Fe^{3+} + LOOH \rightarrow LOO^{\cdot} + Fe^{2+} + H^{+}$

Alkyl radicals ($R'CH_2^{\cdot}$) and aldehydes (R" CHO) form due to degradation and β-scission of alkoxyl radical $LO^{\cdot}$. Free radicals are removed in termination cycle and two radicals combine to form a non-radical compound and finally the propagation cycle is terminated as shown in figure.

Oxidation of Omega-3 Fatty Acids

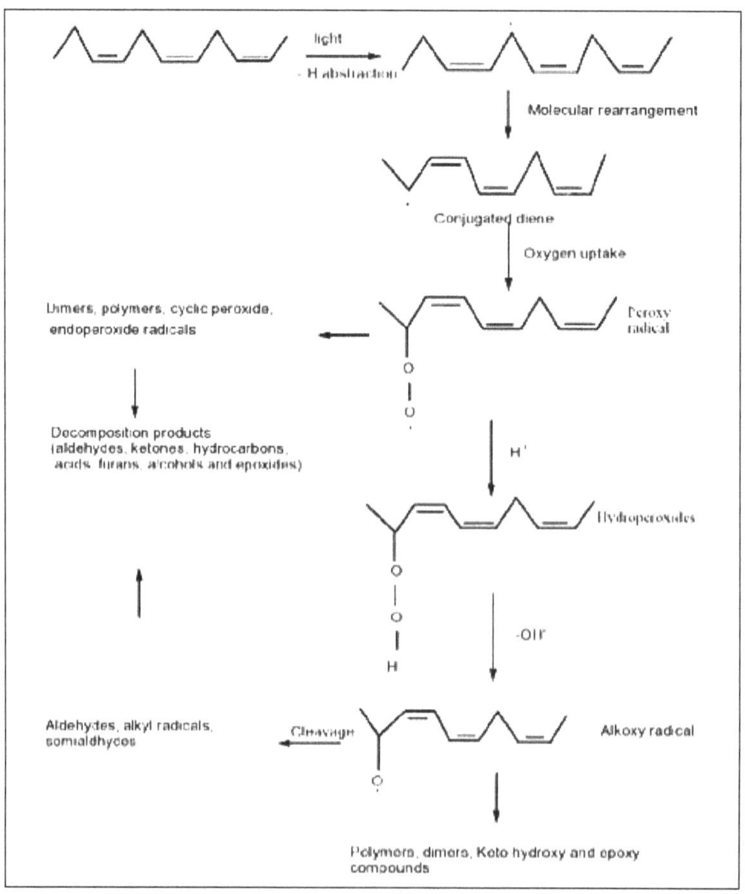

Figure: Autoxidation of unsaturated lipid.

Fish oil without having any antioxidant undergoes rapid autoxidation. Purified menhaden oil when exposed to air at ambient temperature deteriorates rapidly and peroxide value also increase 5 to 12 fold within two days. Normally during fish oil production from fish meal the Peroxide value (PV) ranged from 6-22 mEq O_2 /kg. Different fatty acids have different rate of oxidation, where during oxidation PV increases and omega-3 fatty acids content decreases). Autoxidation of ethyl EPA and DHA is much faster as compared to ethyl linoleate (Lo) and ethyl linolenate (Ln).

Oxidation in fish and fish oil occurs as a result of because of high concentration of polyunsaturated fatty acid exposure to air). These fatty acids are more susceptible to radical initiated triple oxygen oxidation than monounsaturated fatty acids. As singlet oxygen can also react with electron rich double bonds of unsaturated molecules but the rate of triplet oxygen reaction increase as degree of unsaturation increase). Volatile oxidation products formed in fish oil such as 1-penten-3-one, hexanal, 4-(Z)-heptenal, 2,4-(E,E)-heptadienal, nonanal and 2,6-(E,Z)-nonadienal are responsible for off-odors and are good indicators for the oxidation status in fish oil.

EPA and DHA are most polyunsaturated fatty acids (PUFA) in fish oil having five and six double bonds. These are susceptible to oxidation by three phases i.e. initiation, propagation and termination. Free radicals and peroxides are produced during initiation phase when molecular oxygen reacts with unsaturated fatty acids. But in propagation phase some other reactive chemical products form from peroxides such dihydroperoxides because peroxides are unstable and have active methylene group, but in termination phase non-reactive secondary oxidation products such as hydrocarbon, aldehydes and ketones are formed and cause off-flavour. In these PUFA peroxide value is not a good indicator of autoxidation because their hydro-peroxides are not stable and secondary oxidation products play important role in oxidative deterioration in these fatty.

Photo-Oxidation

Photo-oxidation produce aliphatic and aromatic oxidized compounds, such as ketones, aldehydes, carboxylic acids, fatty acids, esters, epoxies, sulfoxides, sulfones, phenols, anhydrides, quinones and alcohols. In the presence of light photosensitizers convert triplet oxygen into singlet oxygen, which is highly reactive and non-radical molecule. It requires low energy to activate the reaction with food as compared to triplet oxygen, so temperature has little effect on the oxidation caused by singlet oxygen than triplet oxygen. When oils are exposed to solar radiation, photo-oxidati on will occur in crude and refine oils. If wave length of solar light is less than 220 nm then unsaturated fatty acids cannot absorb light, but during photo oxidation sensitizer molecules like chlorophyll, porphyrin or dyes such as rose bengal absorb light energy and convert triplet state sensitiser to singlet state sensitiser. Fluorescent light is emitted by sensitiser molecules and to form triplet sensitiser, this triplet sensitiser react by two mechanism In first mechanism sensitiser react with lipid substrate to form complex mixture of free radical and singlet sensitiser then this mixture further react with triplet oxygen and give hydroperoxide similar in structure as form by autoxidation.

- ^{3}Sens + lipid substrate (LH) → Free radical (L$^\cdot$) + singlet sensitiser (SensH)
- Free radical (L$^\cdot$) + singlet sensitiser (SensH) $^3O_2$ → Hydroperoxide (LOOH)

In second mechanism triplet sensitiser also react with triplet oxygen and give singlet oxygen. This singlet oxygen directly reacts with double bond of unsaturated lipids and produce hydroperoxides.

These hydroperoxide are different from autoxidation products and in this case radical scavengers do not interfere because no involvement of free radicals.

- ^{3}Sens + $^3O_2$ (triplet oxygen)→O_2 (singlet oxygen)
- $^1O_2$ (singlet oxygen) + LH → LOOH (Hydroperoxide)

Singlet oxygen produce peroxides by direct reacting with fatty acids without abstracting H atom and proceed chain reaction during lipid peroxidation process. Small amount of singlet oxygen also produce by colliding of peroxyl radicals, illumination of certain compounds in the presence of oxygen by absorbing light and entering to higher energy state.

Enzymatic Oxidation

Hydro-peroxide formed by enzymatic oxidation process are intermediates in the synthesis process of prostagladins and eicosanoids. Enzymatic mechanisms also catalysis the oxidation of lipids in foods by lipoxygenase and cytochromes P450 to produce hydro-peroxides. But oxidation of poly unsaturated fatty acids initiated by lipoxygenases via a free radical mechanisms to form specific hydro-peroxides. Lipoxygenase enzymes molecules contain one heme iron, 1, 4 –cis,cis-pentadiene system that desired in the fatty acid as a substrate for lipoxygenase. Ferric form of lipoxygenase enzyme oxidized 1,4 – cis,cis-pentadiene fatty acid moiety into pentadienyl radical, which react with molecular oxygen to form peroxyl radical and finally produce hydroperoxides. These hydro-peroxides are precursor of further transformation by enzymatic chemical reaction, because lipoxygenase reaction initiates the synthesis of a single fatty acid hydro-peroxide from substrate of fatty acids.

- Fat + Water $^{\text{Lipoxygenase / cytochromes P450}}$ Free fatty acids + Glycerol
- Free fatty acids ^{oxidation} Hydroperoxides degradation Aldehyde, Ketones

Fatty Acids Hydro-Peroxides Isomerization

Isomeric monohydroperoxides formed from autoxidation of methyl linoleate are not stable and isomerizes further into geometrical hydro-peroxide cis-trans and trans-trans isomers. These hydro-peroxides go through rearrangement leading to their isomerization in which their OOH group is relocated and stereochemistry of double bond change by a free radical chain mechanism. Trans, trans-conjugated dienes of hydro-peroxides are produce under thermodynamic control while trans, cis-conjugated dienes of hydro-peroxides are produce under kinetic control. The ratio of trans, cis /trans, trans products form during the oxidation depend on temperature, concentration of fatty acids and solvent. Fatty acids hydro-peroxides are formed by lipoxygenases catalyze dioxygenation of polyunsaturated fatty acids. This reaction is based on initial stereospecific abstraction of a hydrogen atom from a bis-allylic methylene group followed by antarafacial attack by dioxygen at one of the terminal carbon atom of the pentadienyl radical.

Formation of Hydro-Peroxides of Fatty Acids

Hydro-peroxides, as primary oxidation products of lipid are formed through the mechanisms of autoxidation, photo-oxidation and enzymatic oxidation of lipids. Autoxidation is a reaction between oxygen and the unsaturated lipids to form a lipid hydro-peroxide, which then undergoes further reaction with or without the participation of other compounds. This reaction is enhanced by light and metal ions. Another pathway for the formation of hydro-peroxides is photoxidation, where hydro-peroxides are produced from unsaturated fatty acids with the reaction of singlet oxygen. During photoxidation mechanisms each double bond of fatty acid react with singlet oxygen and produces hydro-peroxide that causes flavour deterioration of food by monohydroperoxides hemolytic and heterolytic cleavage mechanisms. Enzyme such as lipoxygenase tightly controls the reaction with molecular oxygen that form mixed products and permit the release of free radicals.

Decomposition of Hydro-Peroxides

Long chain poly unsaturated fatty acids degrade in the presence of oxygen and metals ion that causes deterioration of sensory properties, like taste and odor due to formation of primary oxidation products by decomposition of hydro-peroxides. In the presence of metal catalysts hydro-peroxides decomposed thermally by involvement of hemolytic cleavage of O-O bond of hydro-peroxides to produce alkoxyl and hydroxyl free radicals, reaction of these radicals with lipid substrate propagate the chain reaction and produce hydrocarbons, alcohols and hydro-peroxides. These hydro-peroxides decompose in to α., β-unsaturated aldehydes like 9, 12-dioxo-10(E)-dodecenoic acid, 4-oxo-2(E)-nonenal, 4, 5-epoxy-2(E)-decenal and 4-hydroxy-2(E)-nonenal Decomposition of hydroperoxides of cyclohexene, linoleic, linolenic acids and their methyl esters produce carbonyls. Decomposition of methyl ester hydro-peroxides are slow than of free fatty acids. and produce wide range of carbonyl compounds such as hydrocarbons and furans that contributes to flavour deterioration of foods Lipids hydro-peroxides reduce into dimmers and polymers that can breakdown and produce volatile products such as malonaldehyde. These malonaldehyde as bifunctional breakdown products can interact with amino group of enzymes, proteins and DNA to cause cellular damage.

Free Radical Reactions in Decomposition of Hydroperoxides

Formation of peroxyl or alkoxyl radical is the initial step during decomposition of hydroperoxides through free radicals. This reaction is reversible for peroxyl radicals and irreversible for alkoxyl reaction:

- $LOOH \rightarrow LOO^{\cdot} + H^{\cdot}$
- $LOOH \rightarrow LO^{\cdot} + O\,H$

It relatively difficult to generate peroxyl and alkoxy radicals from hydro-peroxides. Rate of radical reaction is initiated by heat, light and catalyst species. The energy required to dissociate a LOO-H bond is 90 kcal/mol while for LO-OH bond is 44 kcal/mol. Hydrogen ion abstraction occurs with another radical then peroxyl radicals formed from hydroperoxides as shown is the below reaction:

- $LOOH + L^{\cdot} \rightarrow LOO + LH$

Bimolecular reaction decomposes hydroperoxides by forming peroxyl and aloxyl radical.

- $2\,LOOH \rightarrow LO^{\cdot} + LOO^{\cdot} + H_2O$

Trace metals play important role in the decomposition of hydro-peroxides, whereas activity of various metal ions also affected by various variables like pH, complexity of their ligands and solvents. Decomposition of hydro-peroxides through metals ion is high in speed than those of autocatalytic reactions. Higher oxidation process by metal ions produce peroxyl radicals while with lower oxidation process metal ions produce alkoxy radicals.

- $LOOH + M^{(n+1)+} \rightarrow LOO^{\cdot} + H^{+} + M^{n+}$ (peroxyl radicals)
- $LOOH + M^{n+} \rightarrow LO^{\cdot} + OH^{-} + M^{(n+1)+}$ (alkoxy radicals)

References

- Oxidation-of-lipids, omega-in-your-body-oxidation-of-lipids: 1life63.com, Retrieved 15, February 2020
- Oxidation-of-Lipids-in-Foods-307549158: researchgate.net, Retrieved 30, June 2020

5

Inborn Errors of Metabolism

The rare genetic disorders in which the body is not able to metabolize the food properly are known as inborn errors of metabolism. Some of the inborn errors of metabolism are lysosomal storage diseases, glycogen storage diseases and peroxisomal disorders. This chapter discusses in detail these diseases and disorders related to metabolism.

The name 'inborn' error signifies that the symptoms of disorders/defects can be detected in child after birth. He pointed out that there are certain features in common in all these disorders like it occurred with high frequency among the offspring of consanguineous marriages; the symptoms appeared mostly to early days or weeks of neonatal life with familial occurrence in substantial number of cases with overall benign conditions. The original four metabolic disorders have now been amplified to many disorders. Thus, dysregulation of developmental and homeostatic metabolic networks commonly known as 'Inborn errors of metabolism' comprises an important increasing group of genetic disorders. Therefore, it may be defined as an enzymopathy and genetotrophic disease which is genetically based biochemical problem where an enzyme defect obstructs a defined metabolic pathway accumulating the substrate causing disastrous pathologic effect at birth (e.g phenylketonuria) or afterwards in life (e.g. diabetes mellitus). The main four consequences of inborn errors of metabolism are depicted in the following figure.

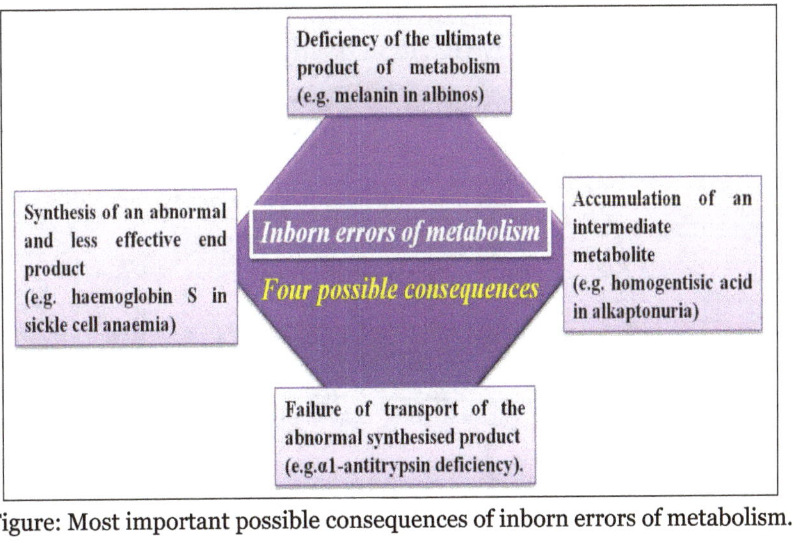

Figure: Most important possible consequences of inborn errors of metabolism.

Assortment of symptoms includes vomiting, diarrhoea, seizures, encephalopathy, hepatomegaly, hypotonia and failure to thrive. The concept of inborn error signifies the aspect of disturbance in the physiological homeostasis. The metabolic pathway is dependent on anabolic (synthetic) and catabolic (degradation) pathway which are balanced. All the pathway-dependent metabolic reactions (endergonic or exergonic) are catalyzed by enzymes. Metabolic abnormality occurs when there is a defect/absence of one particular enzyme of the pathway as given in the figure. These metabolic abnormalities are inherited and leads to most neonatal deaths.

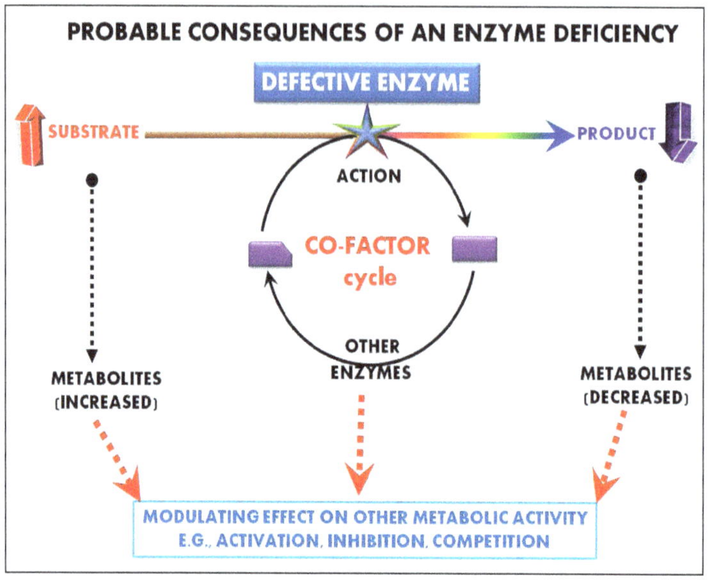

Figure: Simplified representation of metabolic effects of an enzyme deficiency.

The significance of these individually rare yet collectively an important disease can be traced back to the importance of biochemical reactions as it is present in approximately 1 in 2000 births and mostly detectable in childhood. Basically, there are many ways of classifying the metabolic disorders but when the macromolecules are considered, mainly there are four groups like amino acids, carbohydrates, lipids and others as depicted in Figure below. In this introductory section, we are primarily concerned with the lipid metabolic disorders.

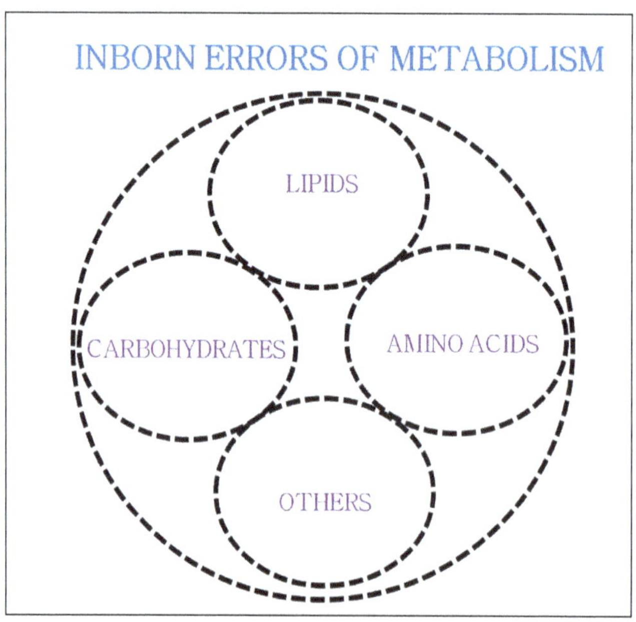

Figure: Major macromolecular system involved in the metabolic syndrome.

New Developments

The therapy of IEM is progressively more promising, even though expensive. Enzyme replacement therapy (ERT) emerge to be safe and effectual for peripheral manifestations in patients with

Fabry disease, Gaucher disease types I and III, mucopolysaccharidosis I (Hurler, Hurler-Scheie, and Scheie syndromes), mucopolysaccharidosis II (Hunter syndrome), mucopolysaccharidosis VI (Maroteaux-Lamy syndrome), and Pompe disease. Efforts are underway to extend enzyme replacement alternative for a number of other disorders. Thus far, ERT has been largely futile in improving central nervous system (CNS) expression of the lysosomal storage diseases, putatively due to complexity in piercing the blood-brain barrier. This has vanguarded to lively clinical trials assessing the safety and efficacy of intrathecal enzyme release in numerous lysosomal storage diseases. Accumulated data indicate that for preventing the progression of CNS symptoms in neuronopathic appearance of lysosomal storage diseases, including some of the lipidose, mucopolysaccharidoses, oligosaccharidoses, sphingolipidoses, and hematopoietic stem cell transplantation may perhaps be efficient under best possible (optimal) conditions. Although, longitudinal natural history data are limited, published guidelines are available to assist with decisions related to the pursuit of transplantation and whether to use bone marrow or umbilical cord blood–derived cells. In general, transplantation yields the best results when performed untimely in the path of the ailment (i.e. in an asymptomatic affected sibling of a child with a lysosomal storage disorder), in centers with experience in performing transplantations to treat inherited metabolic disorders, and in patients healthy enough to tolerate the conditioning and transplantation regimen.

Some evidence indicates that at least in certain disorders, combination ERT and hematopoietic stem cell transplantation together might be superior to hematopoietic stem cell transplantation alone in patients who are appropriate candidates. The accessibility of both ERT and hematopoietic stem cell transplantation has prompted enduring consideration of newborn screening efforts to diagnose and detect lysosomal storage diseases. Although screening for these disorders has not been widely implemented, the potential to treat these disorders is likely to drive further efforts at development.

Manifestations

Although single gene defects typically result in substrate accumulation, the precise causal patho-physiologic mechanisms that lead to clinical symptoms are not wholly lucid. The allocation of accruing material connects with the organs that are affected. Cells of the mononuclear phagocyte system are principally prosperous in lysosomes and accordingly are recurrently exaggerated by lysosomal storage diseases. Glia and neurons are most commonly affected, liable because of the relative paucity of cell turnover in the CNS, yet nonneuronopathic forms of lysosomal storage disease exist. Lysosomal storage diseases may possibly result in a rigorous neurodegenerative phenotype. Milder (typically later-onset or adult-onset) phenotypes have been recognized and are generally associated to the degree of enduring enzyme activity.

Pathophysiology

In general the accumulation of non-degraded substrate is thought to associate to the cellular dysfunction and death that complement lysosomal storage diseases; the precise mechanisms underlying this deterioration are incompletely defined. The pattern of neuronal degeneration in sub-types of lysosomal storage diseases may be surprisingly cell type specific. In some cases, substrate accumulation is also associated with sequestration of important component molecules, leading to a relative deficiency state.

The lysosome dole out as a vital element of the endosomal-lysosomal system. This system is essential for the safeguarding of typical cellular metabolism, operational in conjunction with the chaperone-mediated autophagy and ubiquitin-proteasomal systems. Anomalous function of this system may perhaps lead to ectopic dendritic sprouting (a characteristic quite inimitable to lysosomal storage diseases) and impaired recycling of glutamatergic AMPA receptors. Neuroaxonal spheroid formation is a feature of the ganglioside storage diseases, Niemann-Pick types A and C, and a-mannosidosis, implicating a shared pathology that leads to the production of these compact accumulations of mitochondria and tubulovesicular bodies.

Testing

The constellation of dysmorphic features (coarse facies, macroglossia), bony abnormalities (dysostosis multiplex), cardiac involvement (arrhythmia or cardiomegaly), hepatosplenomegaly, ophthalmologic signs (corneal clouding or macular cherry-red spot), and neurological features might guide to the irrefutable distrust of a lysosomal storage disease. Symptoms are typically gradually progressive rather than episodic, as occurs with other neurometabolic disorders. Neurological symptoms can include ataxia, developmental delay, epilepsy (complex partial or myoclonic), hypotonia, intellectual disability, peripheral neuropathy and/or spasticity.

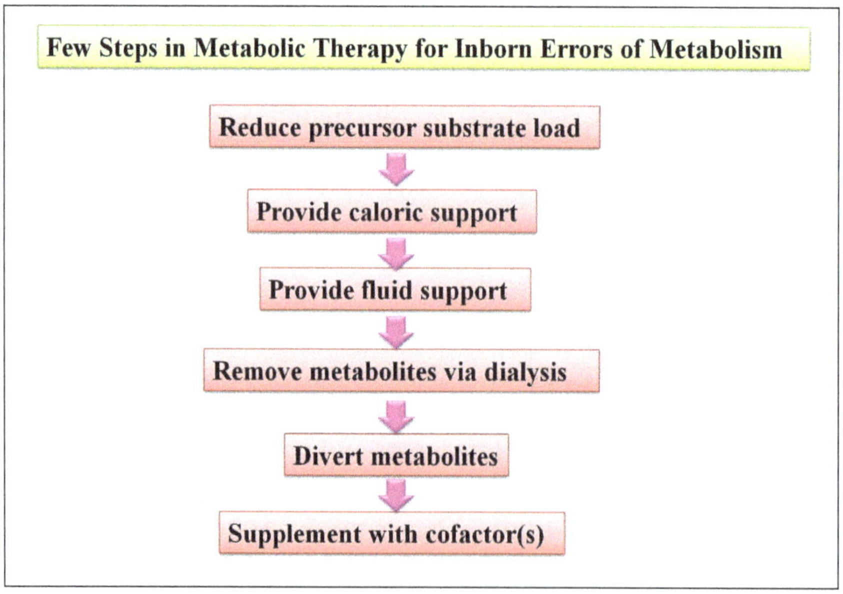

Figure: Major steps in therapy for metabolic diseases.

Screening can be executed via skeletal radiography to gaze for facts of abdominal ultrasonography to identify hepatosplenomegaly, dysostosis multiplex (observed in numerous lysosomal storage diseases), and echocardiography to assess for the cardiac association. Hearing screening results might be irregular in some cases. Ophthalmologic consultation could be useful in recognizing corneal clouding or cherry-red spot. A peripheral blood smear can reveal white blood cell vacuoles (autophagic vacuoles, fingerprint lipid whorls, granular, zebra bodies) that possibly will provide imperative diagnostic clues. Urine can be examined for prominent excretion of oligosaccharides (oligosaccharidoses) and glycosaminoglycans (mucopolysaccharidoses). Blood chitotriosidase (an enzymatic marker of macrophage activation) may be elevated. State-of-the-art testing is most proficiently executed by enzymatic activity measurement in a reference laboratory, characteristically in peripheral white blood cells (even if skin fibroblasts may also be used). For some disorders, enzyme activity can be measured in dried filter paper blood spots, and, occasionally, enzyme activity measurement in other tissues such as muscle can have utility. Urine enzyme activity measurements are seldom helpful, although urine substrate excretion can provide useful information (see screening above). In some cases, confirmatory DNA mutation analysis may be indicated.

Although inborn error of metabolic disorders arises from mostly one enzyme deficiency yet overall as all the signalling pathways are interlinked, therefore it behave as a complex trait. However, other aspect should also be considered like deficiency of cofactor required for enzymatic activity or hyper-activity of an enzyme or problem in transport process leading to excessive accumulation of metabolite. While within a signaling pathways, the concept of upstream and downstream metabolite is also crucial as an accrual of upstream metabolite can divert the pathway of this particular metabolic flux to numerous side pathways. The IEM team at the Cochrane Cystic Fibrosis and Genetic Disorders Group declared 28th February as a Rare Disease Day.

Figure: Symbol of Rare Disease Day.

Regarding the aspect covered in this section can be broadly divided into 4 subgroups which are subsequently described as given in Figure:

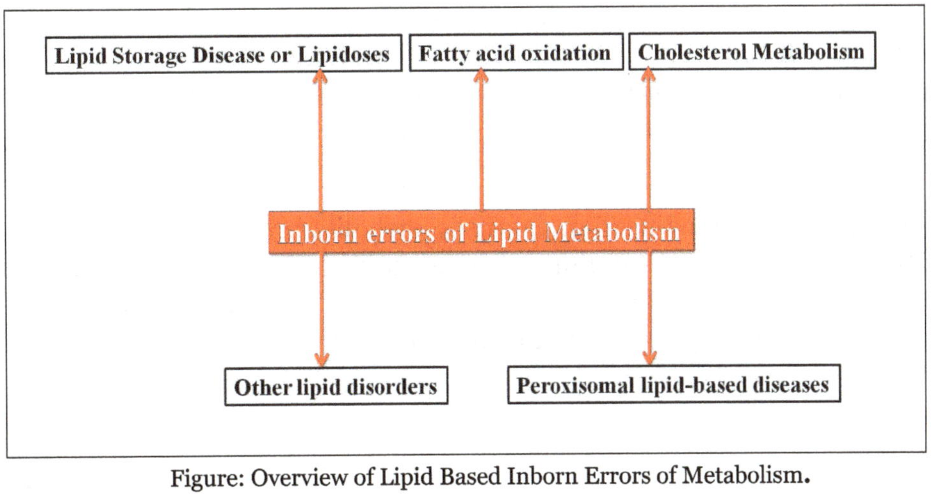

Figure: Overview of Lipid Based Inborn Errors of Metabolism.

Lysosomal Storage Diseases

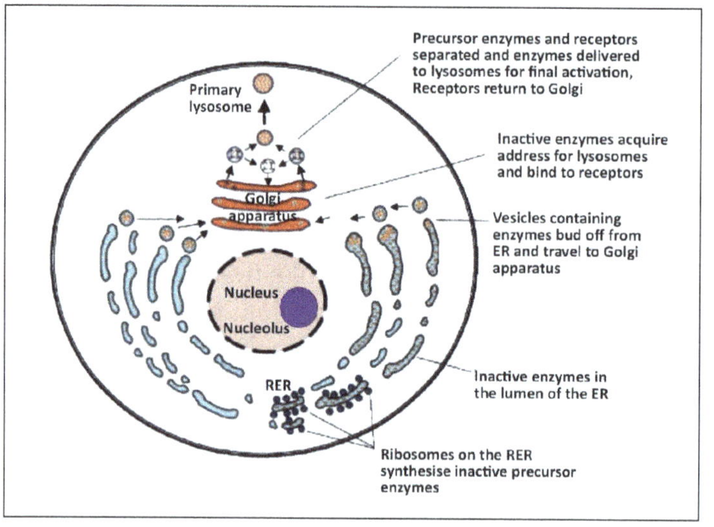

Figure: Functional Depiction of Trans-Golgi Network and Lysosome.

Lysosomal Storage Diseases with Respect to Inborn Errors of Lipid Metabolism

Lysosome, first coined by De Duve et al is the organelle credited for recycling and breakdown of cellular components. The physiological turnover is mainly caused by catabolic enzymes working at low pH within lysosomes. Hindrance in its function will ultimately leads to accrual and storage of components which leads to various clinical abnormalities and death.

Lysosomal storage diseases, in general is collection of hereditary metabolic disorders caused by flaw in protein responsible for proper functioning of lysosomes. Most of the lysosomal storage diseases are fatal. The major abnormality includes ocular pathology, organomegaly, connective tissue pathologies and mental retardation. Neurological problems of the lipidosis mainly are brain degeneration, ataxia, pain in the arms and legs, seizures, eye paralysis, learning troubles, swallowing food difficulties, spasticity, and failure of muscle tone, slurred speech, and hypersensitivity to touch and clouding of the cornea. It is now accepted that these disorders are not simply an outcome of storage, but an end result of perturbation of multifaceted cell signaling pathways which leads to specific structural and biochemical changes which can provide hint for therapy but still there exist challenges for treating the effects on central nervous system.

The main types of enzymes includes enzyme responsible for catabolism like hydrolase, post-translational lysosomal enzymes modification defects, metabolite proteins important for appropriate intracellular trafficking mainly in the trans-golgi network which involves lysosomes and activator proteins. The enzymes specifically responsible with respect to lipid metabolism are discussed in this section. They are caused by defect in proteins that are critical to the function of lysosomes.

Lipidoses

Lipidoses (or lipid storage diseases) are a collection of hereditary metabolic inborn errors in which detrimental quantity of lipids (fats) mount up in several of the body's cells and tissues. Lipidoses are inborn metabolic errors, leading to the accumulation of various lipids in the nervous system and in the reticulo-endothelial system. These metabolic disorders can occur because of mainly two reasons, first is that enough enzyme production or improper functioning of enzymes. Therefore, unnecessary storage of fats can cause everlasting cellular and tissue harm, chiefly in the peripheral nervous system, brain, liver, bone marrow and spleen.

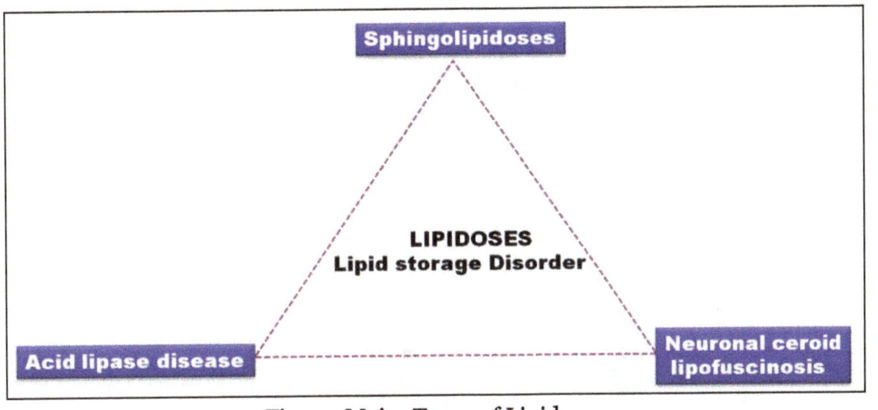

Figure: Major Types of Lipidoses.

Lipids are generally distinct as any naturally occurring fat-soluble (lipophilic), such as oils, fats, steroids, waxes, and others. Lipids are vital parts of the membranes found within and between each cell and in the myelin sheath that coats and guards the nerves. Lysosomes are the main organelle which metabolize and converts lipids and proteins into smaller parts for energy production. Lipid storage diseases are related to diverse disorders which are hereditary mostly occurs due to deficiency of lysosomal enzymes.

Inheritance Pattern of Lipidoses

Lipid storage diseases are hereditary in which defective gene regulating a specific protein carried by one or both parents. These are inherited in two ways:

- Autosomal recessive inheritance happens when both parents transmit and convey a copy of the defective gene, but neither parent is influenced by the disorder. Only 25% of children can be affected by an autosomal recessive inheritance.

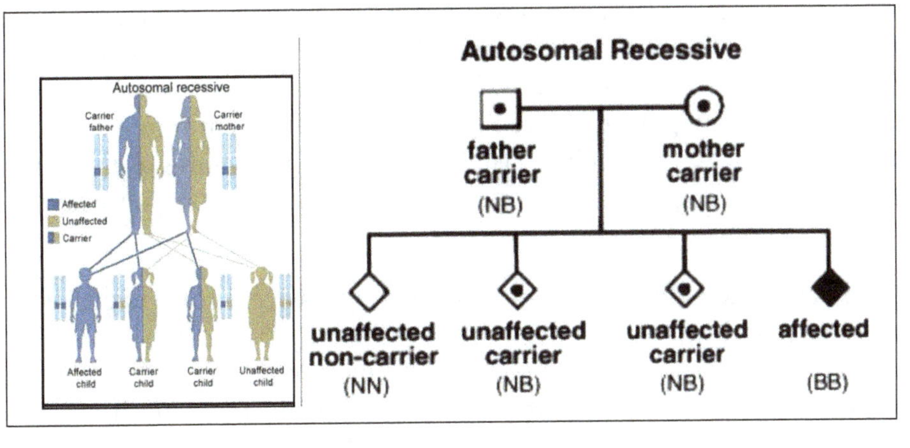

- X-linked (or sex-linked) recessive inheritance happens on the X-chromosomes, the mother carries the targeted gene that decide the child's gender and passes it to her son with 50 percent possibility in carriers of disorder inheritance.

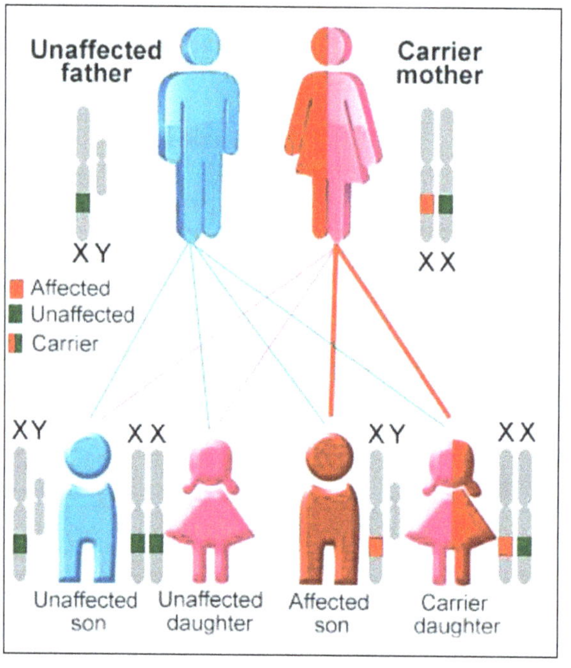

Types of Lipidoses

Sphingolipidoses

Sphingolipidoses are the sphingolipid storage diseases essentially an autosomal recessive disease except Fabry disease which is X-linked. Broadly, Sphingophospholipids and Glycosphingolipids are the main groups of sphingolipidoses. Ceramides are the precursors of both phosphorylatred and glycosylated sphingolipids.

Sphingophospholipidoses

The most important sphingophospholipid in human is sphingomyelin which is an important constituent of the myelin of nerve fibres and have a structure in which an alcohol group at carbon 1 of sphingosine is esterified to phosphorylcholine. Ceramide is the immediate precursor of sphingomyelin and it is the major constituent of skin and regulates skin`s water permeability. Sphingomyelin is degraded by Sphingomyelinase, a lysosomal enzyme that removes phosphorylcholine and leaving behind ceramide. Ceramidase cleaves ceramide into sphingosine and free fatty acid. Ceramide and sphingosine regulates many signal transduction pathways mostly PKCdependent pathways and also promotes apoptosis.

Niemann-Pick Disease is actually a group of autosomal recessive disorders caused by its inability to degrade sphingomyelin by the deficient sphingomyelinase enzyme (type A and B) or lack of the NPC1 or NPC2 proteins (type C and D). Sphingomyelinase is a type of phospholipase C and the liver and spleen are the primary sites of deposition of lipid while the deposition of sphingomyelin in the central nervous system leads to progressive neurodegeneration.

Glycosphingolipidoses

Transformed cells with dysregulated cell growth and division leads to a dramatic change in the glycosphingolipid composition of the plasma membrane. They are antigenic (blood group antigens or embryonic antigens specific for particular stages of fetal development and tumor antigens).

Glycosphingolipids consists of neutral and acidic types. Further within acidic glycosphingolipids, there are two types: Gangliosides and Sulfatides. The main features of sphingolipidoses diseases are as follows:

G_{M1} Gangliosidoses

In this there is an accumulation of gangliosides (G_{M1}) and keratin sulphate:

- Neurologic deterioration.
- Hepatosplenomegaly.
- Skeletal deformaties.
- Cherry-red macula.

Tay-Sachs Disease

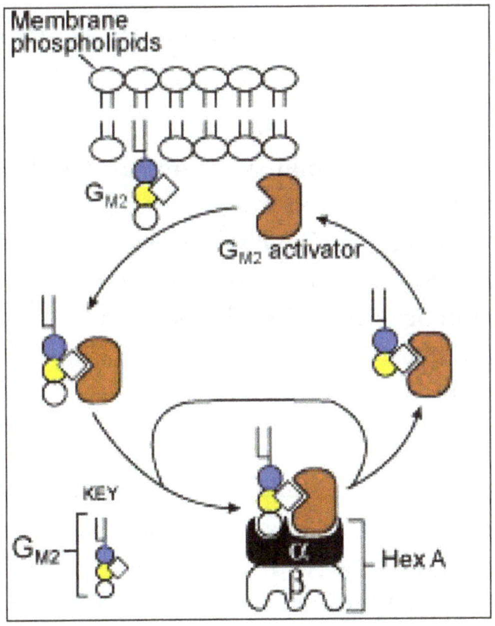

Figure: Model for G_{M2} ganglioside metabolism.

Under normal conditions, β-hexosaminidase works in the lysosome of nerve cells to breakdown unwanted ganglioside G_{M2}, a component of the nerve cell membrane. This requires three components; an α-subunit. In Tay Sachs disease, the alpha subunit of hexosaminidase malfunctions, leading to a toxic build-up of the G_{M2} ganglioside in the lysosome.

Tay-Sachs disease, a heritable metabolic disorder commonly associated with Ashkenazi Jews, has also been found in the French Canadians of Southeastern Quebec, the Cajuns of Southwest Louisiana, and other populations throughout the world. The severity of expression and the age at onset of Tay-Sachs varies from infantile and juvenile forms that exhibit paralysis, dementia, blindness and early death to a chronic adult form that exhibits neuron dysfunction and psychosis.

Tay-Sachs is an autosomal recessive disease caused by mutations in both alleles of a gene (*HEXA*) on chromosome 15. HEXA codes for the alpha subunit of the enzyme β-hexosaminidase A. This enzyme is found in lysosomes, organelles that break down large molecules for recycling by the cell. Normally, β-hexosaminidase A helps to degrade a lipid called GM2 ganglioside, but in Tay-Sachs individuals, the enzyme is absent or present only in very reduced amounts, allowing excessive accumulation of the GM2 ganglioside in neurons. The progressive neurodegeneration seen in the varied forms of Tay-Sachs depends upon the speed and degree of G_{M2} ganglioside accumulation, which in turn is dependent upon the level of functional β-hexosaminidase A present in the body.

A mouse model has been developed for Tay-Sachs, although its usefulness is limited since Tay-Sachs mice possess a minor alternative pathway for breaking down GM2 ganglioside. Treatment of the late onset form of Tay-Sachs with a ganglioside synthesis inhibitor shows promise. The effectiveness this and other treatments on individuals with the infantile (the most common) form of the disease is extremely limited since the extent of neurological damage prior to birth is unknown. The difficulty in reversing such damage will make it hard to develop an effective treatment for the infantile form of the disease. It is hoped, however, that the latter onset forms of Tay-Sachs may prove responsive to treatment, and such treatment combined with the DNA and enzymatic screening programs currently in use will lead to the eventual control of this disease.

Gaucher's Disease

Gaucher disease is categorized as a lysosomal storage disorder (LSD). Lysosomes are the major digestive units in cells. Enzymes within lysosomes break down or "digest" nutrients, including certain complex carbohydrates and fats. In Gaucher disease certain sugar (glucose) containing fat, known as glycolipids, abnormally accumulate in the body because of the lack of the enzyme, glucocerebrosidase. This accumulation or "storage" of lipids leads to the various symptoms or physical findings associated with a lysosomal storage disease. Gaucher disease is the second most common type of lysosomal storage disorder.

Signs and Symptoms

Researchers have identified three distinct forms of Gaucher disease separated by the absence (type 1) or presence and extent (type 2 or type 3) of neurological complications. Additional forms of Gaucher disease include perinatal-lethal form and cardiovascular form. The specific symptoms present in individuals with Gaucher disease vary greatly from person to person. Some individuals exhibit few or no symptoms (asymptomatic); others experience chronic, and sometimes severe, complications.

Gaucher disease type 1 is also known as non-neuronopathic, because it does not involve the central nervous system (brain and spinal cord). Type 1 Gaucher disease is the most common form of the condition. Most individuals with Gaucher disease type 1 experience easy bruising due to low levels of blood clotting cells known as platelets (thrombocytopenia), chronic fatigue due to low levels of circulating red blood cells (anemia), and an abnormally enlarged liver and/or spleen (hepatosplenomegaly). Affected individuals may also experience lack of blood supply (infarction) to various bones of the body resulting in dull or intense bone pain (bone crises), degeneration (avascular necrosis) and deformity of affected bones, and thinning and weakening of bones (osteoporosis). Such skeletal abnormalities result in an increased susceptibility to fractures. In rare cases, affected individuals may also experience involvement of the lungs and/or kidneys.

Gaucher disease type 2, also known as acute neuronopathic Gaucher disease, occurs in newborns and infants and is characterized by neurological complications due to the abnormal accumulation of glucocerebroside in the brain. Enlargement of the spleen (splenomegaly) is often the first symptom and may become apparent before six months of age. Enlargement of the liver (hepatomegaly) is not always evident. Affected infants may lose previously acquired motor skills and exhibit low muscle tone (hypotonia), involuntary muscle spasms (spasticity) that result in slow, stiff movements of the arms and legs, and crossed eyes (strabismus). In addition, affected infants may experience difficulty swallowing (dysphagia), which may result in feeding difficulties; abnormal positioning or bending of the neck (retroflexion); and failure to gain weight and grow at the expected rate (failure to thrive) and high-pitched breathing (stridor) due to contraction of the muscles of the voice box (laryngeal spasm). Anemia and thrombocytopenia may also occur. Gaucher disease type 2 often progresses to life-threatening complications such as respiratory distress or the entrance of food into the respiratory passages (aspiration pneumonia). Severely affected newborns may show skin abnormalities (collodion skin or ichthvosiform changes) and generalized swelling (hydrops), with death in the first few weeks of life. Other children with Gaucher disease type 2 have greatly reduced lifespans, with death usually occurring between 1 and 3 years of life.

Gaucher disease type 3, also known as chronic neuronopathic Gaucher disease, occurs during the first decade of life. In addition to the blood and bone abnormalities, affected individuals develop neurological complications that develop and progress slower than in Gaucher disease type 2. Associated neurological complications include mental deterioration; an inability to coordinate voluntary movements (ataxia); and brief, shock-like muscle spasms of the arms, legs or entire body (myoclonic seizures). Some individuals with Gaucher disease type 3 may have difficulty moving their eyes from side-to-side (horizontal gaze palsy). Patients with Type 3 Gaucher disease can also have a vertical gaze palsy that usually occurs later than the horizontal gaze paresis. A significant proportion of patients also develop pulmonary (lung) disease (interstitial lung disease). There can be wide variability in presentation and clinical course among patients with type 3 Gaucher disease. Some affected patients may live into their teens and early 20's, while others have lived for much longer (30's and 40's). With increasing difficulties, affected individuals may require assistance to fulfill the task of daily living (for example, with eating, bathing, and ambulation).

The perinatal-lethal form or fetal/neonatal Gaucher disease occurs in less than 5% of patients. This type is very severe and associated with death before 3 months of age or even in the womb. The fetus/newborn may present with widespread swelling of the skin (edema or anasarca) leading to fluid buildup in the heart, skin, or lungs (hydrops fetalis). Other symptoms include bleeding within the skull (intracranial hemorrhage), scaling of the skin (non-bullous ichthyosiform erythrodema) with a reddish appearance, and contraction of the joints in fixed, bent position (arthrogryposis multiplex congenita).

The cardiovascular form is characterized by CNS involvement, such as having difficulty initiating eye movement in desired directions (oculomotor apraxia). Other symptoms include calcification of the mitral and aortic valve, corneal opacity, and mild splenomegaly. Calcium deposits on the heart can reduce blood flow to these valves, and can increase blood pressure. Supranuclear ophthalmoplegia can also be present, which causes problems with balance, walking, and thinking. Cardiac-related complications and associated neurologic problems lead to a reduced lifespan, although this can extend into young adulthood in some patients.

Causes

- Gaucher disease is caused by changes (mutations) in the GBA gene.
- All three forms of Gaucher disease are inherited in an autosomal recessive pattern. Human traits, including the classic genetic diseases, are the product of the interaction of two genes, one received from the father and one from the mother.
- Recessive genetic disorders occur when an individual inherits an abnormal gene from each parent. If an individual receives from each parent one normal gene and one abnormal gene for the disease, the person will be a carrier for the disease, but usually will not show symptoms. The risk for two carrier parents to both pass the abnormal gene and, therefore, have an affected child is 25% with each pregnancy. The risk to have a child who is a carrier, like the parents, is 50% with each pregnancy. The chance for a child to receive normal genes from both parents is 25%. The risk is the same for males and females.

Sandhoff Disease

Sandhoff disease is a rare inherited disorder that progressively destroys nerve cells (neurons) in the brain and spinal cord.

The most common and severe form of Sandhoff disease becomes apparent in infancy. Infants with this disorder typically appear normal until the age of 3 to 6 months, when their development slows and muscles used for movement weaken. Affected infants lose motor skills such as turning over, sitting, and crawling. They also develop an exaggerated startle reaction to loud noises. As the disease progresses, children with Sandhoff disease experience seizures, vision and hearing loss, intellectual disability, and paralysis. An eye abnormality called a cherry-red spot, which can be identified with an eye examination, is characteristic of this disorder. Some affected children also have enlarged organs (organomegaly) or bone abnormalities. Children with the severe infantile form of Sandhoff disease usually live only into early childhood.

Other forms of Sandhoff disease are very rare. Signs and symptoms can begin in childhood, adolescence, or adulthood and are usually milder than those seen with the infantile form. Characteristic features include muscle weakness, loss of muscle coordination (ataxia) and other problems with movement, speech problems, and mental illness. These signs and symptoms vary widely among people with late-onset forms of Sandhoff disease.

Causes

Mutations in the HEXB gene cause Sandhoff disease. The HEXB gene provides instructions for making a protein that is part of two critical enzymes in the nervous system, beta-hexosaminidase A and beta-hexosaminidase B. These enzymes are located in lysosomes, which are structures in cells that break down toxic substances and act as recycling centers. Within lysosomes, these enzymes break down fatty substances, complex sugars, and molecules that are linked to sugars. In particular, beta-hexosaminidase A helps break down a fatty substance called G_{M2} ganglioside.

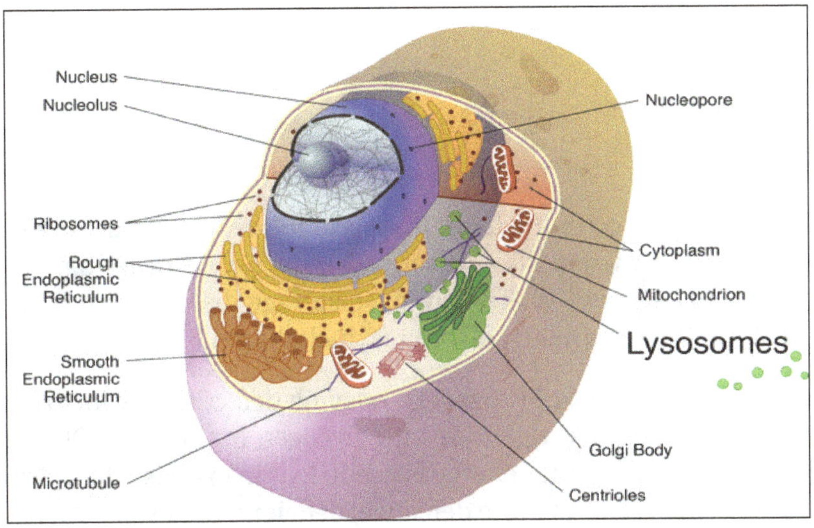

Mutations in the HEXB gene disrupt the activity of beta-hexosaminidase A and beta-hexosaminidase B, which prevents these enzymes from breaking down G_{M2} ganglioside and other molecules. As a result, these compounds can accumulate to toxic levels, particularly in neurons of the brain and spinal cord. A buildup of G_{M2} ganglioside leads to the progressive destruction of these neurons, which causes many of the signs and symptoms of Sandhoff disease. Because Sandhoff disease impairs the function of lysosomal enzymes and involves the buildup of G_{M2} ganglioside, this condition is sometimes referred to as a lysosomal storage disorder or a G_{M2}-gangliosidosis.

Glycogen Storage Diseases

Glycogen storage diseases (GSDs) are a group of inherited genetic disorders that cause glycogen to be improperly stored in the body. Children with glycogen storage diseases have a buildup of abnormal amounts or types of glycogen in their tissues.

Glycogen is the storage form of glucose in our bodies. Glucose is a simple sugar, which is a form of carbohydrate. It is found in many foods and is the main source of energy in our bodies.

Types of Glycogen Storage Disease

The main types of glycogen storage diseases in children are categorized by number and name. They include:

- Type I (Von Gierke disease) – this is the most common type of glycogen storage disease, and accounts for 90% of all glycogen storage disease cases.
- Type II (Pompe's disease, acid maltase deficiency).

- Type III (Cori's disease).
- Type IV (Andersen's disease).
- Type V (McArdle's disease).
- Type VI (Hers' disease).
- Type VII (Tarui's disease).
- Type VIII.

Since glycogen is primarily stored in the liver or muscle tissue, glycogen storage diseases usually affect functioning of the liver, the muscles, or both. The glycogen storage diseases that mainly affect the liver are types I, III, IV, and VI. The glycogen storage diseases that mainly affect muscles are types V and VII. Type II affects nearly all organs, including the heart.

Glycogen storage diseases are caused by a genetic enzyme defect that is inherited from both parents. Normally, enzymes help convert glucose into glycogen for storage. Other enzymes convert the glycogen back to glucose when quick energy is needed, as in exercise. In a person with glycogen storage diseases, some of these enzymes are defective, deficient, or absent. This causes the buildup of abnormal amounts and types of glycogen in liver and/or muscle tissues. Since glycogen storage diseases are hereditary, the primary risk factor for is having a family member with this disease.

Glycogen Storage Disease Symptoms

Glycogen storage disease symptoms in pediatric patients depend on its type. The following is a list of common glycogen storage disease symptoms:

- Low blood sugar.
- Enlarged liver.
- Slow growth.
- Muscle cramps.

Symptoms of specific types of glycogen storage diseases include:

Type I - Von Gierke Disease

- Enlarged liver and kidneys.
- Low blood sugar.
- High levels of lactate, fats, and uric acid in the blood.
- Impaired growth and delayed puberty.
- Bone thinning from osteoporosis.
- Increased mouth ulcers and infection.

Type II - Pompe's Disease

- Enlarged liver and heart.
- In severe cases, muscle weakness and heart problems develop.
- In severe cases, infants may suffer heart failure by the age of 18 months.
- Milder forms of type II may not cause heart problems.

Type III - Cori's Disease

- Swollen abdomen due to an enlarged liver.
- Growth delay during childhood.
- Low blood sugar.
- Elevated fat levels in blood.
- Possible muscle weakness.

Type IV - Anderson's Disease

- Growth delay in childhood.
- Enlarged liver.
- Progressive cirrhosis of the liver (which may lead to liver failure).
- May affect muscles and heart in late-onset type.

Type V - McArdle's Disease

- Muscle cramps during exercise.
- Extreme fatigue after exercise.
- Burgundy-colored urine after exercise.

Types VI, IX - Hers' Disease

- Liver enlargement occurs, but diminishes with age.
- Low blood sugar.

Type VII- Tarui's Disease

- Muscle cramps with exercise.
- Anemia.

Type VIII

- Muscle weakness.

- Anemia.
- Increased levels of uric acid.

Glycogen Storage Disease Diagnosis

Glycogen storage disease diagnosis usually occurs in infancy or childhood as a result of the above symptoms. If your child's doctor suspects a glycogen storage diseases, he or she will ask about your child's symptoms and medical history, then perform a physical exam. The doctor will perform tests to rule out or confirm the diagnosis. These tests may include:

- Biopsy of the affected organs.
- Blood tests and urine tests.
- MRI scan – A test that uses magnetic waves to make pictures of the inside of the body.

Glycogen Storage Disease Treatment

Glycogen storage disease treatment will depend on the type of disease and the symptoms. The following general treatment guidelines apply to people who have glycogen storage diseases that affect the liver, or types I, III, IV, and VI. Your child's doctor will develop a treatment regimen based on your child's specific symptoms. The goal of treatment is to maintain normal blood glucose levels. This may be done with:

- A nasogastric infusion of glucose in infants and children under age two.
- Dietary changes, including:
    - In children over age two, frequent small carbohydrate feedings are given throughout the day. This may include uncooked cornstarch. (Uncooked cornstarch provides a steady slow-release form of glucose).
    - Elimination of foods that are high in fructose or lactose (type I only).
    - Allopurinol (Aloprim, Zyloprim) may be prescribed to reduce uric acid levels in the blood. This is done to prevent gout and kidney stones.
    - Type IV is sometimes treated with liver transplantation.

This next group of glycogen storage disease treatment guidelines applies to people who have glycogen storage diseases that affect the muscles, or types V and VII. Your child's doctor will develop a treatment regimen based on your child's specific symptoms.

The goal of treatment is to avoid muscle fatigue and/or cramps induced by exercise. This is done by:

- Regulating or limiting strenuous exercise to avoid fatigue symptoms.
- Improving exercise tolerance by oral intake of glucose or fructose (fructose must be avoided in people with type I), or an injection of glucagon.
- Eating a high protein diet.

There is no way to prevent glycogen storage diseases. However, early treatment can help control the disease once a person has it. If you have a glycogen storage disease or a family history of the disorder, you can talk to a genetic counsellor when deciding to have children.

Peroxisomal Disorders

The peroxisomal disorders represent defects in peroxisomal functions leading to genetic diseases in man. The peroxisomal enzyme deficit group comprises seven different disorders of which phytanoyl-CoA hydroxylase (adult Refsum disease) and D-bifunctional protein deficiencies are the most recurrent. X-linked adrenoleukodystrophy is the only disease of peroxisomal substrate transport deficiency group. The therapeutic interventions are typically restricted to X-linked adrenoleukodystrophy. The most frequent adult-onset peroxisomal disorders are X-linked adrenoleukodystrophy and Refsum disease, recognized biochemically by elevated phytanate and a very long chain fatty acids respectively. Nonetheless, only ~85% of female carriers of adrenoleukodystrophy have unusual plasma concentrations and it ought to be assume in mind that distinctly unlike phenotypes of adrenoleukodystrophy can arise in one family.

2-methyl-CoA racemase deficiency presents an instance of a lately described adult-onset IEM. The clinical features are analogous to the Refsum disease and typify biochemically by the elevated concentrations of plasma pristanate. Quantifiable features include peripheral neuropathy, pigmentary retinopathy and seizures. Acute and sub-acute appearance crop up by fever, pregnancy and rapid weight loss.

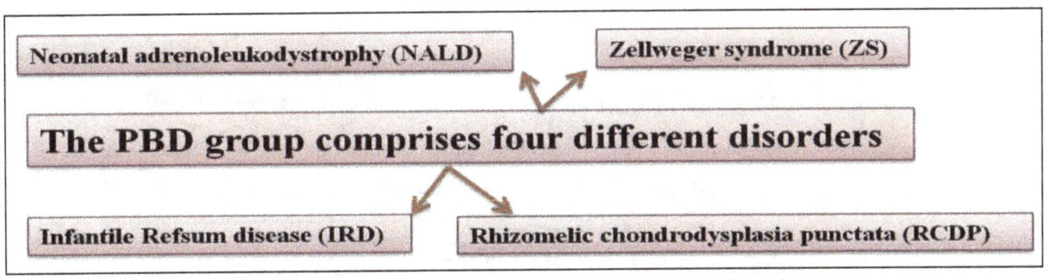

Figure: Examples of Peroxisomal biogenesis disorders.

Cerebrohepatorenal syndrome also known as Zellweger syndrome is a congenital and rare disorder illustrated by the absence or reduction of peroxisomes in the cells of the brain, kidneys and liver that distress the aptitude to βoxidize very-long chain fatty acids in the peroxisomes of the cell, owing to a genetic disorder in one of the numerous genes involved with peroxisome biogenesis. A number of peroxins are allied with Zellweger syndrome, including PEX1, PEX2, PEX3, PEX5, PEX6, PEX12, PEX14, and PEX26. The disorder is also known as the Zellweger spectrum occurs in one of three peroxisome biogenesis disorders. The further two diseases are infantile Refsum disease and neonatal adrenoleukodystrophy (NALD).

β-Oxidation of Very Long Chain Fatty Acid Oxidation

A very long chain fatty acids in which carbon atoms are 22 or more than that undergoes β-oxidation in the peroxisomes and the shortened fatty acids that undergoes β-oxidation in mitochondria.

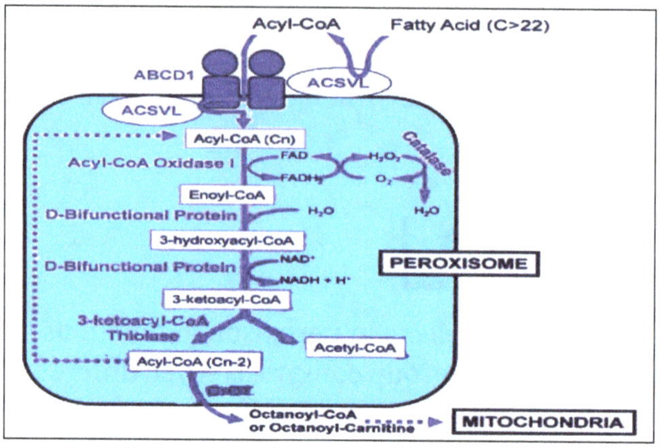

Figure: β-oxidation of very long chain fatty acid in peroxisomes.

In contrast to mitochondrial β-oxidation, the preliminary dehydrogenation in peroxisomes is catalyzed by an FAD-containing acyl CoA oxidase. The $FADH_2$ produced is oxidized by molecular oxygen, which is abridged to H_2O_2; thus, no ATP is generated by this step. The H_2O_2 is reduced to H_2O by catalase [Genetic defects either in the ability to target matrix proteins to peroxisomes (resulting in Zellweger syndrome —a peroxisomal biogenesis disorder) or in the ability to transport VLCFA across the peroxisomal membrane (resulting in Xlinked adrenoleukodystrophy), lead to accumulation of VLCFA in the blood and tissues.]

Omega Oxidation of Fatty Acids

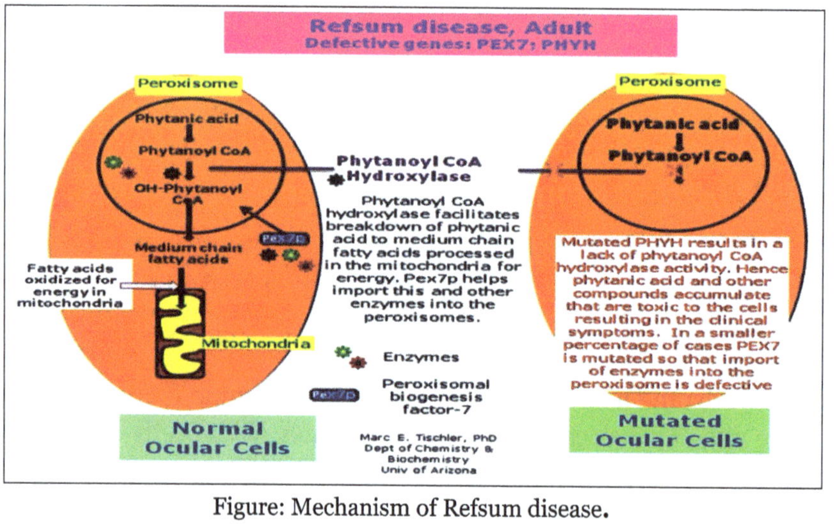

Figure: Mechanism of Refsum disease.

Phytanic acid (Branched-chain, 20 carbon fatty acid): This fatty acid is not a substrate for acyl CoA dehydrogenase because of the methyl group on its β carbon. Instead, it is hydroxylated at the α-carbon by phytanoyl CoA α-hydroxylase (PhyH), carbon 1 is released as CO_2, and the product, 19 carbon pristanic acid, is activated to its CoA derivative and undergoes β-oxidation. Refsum disease is an autosomal and rare recessive disorder caused by the absence of peroxisomal PhyH. This results in the accretion of phytanic acid in the tissue and plasma. The symptoms are primarily neurologic, and the treatment involves dietary restriction to halt disease progression. [ω-Oxidation (at the methyl terminus) also is known, and generates dicarboxylic acids. Normally a minor pathway of the ER, it's up-regulation is seen with conditions such as MCAD deficiency that limit fatty acid β-oxidation.]

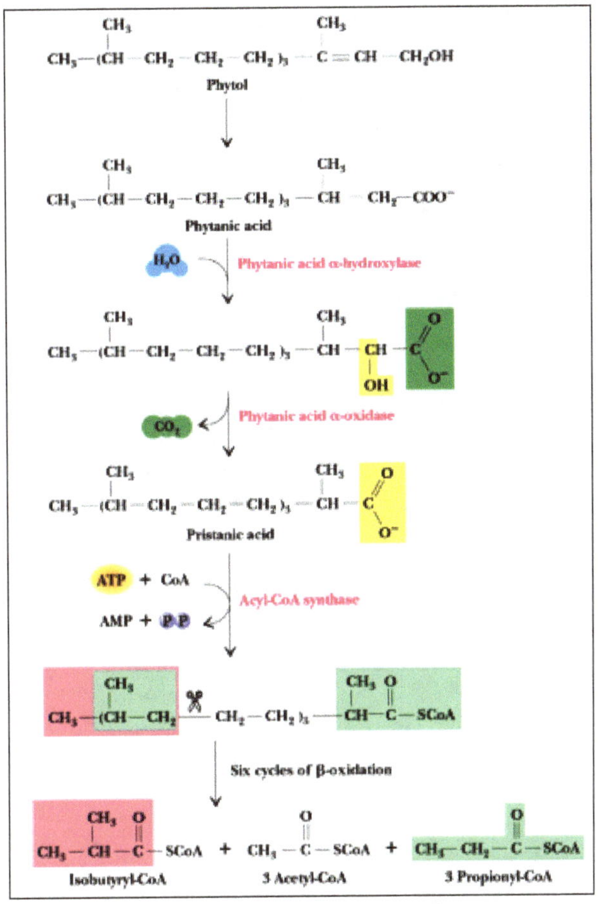

Figure: Oxidation of Phytol.

Phytanic acid α oxidase oxidizes phytanic acid to yield carbon dioxide and pristanic acid (odd chain fatty acid) that can be consequently oxidized by β-oxidation. This process rivets hydroxylation of the α carbon, removal of the terminal carboxyl group and concomitant alteration of the α hydroxyl group to a terminal carboxyl group, and linkage of CoA to the terminal carboxyl group. This branched substrate will operate in the β-oxidation procedure, eventually yielding Acetyl CoA, Propionyl-CoA, and 2-methyl propionyl CoA (Iso-butyryl CoA, in the case of phytanic acid).

Inborn Errors of Cholesterol Biosynthesis

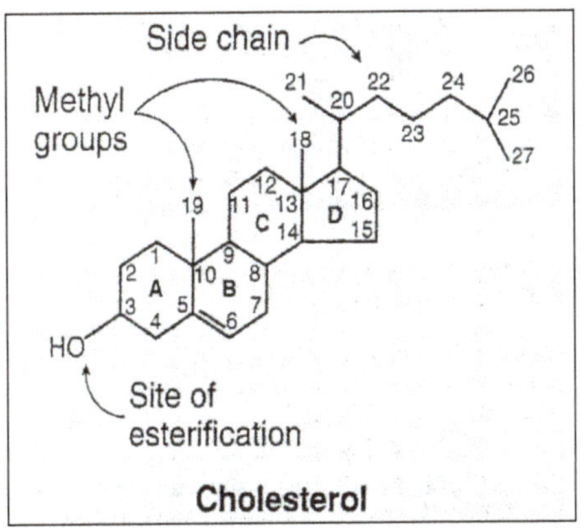

Figure: Structure of cholesterol.

Cholesterol have an important function as it is a part of membrane component as well as it is a precursor to bile acids, vitamin-D and steroid hormones. The cholesterol biosynthesis disorders have lately emerged as imperative IEM which have taught us a lot of new genetic and biochemical edifications. Despite the fact that most of the metabolic diseases are exemplified by exclusively or largely post-natal biochemical deficiencies or toxicities, cholesterol biosynthesis disorders are prominent for their relentless effects on pre-natal development.

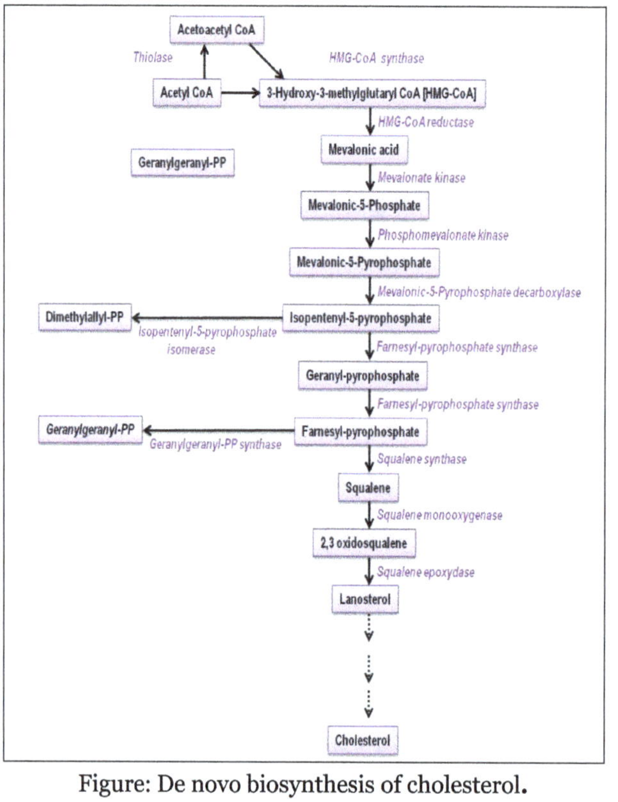

Figure: De novo biosynthesis of cholesterol.

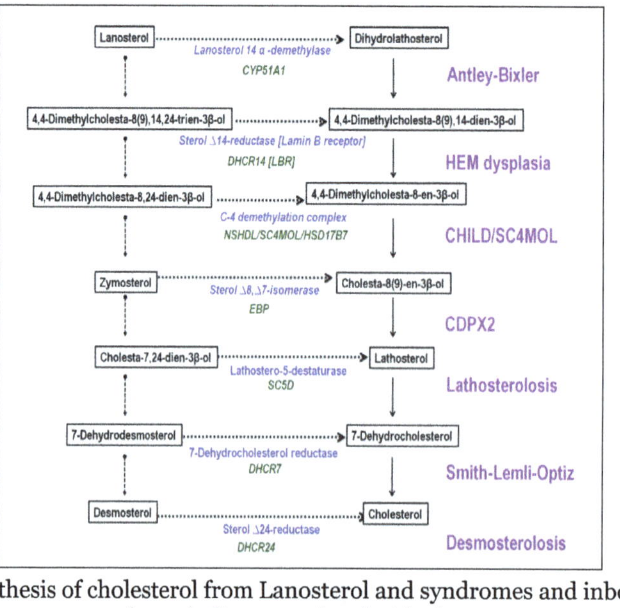

Figure: Synthesis of cholesterol from Lanosterol and syndromes and inborn errors of metabolism associated with them.

Familial Hypercholesterolemia

Familial hypercholesterolemia also known as spelled familial hypercholesterolaemia (FH or SFH) is a genetic disorder portrayed by high cholesterol levels, exclusively very high levels of LDL (low density lipoprotein or bad cholesterol) in the blood and untimely cardiovascular illness. Individuals with FH or SFH displayed different body biochemistry and their elevated cholesterol levels are less responsive to the control methods of cholesterol frequently more efficient in people without FH or SFH (e.g. dietary modification and statin tablets). However, the high doses of statin as a treatment regimen are usually effective.

Smith-Lemli-Opitz Syndrome

Smith-Lemli-Opitz syndrome (SLOS) is a manifold congenital anomalies (MCA) / mental retardation (MR) syndrome caused by a deficiency in the synthesis of cholesterol. SLOS is an autosomal recessive genetic state caused by the deficiency of 3 β-hydroxysterol-δ 7-reductase (7-dehydrocholesterol-δ 7-reductase [DHCR7] EC 1.3.1.21), the ultimate enzyme in the sterol biosynthetic pathway that converts 7-dehydrocholesterol to cholesterol.

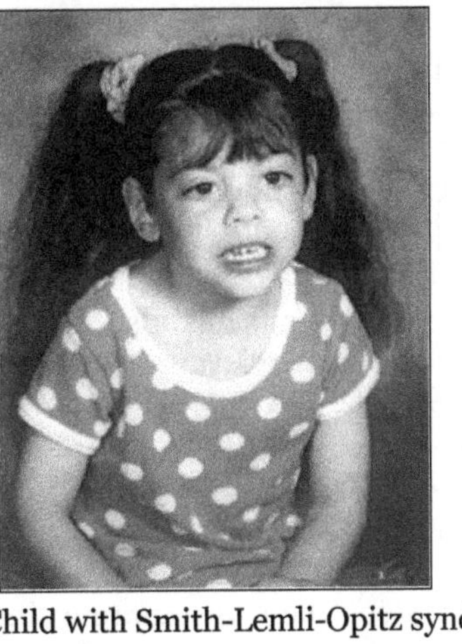

Figure: Child with Smith-Lemli-Opitz syndrome.

References

- Gaucher-disease, rare-diseases: rarediseases.org, Retrieved 18, February 2020
- Glycogen-storage-diseases, liver-disease-states, liver, education, transplant: chp.edu, Retrieved 24, May 2020

Permissions

All chapters in this book are published with permission under the Creative Commons Attribution Share Alike License or equivalent. Every chapter published in this book has been scrutinized by our experts. Their significance has been extensively debated. The topics covered herein carry significant information for a comprehensive understanding. They may even be implemented as practical applications or may be referred to as a beginning point for further studies.

We would like to thank the editorial team for lending their expertise to make the book truly unique. They have played a crucial role in the development of this book. Without their invaluable contributions this book wouldn't have been possible. They have made vital efforts to compile up to date information on the varied aspects of this subject to make this book a valuable addition to the collection of many professionals and students.

This book was conceptualized with the vision of imparting up-to-date and integrated information in this field. To ensure the same, a matchless editorial board was set up. Every individual on the board went through rigorous rounds of assessment to prove their worth. After which they invested a large part of their time researching and compiling the most relevant data for our readers.

The editorial board has been involved in producing this book since its inception. They have spent rigorous hours researching and exploring the diverse topics which have resulted in the successful publishing of this book. They have passed on their knowledge of decades through this book. To expedite this challenging task, the publisher supported the team at every step. A small team of assistant editors was also appointed to further simplify the editing procedure and attain best results for the readers.

Apart from the editorial board, the designing team has also invested a significant amount of their time in understanding the subject and creating the most relevant covers. They scrutinized every image to scout for the most suitable representation of the subject and create an appropriate cover for the book.

The publishing team has been an ardent support to the editorial, designing and production team. Their endless efforts to recruit the best for this project, has resulted in the accomplishment of this book. They are a veteran in the field of academics and their pool of knowledge is as vast as their experience in printing. Their expertise and guidance has proved useful at every step. Their uncompromising quality standards have made this book an exceptional effort. Their encouragement from time to time has been an inspiration for everyone.

The publisher and the editorial board hope that this book will prove to be a valuable piece of knowledge for students, practitioners and scholars across the globe.

Index

www.ingramcontent.com/pod-product-compliance
Lightning Source LLC
LaVergne TN
LVHW080729210926
848485LV00003B/540

* 9 7 8 1 6 3 9 8 7 3 4 4 9 *